# Crystallization Process and Simulation Calculation

# Crystallization Process and Simulation Calculation

Editors

**Mingyang Chen**
**Jinbo Ouyang**
**Dandan Han**

Basel • Beijing • Wuhan • Barcelona • Belgrade • Novi Sad • Cluj • Manchester

*Editors*

Mingyang Chen
Tianjin University
Tianjin
China

Jinbo Ouyang
East China University of
Technology
Nanchang
China

Dandan Han
Tianjin University
Tianjin
China

*Editorial Office*
MDPI
St. Alban-Anlage 66
4052 Basel, Switzerland

This is a reprint of articles from the Special Issue published online in the open access journal *Crystals* (ISSN 2073-4352) (available at: https://www.mdpi.com/journal/crystals/special_issues/crystallization_process2).

For citation purposes, cite each article independently as indicated on the article page online and as indicated below:

Lastname, A.A.; Lastname, B.B. Article Title. *Journal Name* **Year**, *Volume Number*, Page Range.

**ISBN 978-3-0365-9154-4 (Hbk)**
**ISBN 978-3-0365-9155-1 (PDF)**
doi.org/10.3390/books978-3-0365-9155-1

© 2023 by the authors. Articles in this book are Open Access and distributed under the Creative Commons Attribution (CC BY) license. The book as a whole is distributed by MDPI under the terms and conditions of the Creative Commons Attribution-NonCommercial-NoDerivs (CC BY-NC-ND) license.

# Contents

**Preface** . . . . . . . . . . . . . . . . . . . . . . . . . . . . . . . . . . . . . . . . . . . . . . . . . . . . . . . . . . . . . . . . . . . . . . . vii

**Xiaowei Wang, Kangli Li, Xueyou Qin, Mingxuan Li, Yanbo Liu, Yanlong An, et al.**
Research on Mesoscale Nucleation and Growth Processes in Solution Crystallization: A Review
Reprinted from: *Crystals* **2022**, *12*, 1234, doi:10.3390/cryst12091234 . . . . . . . . . . . . . . . . . . . 1

**Qiuyan Ran, Mengwei Wang, Wenjie Kuang, Jinbo Ouyang, Dandan Han, Zhenguo Gao, et al.**
Advances of Combinative Nanocrystal Preparation Technology for Improving the Insoluble Drug Solubility and Bioavailability
Reprinted from: *Crystals* **2022**, *12*, 1200, doi:10.3390/cryst12091200 . . . . . . . . . . . . . . . . . . . 23

**Xinyuan Zhang, Pingping Cui, Qiuxiang Yin and Ling Zhou**
Measurement and Correlation of the Solubility of Florfenicol in Four Binary Solvent Mixtures from $T$ = (278.15 to 318.15) K
Reprinted from: *Crystals* **2022**, *12*, 1176, doi:10.3390/cryst12081176 . . . . . . . . . . . . . . . . . . . 45

**Yan Huo, Xin Li and Binbin Tu**
Image Measurement of Crystal Size Growth during Cooling Crystallization Using High-Speed Imaging and a U-Net Network
Reprinted from: *Crystals* **2022**, *12*, 1690, doi:10.3390/cryst12121690 . . . . . . . . . . . . . . . . . . . 67

**Kirils Surovovs, Maksims Surovovs, Andrejs Sabanskis, Jānis Virbulis, Kaspars Dadzis, Robert Menzel, et al.**
Numerical Simulation of Species Segregation and 2D Distribution in the Floating Zone Silicon Crystals
Reprinted from: *Crystals* **2022**, *12*, 1718, doi:10.3390/cryst12121718 . . . . . . . . . . . . . . . . . . . 79

**Felix Sturm, Matthias Trempa, Gordian Schuster, Rainer Hegermann, Philipp Goetz, Rolf Wagner, et al.**
Long-Term Stability of Novel Crucible Systems for the Growth of Oxygen-Free Czochralski Silicon Crystals
Reprinted from: *Crystals* **2023**, *13*, 14, doi:10.3390/cryst13010014 . . . . . . . . . . . . . . . . . . . 99

**Raghda Hamdi and Mohamed Mouldi Tlili**
Influence of Foreign Salts and Antiscalants on Calcium Carbonate Crystallization
Reprinted from: *Crystals* **2023**, *13*, 516, doi:10.3390/cryst13030516 . . . . . . . . . . . . . . . . . . . 113

**Atef Korchef, Salwa Abouda and Imen Souid**
Optimizing Struvite Crystallization at High Stirring Rates
Reprinted from: *Crystals* **2023**, *13*, 711, doi:10.3390/cryst13040711 . . . . . . . . . . . . . . . . . . . 129

**Junkai Wang, Laishi Li, Yusheng Wu and Yuzheng Wang**
The Influence of Hydrothermal Temperature on Alumina Hydrate and Ammonioalunite Synthesis by Reaction Crystallization
Reprinted from: *Crystals* **2023**, *13*, 763, doi:10.3390/cryst13050763 . . . . . . . . . . . . . . . . . . . 145

**Kai Guo, Wenchong Cheng, Haiyuan Liu, Wenhao She, Yinpeng Wan, Heng Wang, et al.**
Sn-Doped Hydrated $V_2O_5$ Cathode Material with Enhanced Rate and Cycling Properties for Zinc-Ion Batteries
Reprinted from: *Crystals* **2022**, *12*, 1617, doi:10.3390/cryst12111617 . . . . . . . . . . . . . . . . . . . 157

**Jia-Jing Luo, Xiang-Xin Cao, Qi-Wei Chen, Ying Qin, Zhen-Wei Zhang, Lian-Qiang Wei, et al.**
2D Layer Structure in Two New Cu(II) Crystals: Structural Evolvement and Properties
Reprinted from: *Crystals* **2022**, *12*, 585, doi:10.3390/cryst12050585 . . . . . . . . . . . . . . . . . . . 167

Shengtao Zhang, Hao Fu, Guofeng Fan, Tie Li, Jindou Han and Lili Zhao  
**Study on Deposition Conditions in Coupled Polysilicon CVD Furnaces by Simulations**  
Reprinted from: *Crystals* **2022**, *12*, 1129, doi:10.3390/cryst12081129 . . . . . . . . . . . . . . . . . **175**

# Preface

Crystallization is a key operating unit for the effective separation and purification of solid products. To enhance product quality and lower production costs, comprehending and controlling the crystallization process is now crucial. The research that focuses on the basic principles of nucleation, growth, breakage, and agglomeration in crystallization, the choice of crystallization methods like continuous crystallization, and the operation technology and control methods in the crystallization process have all made significant advancements and great progress in recent years. As a complex physical and chemical process involving many parameters and variables, the characteristics of strong coupling, nonlinearity, and large lagging in the crystallization process are very prominent, which greatly increases the complexity of process research and design as well as the challenge of quantifying process stability and reliability. Based on this, in recent years, accurate inline or online monitoring measurement and process analysis technical tools have become the focus of breakthroughs, which have achieved rapid development and application in guiding crystallization process regulation and optimizing crystallization processes. At the same time, against the background of digitalization, automation, and intelligence in the crystallization process, it has continuously promoted the development of the simulation calculation field by constructing mathematical models to describe and study complex processes. Especially in simulation technology of molecular dynamics and hydrodynamics, through mathematical modeling and multi-physical field simulation, dynamic data of multi-scale and cross-time and space can be provided, which has become an effective tool for in-depth research. The application of artificial intelligence technology such as machine learning will also become an important direction of crystallization process research in the future.

Therefore, this Special Issue includes 12 excellent works, covering a broad spectrum of topics from the study of crystallization process research and simulation calculation. The topics include, but are not limited to, the following:

Crystallization process analysis and control;
Secondary crystallization process;
Crystallization process strengthening;
Cocrystallization;
Continuous crystallization;
Dynamic simulation.

We hope that this Special Issue can serve as an open platform for scientists, graduate students and engineers working in the field of crystallization to exchange and publish the latest research results.

We deeply appreciate the outstanding research work of all authors. We are also very grateful to Mr. Mars Tan from the Editorial Office for his continuous assistance and support, which enabled our Special Issue to be successful.

Mingyang Chen, Jinbo Ouyang, and Dandan Han
*Editors*

*Review*

# Research on Mesoscale Nucleation and Growth Processes in Solution Crystallization: A Review

Xiaowei Wang [1,2,†], Kangli Li [1,3,†], Xueyou Qin [1,2], Mingxuan Li [1,2], Yanbo Liu [1,2], Yanlong An [4], Wulong Yang [4], Mingyang Chen [1,2,\*], Jinbo Ouyang [5,\*] and Junbo Gong [1,2]

1. State Key Laboratory of Chemical Engineering, School of Chemical Engineering and Technology, Tianjin University, Tianjin 300072, China
2. Haihe Laboratory of Sustainable Chemical Transformations, Tianjin 300072, China
3. Institute of Shaoxing, Tianjin University, Shaoxing 312300, China
4. Zhejiang Huakang Pharmaceutical Co., Ltd., Quzhou 324302, China
5. School of Chemistry, Biology and Materials Science, East China University of Technology, Nanchang 330013, China
\* Correspondence: chenmingyang@tju.edu.cn (M.C.); oyjb1001@163.com (J.O.)
† These authors contributed equally to this work.

**Abstract:** In recent studies, the existence of mesoscale precursors has been confirmed in crystallization. Different from the classical crystallization theory, which only considers the sequential attachment of basic monomers (atoms, ions, or molecules), the nonclassical crystallization process involving precursors such as prenucleation clusters, nanoparticles, and mesocrystals is more complicated. The mesoscale structure is important for the quantitative description and directional regulation of the solution crystallization process. It is necessary to explore the mechanism by the mesoscale scientific research methods on the base of traditional chemical engineering and process system engineering research methods. Therefore, the paper reviews several representative nonclassical nucleation and growth theories, mainly including two-step nucleation theory, prenucleation clusters theory, particle agglomeration theory, amorphous precursor growth theory, particle attachment growth theory and mesocrystal growth theory. Then, the mesoscale structure and its spatiotemporal dynamic behavior are discussed, and the application of the EMMS model in the nucleation and growth process is analyzed. Finally, we put forward our views on the prospect of the paradigms and theoretical innovations of using mesoscale methods in crystal nucleation and growth.

**Keywords:** nucleation; growth; mesoscale; precursor; EMMS model

## 1. Introduction

Crystallization has been widely used as an important separation and purification method in chemical, food and pharmaceutical engineering [1–3]. Solution crystallization is the most widely used crystallization method due to its high efficiency and low pollution [4]. Solution crystallization is the transformation of a substance from liquid state to crystalline state and its process undergoes two steps, which are nucleation and crystal growth. Nucleation is the generation of a number of crystal nuclei in the supersaturated solution and crystal growth is the further growth of crystal nuclei. Nucleation and growth of crystals are the most important scientific questions in the crystallization field, which have been attracting the extensive attention of researchers [5].

There are few theories describing the nucleation and growth processes of crystals. However, most of the existing theories are based on simple assumptions, which lead to significant limitations in the application of their models [6,7]. Under the classical nucleation and growth theory, solute molecules in supersaturated solutions form clusters through motion and collisions, and nuclei are formed when the clusters exceed a certain critical size. The crystal nuclei grow by adding monomers such as molecules, atoms and ions to the surface one by one, and finally form a crystal with an ordered structure. Both system free-energy changes and reaction kinetics lead to diversification of crystallization pathways [8,9]. With the further study on crystallization and the development of process analytical technology (PAT), it is possible to directly observe the nucleation and growth process of crystals at the molecular and nanoscale. According to experimental observations, the classical crystallization theory cannot apply in all systems. Then, nonclassical nucleation theories, such as two-step nucleation theory, prenucleation clusters theory, and particle agglomeration theory have been proposed. The most typical difference between these theories and classical nucleation theories is that molecules undergo intermediate states such as prenucleation clusters, metastable crystalline states and amorphous states before nucleation. In addition, nonclassical growth theories have also been proposed, which explain that the growth units from molecules, atoms and ions to precursors such as oligomers, solute-rich droplets, mesocrystals and nanoparticles [10]. Nonclassical crystallization pathways have been observed in the nucleation and growth of protein crystals, inorganic nanomaterials and organic crystals [11,12]. In a word, the exact mechanism of crystal nucleation and growth is still unclear, and there are still lots of difficulties to explore in the boundary between micro- and macro-scales.

In the process of crystal nucleation and growth, the precursors between molecules and crystals are typical mesoscale structures. Mesoscale is a relative concept, which refers to a complex structural scale between the individual microscale and the whole scale composed of them [13]. Mesoscale is universal in different levels, but its specific forms are different. The physical mechanisms in different mesoscale structures have a commonality that the competition and coordination among the dominant mechanisms tend to minimize the energy consumed in the system. The commonality is the core of mesoscience, based on the principle of the energy minimization multi-scale (EMMS) [14]. The EMMS principle of mesoscience provides an important research idea and method for the study of crystal nucleation and growth. The traditional crystallization theory mainly focuses on the crystal nucleation and growth at the macroscopic and microscopic scales, but the mesoscale structure is less involved. Therefore, it is the focus of subsequent research work to fully consider the mesoscale structure of the nucleation and growth processes, and to establish the stability conditions of the mesoscale structure to achieve the quantitative description of its structure and properties [15].

In this paper, we review the basic principles and concepts of classical nucleation and growth theory and summarize the recent progress of mesoscale structure research in nonclassical nucleation and growth theory. In addition, we also analyze the application of the EMMS model in the nucleation and growth process and describe the prospect of the mesoscale research paradigm and theoretical development of crystal nucleation and growth in solution crystallization.

## 2. Mesoscale Structure during Nucleation

The spatio-temporal dynamic structure of nucleation precursors is a typical mesoscale problem, and it is also a key factor that could affect the nucleation process in solution crystallization. Various nucleation theories mainly involve nucleation precursors with different mesoscale structures, such as ordered clusters, disordered clusters, prenucleation clusters, and nanoparticles.

### 2.1. Ordered Clusters in Classical Nucleation Theory

The classical nucleation theory (CNT) stemmed from the job of Gibbs [16] in 1877, then Farkas, Volmer and Weber [17] improved it by quantitative analysis and modification, and it was formally proposed by Becker and Döring [18] in 1935. CNT is based on the process of condensation of supersaturated vapor into liquid and can be applied to liquid–solid equilibrium systems, such as solution crystallization. CNT considers that when primary monomers such as molecules, atoms or ions collide with each other, they will follow the dynamic equilibrium of multi-stage adsorption and desorption to form ordered clusters [19,20]. The ordered clusters can grow into crystal embryos, which can establish thermodynamic equilibrium with the solution during the growth process and form crystal nuclei. The crystal nuclei may continue to grow into crystals, and the specific process is shown in Figure 1a.

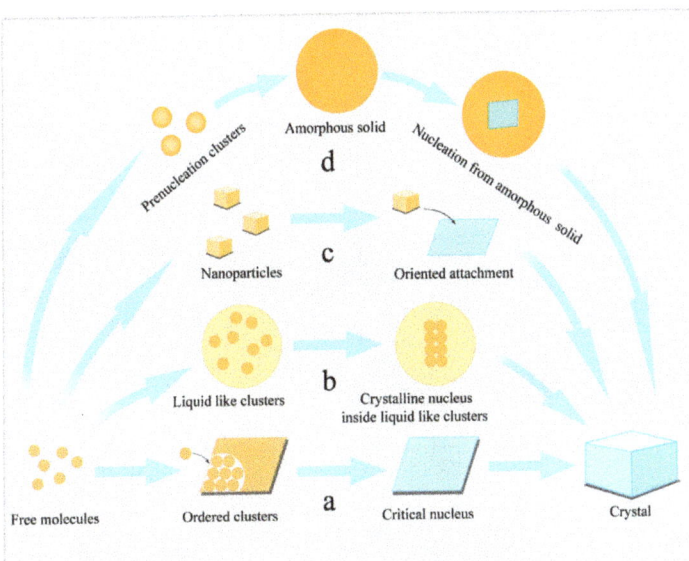

**Figure 1.** Schematic diagram of (**a**) classical nucleation theory, (**b**) two-step nucleation theory, (**c**) particle agglomeration theory, (**d**) prenucleation clusters theory.

The structure and behavior of critical ordered clusters are the highlight of research in CNT. The prediction method of critical size has also been reported in many papers. Joseph L. Katz [21] used CNT and Reiss theory to predict critical ordered clusters size of nonane, which consisted of eight and nine molecules. Joel D [22] investigated the nucleation process of AgBr in water using molecular dynamics methods and found that for $Ag_6Br_6$ and $Ag_9Br_9$, the critical ordered clusters consisted of 5–6 monomers. However, the critical ordered clusters of $Ag_{18}Br_{18}$ consisted of three monomers because nucleation occurred immediately. Adamski [23] further found that critical ordered clusters formed by $10^{-15}$ g barium salts consisted of one million monomers. Moreover, CNT with critical ordered clusters has been confirmed by relevant experiments. S. T [24] observed the structure of ordered clusters in apoferritin crystallization by atomic force microscopy (AFM) (Figure 2). As shown in Figure 2, the number of molecules constituting these clusters was different and they were all monolayers. These ordered clusters would dissolve or continue to grow after the remaining adsorbed for 2 to 30 min, which demonstrated the formation of critical ordered clusters during nucleation.

In CNT, the molecular self-assembly process from the molecular scale to the crystal scale is relatively ordered and involves fewer mesoscale structures. CNT has been successfully used to qualitatively analyzed the crystallization process and has achieved relatively accurate results in crystal engineering design and control. However, CNT cannot be applicable to nucleation processes with mesoscale structures such as prenucleation clusters.

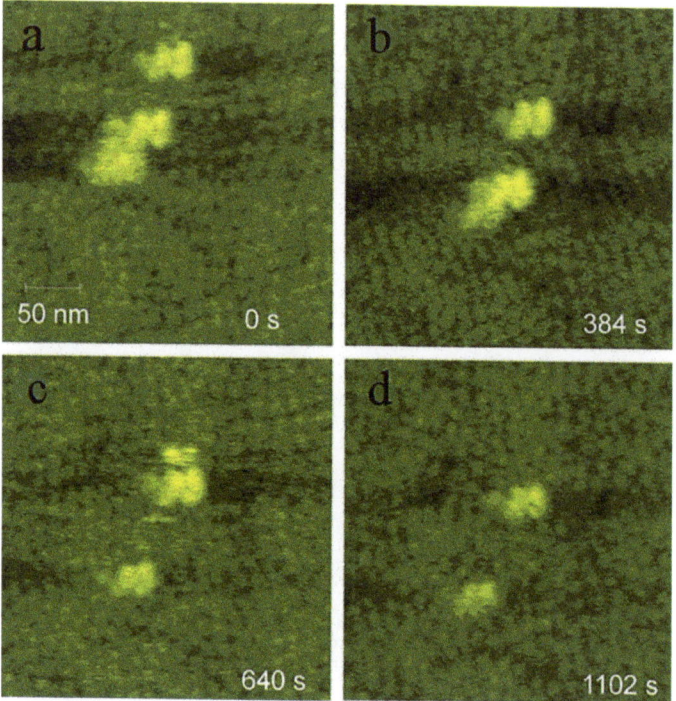

**Figure 2.** AFM image of clusters evolution in protein solution with C = 0.23 mg mL$^{-1}$ and σ = 2.3. (**a**) 0 s, (**b**) 384 s, (**c**) 640 s, (**d**) 1102 s. C as the actual protein concentration and σ as the supersaturation of the solution with respect to the crystalline phase. The figure is reproduced with copyright permission of ref. [24].

*2.2. Disordered Clusters in Two-Step Nucleation Theory*

As shown in Figure 1b, the two-step nucleation theory indicates that solute molecules first aggregate into dense disordered nucleation clusters in supersaturated solutions. When dense disordered clusters reach a certain level, their interior will reorganize into an ordered crystal nucleus and eventually form crystals [25,26]. The two-step nucleation theory was originally proposed to explain the process of protein nucleation, but now, this theory has also been applied to small molecule materials and colloids [27,28].

Disordered clusters are mesoscale structures in the two-step nucleation theory. Since disordered clusters are microstructures with spatio-temporal dynamic changes, it is very difficult to observe and analyze them with instruments. Therefore, simulation methods are often used for this research. The mechanism of two-step nucleation was initially proposed by Wolde and Frenkel [29], who used the Monte Carlo method to study uniform nucleation in the Lennard–Jones system. Their simulations showed simultaneous changes in density and structural order parameters away from the liquid–liquid critical point, but high-density fluctuations were observed at the critical point, which made a significant change in the crystallization nucleation process. Disordered droplets were first formed, and then crystal nuclei were generated inside droplets larger than a certain critical size. Nathan Duffl [30] proposed the Potts lattice gas model to study the nucleation in solution, and the simulation results showed that the nucleation pathway would change from one-step to two-step nucleation as the temperature approaches the melting temperature, which conformed with the two-step nucleation mechanism. Myerson [31] treated glycine solution with near-infrared laser and obtained different crystal forms of glycine by using linear polarization and circular polarization pulses, respectively. The result suggested that molecules underwent a high-density intermediate state before aggregating to form a nucleus, confirming the two-step nucleation mechanism of glycine. They later confirmed the two-step nucleation process of glycine using small-angle X-ray scattering (SAXS) [32].

The most important mechanism in the two-step nucleation theory was proposed by Vekilov et al. [33]. They demonstrated that in supersaturated solutions, dense disordered clusters were formed first, and then crystal nuclei were formed from the dense disordered clusters, as shown in Figure 3. The second step determined the crystal nucleation rate. Two-step nucleation theory successfully explains some difficult questions of crystal nucleation: the actual nucleation rate is much lower than that predicted by CNT, and what is the effect of heterogeneous matrices on polymorphic selection?

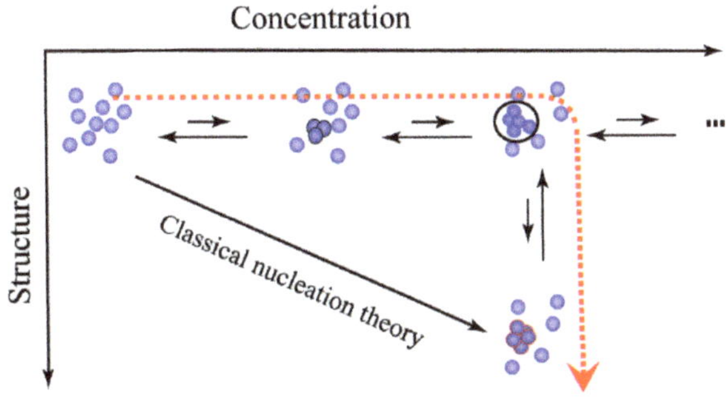

**Figure 3.** Microscopic views on the (Concentration, Structure) plane during the two-step nucleation process and classical nucleation process. The figure is reproduced with copyright permission of ref. [33].

In general, the two-step nucleation theory first considers mesoscale disordered clusters, and preliminarily describes the mesoscale structure and behavior from the molecular scale to the crystal scale, which is a supplement to the relatively ordered CNT. Moreover, the two-step nucleation theory can perfectly explain the nucleation process of proteins, some organic small molecules, inorganic molecular systems, colloidal systems, and biological minerals [34–38].

*2.3. Prenucleation Clusters in Prenucleation Clusters Theory*

In recent years, much evidence has shown that inorganic systems, such as calcium carbonate and calcium phosphate, have stable aggregates in unsaturated and supersaturated solutions, also known as prenucleation clusters [39–41]. The discovery of these prenucleation clusters contradicts the hypothesis of ordered clusters formed by monomer aggregation in CNT, and thus the mechanism of prenucleation clusters is a nonclassical nucleation mechanism. Except in inorganic systems, studies have also reported the existence of prenucleation clusters in some amino acid systems [42]. As shown in Figure 1d, prenucleation clusters theory suggests that stable prenucleation clusters do not form crystal nuclei through the gradual accumulation of monomers but form larger amorphous solid through collisional agglomeration between prenucleation clusters. Subsequently, the internal molecular structure of the amorphous solid reorganizes to crystal nuclei, which grow into macroscopic crystals.

Prenucleation clusters are mesoscale structures in the nucleation process described by prenucleation clusters theory. Prenucleation clusters are thermodynamically stable clusters composed of atoms, molecules or ions, and there is no phase boundary between them and the surrounding solution. Meanwhile, prenucleation clusters are molecular precursors for solution nucleation and can participate in the phase separation process. Compared with the disordered cluster structure described by the two-step nucleation theory, the key difference of this mesoscale structure is the formation of stable and ordered solute self-associations [43,44].

Zhang [45] proposed that at high magnesium concentrations, amorphous calcium carbonate coexisted with stable prenucleation clusters. These prenucleation clusters consisting of ions and water molecules played an important role in the crystallization of calcium carbonate. Habraken [46] used cryo-TEM and in-situ AFM to detect calcium phosphate prenucleation clusters, which were confirmed to be calcium triphosphate clusters $[Ca_2(HPO_4)_3]^{-2}$, and the prenucleation clusters aggregated and absorbed additional calcium ions to form amorphous calcium phosphate. Thi Thanh [47] demonstrated that prenucleation clusters were formed during the crystallization process of L-glutamic acid by TEM and SAED. As shown in Figure 4, when the crystalline system passed through the metastable region, prenucleation clusters formed by spontaneous aggregation from solution. The prenucleation clusters may grow further into larger amorphous solid through agglomeration or Ostwald ripening mechanisms. After this stage, the amorphous solid may undergo shrinkage and reorganization, which led to compaction into nuclei.

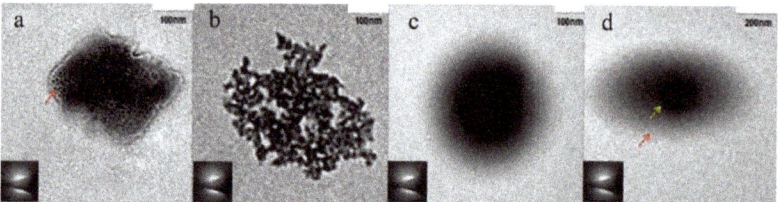

**Figure 4.** (**a**–**d**) show the TEM images of the evolution of the amorphous intermediate in the initial stage of L-glutamic acid phase separation. Red arrow in (**a**) indicates smaller units with a diameter of 3 nm to 5 nm in diameter. Yellow arrow in (**d**) indicates compaction process. The figure is reproduced with copyright permission of ref. [47].

Overall, prenucleation clusters theory further considers prenucleation clusters generated by self-association of solute molecules, and the nucleation pathway for collisional aggregation of the aggregate cluster is formed. The theory further complements the mesoscale structure and behavior from the molecular scale to the crystal scale.

*2.4. Particles in Particle Agglomeration Theory*

Particle agglomeration nucleation is another nonclassical nucleation theory. When the supersaturation of the system is low and the number of nanoparticles generated before nucleation is large, the agglomeration nucleation of nanoparticles often occurs. The specific process is shown in Figure 1c. The substance first forms fine nanoparticles, and then nanoparticles aggregate to form larger and a stable crystal nucleus [48,49].

In recent years, particle agglomeration theory has gradually attracted the attention of researchers. However, due to the limitations of conventional characterization methods, there is rare understanding of the theory. The main challenge in studying agglomeration nucleation processes is that few PAT are available to monitor the process in real time. Fortunately, researchers have made many efforts to solve this challenge in terms of both computation and experiment. As shown in Figure 5, Baumgartner [50] used Cryo-TEM to observe the presence of particle agglomeration consisting of $Fe^{2+}$ and $Fe^{3+}$ with the size of about 2 nm before magnetite nucleation. This is in contrast to the formation of the amorphous phase through the attachment of prenucleation clusters in calcium carbonate and phosphate. Mirabello [51] demonstrated that the generation of magnetite was a nanoparticle self-assembly process. The crystal nuclei of magnetite were produced by the agglomeration of metastable primary particles and dehydration. Theoretical calculations showed that the agglomerated nucleation of the initial particles had a lower energy barrier than CNT. The Pt 3D structure reconstructed by Jungwon using electron tomography showed that the metal nuclei consisted of multiple crystalline domains, which resulted from the agglomeration of smaller nanoparticles during nucleation [52]. In addition, PATs have detected in real time the aggregation of nanoparticles in prenucleation solutions.

Polte [53] used in-situ small-angle X-ray scattering to monitor the formation of numerous particles with the size of about 0.8 nm in solution during the growth of gold nanocrystals. Subsequently, the number of nanoparticles decreased rapidly, and the size of the particles gradually became larger, which confirmed that the nucleation process of gold nanocrystals was formed by particle agglomeration.

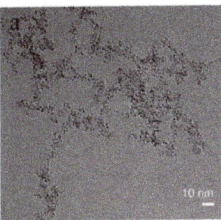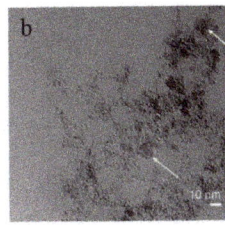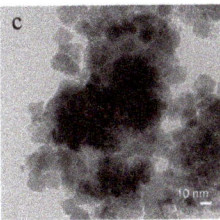

**Figure 5.** TEM images of magnetite particles aggregated to form nuclei: (**a**) 2 min, (**b**) 6 min, and (**c**) 82 min. Arrows in (**b**) indicate early formed crystalline magnetite nanoparticles. The figure is reproduced with copyright permission of ref. [50].

Compared with the prenucleation cluster theory, the particle agglomeration nucleation theory considers the existence of precursors such as nanoparticles and colloidal particles for the first time, and more specifically considers the nucleation modes at the mesoscale, such as directional attachment, self-assembly, and polymerization from a kinetic perspective, which further enriches the mesoscale structure and behavior.

## 3. Mesoscale Structure during Growth

With the development of microscopic research methods, researchers have gone deeper into the crystal growth theory. The growth unit in the crystal growth process has expanded from atomic, molecular and ions to mesoscale intermediate states such as amorphous precursors and nanoparticles. The intermediate states are essential for crystal growth.

### 3.1. Units in Classical Growth Theory

Since the early 20th century, researchers have tried various experimental methods and numerical analysis methods to reveal the crystal growth mechanism at the atomic, molecular and ionic levels, and summarized many classical growth theories, such as diffusion control growth theory, two-dimensional nucleation growth theory, and dislocation growth theory. The classical crystal growth theory considers that the basic unit of crystal growth is atoms, molecules or ions, and during the growth process the units are transferred and attached from solution to solid crystal surface one by one.

The diffusion control growth theory means that the interface structure is a rough interface, which is equivalent to accepting molecules deposited from the liquid phase, and the advancement of the interface is mainly due to the random and continuous attachment of molecules on the interface [54]. Molecules diffused from the liquid phase are easily attached to the crystal, and the crystal grows much easier than the smooth surface. The rate at which solute molecules are transported from the supersaturated solution to the crystal surface determines the crystal growth rate [55]. Wang investigated the effect of protein on calcium phosphate growth based on diffusion-controlled growth theory. The addition of protein increased the viscosity of the solution and increased the diffusion resistance of calcium and phosphorus ions, leading to a decline in growth rate of calcium phosphate [56].

Two-dimensional nucleation growth theory means that after the nucleus is formed, the atoms in solution are attracted to a certain lattice surface of the nucleus at first and aggregate with each other to form a stable atomic layer with sufficient area to form a two-dimensional crystal island raised on the lattice surface, which is called a two-dimensional nucleus. Then the atoms in solution can spontaneously accumulate outwards at the concave corners of the platform's periphery until a complete layer is formed. The above process is repeated continuously on the new lattice plane, so that it grows into a complete crystal layer by layer, as shown in Figure 6 In the 1920s, Volmer [57] first proposed the two-dimensional nucleation growth theory. Burton, Cabrera and Frank [58] elaborated on the mechanism of two-dimensional nucleation, concluding that the formation of a nucleus exceeding the critical radius required large activation energy, so it is unfeasible to grow through two-dimensional nucleation at low supersaturation. Mallink [59] used atomic force microscopy (AFM) to study the growth mechanisms of more than 30 types of thaumatin, peroxidases, egg lysozyme and chromosomal mosaic viruses. The results showed that the crystal growth was consistent with two-dimensional nucleation growth under the condition of low or high supersaturation. This result is contrary to the conclusion that two-dimensional nucleation at low supersaturation was difficult to achieve. Furthermore, Nozawa [60] demonstrated by single-particle resolution that the dominant growth mechanism in colloidal crystallization was the two-dimensional nucleation growth mechanism.

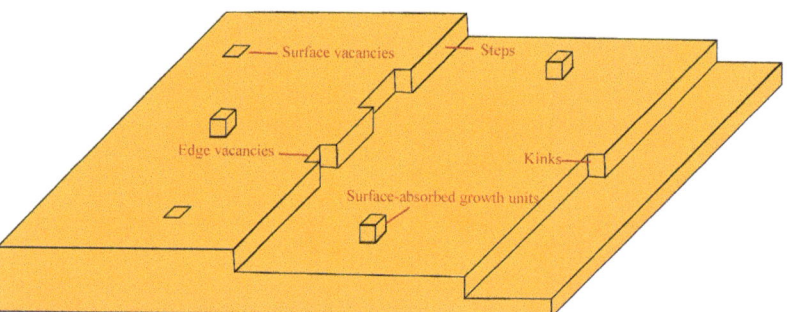

**Figure 6.** Schematic diagram of two-dimensional nucleation growth theory.

In the actual crystal growth, some researchers have observed that crystals can often grow far below the critical supersaturation required for two-dimensional nucleation growth, and this growth mechanism is dislocation growth theory. The screw dislocation mechanism points out that the crystal is not complete, and there are screw dislocation outcrops on the crystal surface, which can be used as a step source for crystal growth. During the crystal growth process, the growth steps spirally diffuse around the screw dislocation lines and never disappear [61]. Therefore, crystal growth no longer requires two-dimensional nucleation and can grow at low supersaturation. Chernov [62] combined the basic parameters of the actual step motion and the bulk diffusion model into the basic idea of screw dislocation growth, which better generalized the screw dislocation growth theory. At first, the contribution of defects other than screw dislocations to crystal growth was not considered in crystal growth until Bauser [63] observed that edge dislocations can also provide a step source for crystal growth. Min [64] analyzed the distortion of the lattice planes caused by different types of dislocations and the molecular configuration adjacent to the surface outcrop, and proposed that if the surface intersects with the dislocation line and was not in the crystal band with the dislocation's Berger vector as the axis, the never-disappearing step will exist at the outcrop of the surface dislocation, regardless of the orientation of the dislocation line. Therefore, whether it is the screw dislocation, the edge dislocation, or the mixed dislocation, their contributions to crystal growth are exactly the same, and they all provide a never-disappearing step for crystal growth, which makes the dislocation growth theory continuously improve.

In the classical growth theory, crystal growth is the process in which particles such as atoms, molecules, and ions in supersaturated solution deposit and grow on the formed crystal nucleus, which involves fewer mesoscale structures. With further exploration, many researchers have found that the growth of crystals with special morphologies involves mesoscale structures such as clusters and nanoparticles, which cannot be explained by classical crystal growth theories.

## 3.2. Amorphous Precursors in Amorphous Precursor Growth Theory

Since amorphous precursors have good plasticity and can form crystals with various morphologies, the formation and transformation of amorphous precursors are often the first step in the nonclassical crystallization process, as shown in Figure 7a. Amorphous particles and liquid droplets are common amorphous precursors with mesoscale structures whose size and structure are between those of solutes and crystals. According to Ostwald ripening principle, the more unstable the formed substance and the lower the degree of modification, the lower the activation energy barrier required. Therefore, in highly supersaturated solutions, amorphous precursors are preferentially formed, and the higher the degree of supersaturation, the more amorphous clusters are formed.

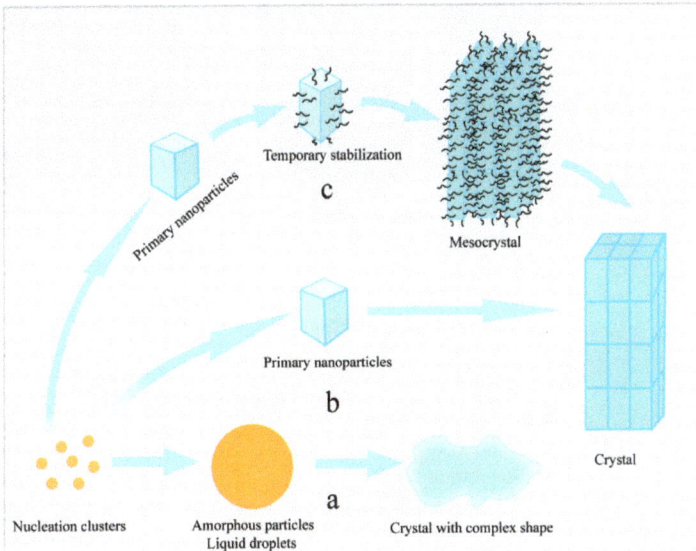

**Figure 7.** Schematic diagram of (**a**) amorphous precursor growth theory, (**b**) oriented attachment growth theory, (**c**) mesocrystal growth theory.

Amorphous precursors have transient characteristics and instability, making them difficult to detect [65]. However, by applying corresponding analytical methods, the changes of ions and supersaturation in the solution can be monitored in situ, and the redissolution and mesoscopic transformation of amorphous precursors can be distinguished. In addition, amorphous precursors can also be characterized by transmission electron microscopy, scanning electron microscopy, selected-area electron diffraction, and X-ray diffraction. Yang [66] observed the nonclassical formation process of nickel nanocrystals using high-resolution TEM. At the beginning of the reaction, amorphous particles were formed due to the induction of surface interactions, and crystalline domains appeared and gradually grew into nanocrystals. This amorphous-precursor-mediated nonclassical growth mechanism had significant meaning for the formation of nanocrystals. Vuk [67] demonstrated that the growth patterns on different crystal planes were different during the growth of hydroxyapatite (HAp). Figure 8 showed that the crystal growth on the (100) crystal plane was the classical growth process, in which atoms and ions were continuously accumulated to form crystals. The crystal growth on the (001) crystal plane was the nonclassical growth process, and the spherical amorphous nanoparticles were continuously accumulated to form single crystals.

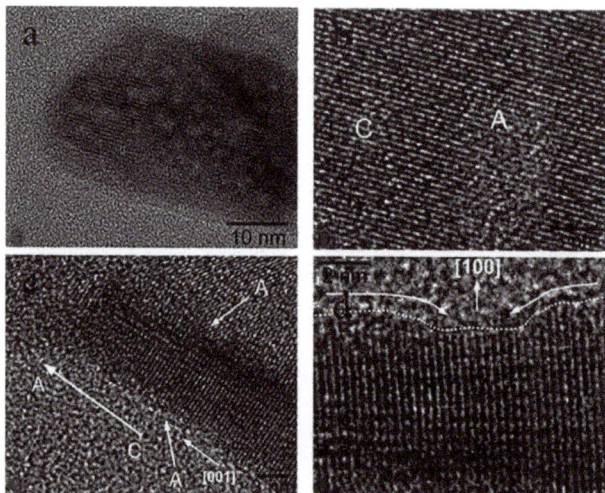

**Figure 8.** TEM images: (**a**) amorphous spherical units attached to grow rod-shaped HAp nanoparticles; (**b**) the interface between amorphous spheroid unit and nanoparticle; (**c**) [001] crystallographic direction and the amorphous coating around the crystalline particle interior; and (**d**) steps formed at the growth front of an HAp nanoparticle in the (**a**). Arrows in (**c**) indicate the growth direction. The figure is reproduced with copyright permission of ref. [67].

Droplets are disordered and liquid amorphous precursors, around 100 nm in size. It appears as small droplets under the optical microscope, and then grows into crystals through aggregation and fusion. Liquid precursors do not have a specific ionic composition but can be neutralized by inclusion of ternary counterions and additives themselves. The composition of the liquid precursors is intimately associated with the reaction conditions, such as the concentration of the reagents, the reaction temperature, the order of adding the reagents and the way of mixing. For example, $CaCO_3$ liquid precursors present two forms of over-carbonated and under-carbonated, which means that they need to absorb specific ions from the external environment to complete the crystallization. DiMasi [68] discovered that carbonate transport was the decisive step in under-carbonated $CaCO_3$ liquid precursor transformation.

*3.3. Nanoparticles in Particle Attachment Growth Theory*

The particle attachment growth theory is a well-recognized nonclassical growth theory. Different from the classical growth theory, the particle attached growth theory extends the traditional crystal growth unit from atoms, ions or molecules to clusters and nanoparticles with mesoscale structures. It is pointed out that the crystal growth is the alignment of specific crystal planes through lattice matching between nanoparticles to obtain a highly ordered assembled superstructure, and finally fuse to grow into complete crystals. The particle attachment growth theory mainly includes oriented attachment and random attachment.

Oriented attachment growth is a process of self-assembly growth between adjacent nanoparticles. As shown in Figure 7b, during the crystal growth process, nanoparticles are continuously contacted, separated and repeatedly aligned before attachment to match the lattice orientation, and then align along in the same direction to form a larger crystal. Nanoparticles tend to attach through the crystal face with the highest surface energy or surface area, minimizing the total energy between the two particles. Penn and Banfield [69,70] used high-resolution TEM to observe the assembly and growth of smaller $TiO_2$ nanoparticles in the process of hydrothermal synthesis of $TiO_2$ nanocrystals, revealing this nonclassical oriented attachment growth mechanism for the first time. Subsequently, numerous studies have found that inorganic crystals such as $Co_3O_4$ [71], $CaCO_3$ [72], ZnS [73,74] generally have an oriented attachment growth during the growth process. Cho [75] considered that the electric dipole moments in the {100}, {110}, and {111} orientations of octahedral PbSe nanocrystals could facilitate the occurrence of oriented attachment growth. Pacholski [76] reported that hemispherical ZnO nanoparticles formed high-quality single-crystal ZnO nanorods by the oriented attachment growth mechanism.

In recent years, the development of in-situ TEM has further revealed the mechanism of the oriented attachment growth process. Cao [77] investigated the attachment growth process of ZnO nanoparticles in ethanol solution by a series of methods such as transmission electron microscopy. As shown in Figure 9, ZnO nuclei were formed first, and the nuclei gradually grew into ZnO nanoparticles, and then the surface of the nanoparticles became rough, and the nanoparticles with rough surfaces underwent oriented attachment to form complete ZnO crystals. Zhu [78] used in-situ liquid cell TEM to observe that ligand interactions on the surface of gold nanoparticles induce the growth of nanoparticles by oriented attachment. The organic ligands on the Au surface after the addition of citric acid could guide the rotation of the particles, and fusion occurred only when the {111} facets of the two particles were completely aligned. First-principles calculations confirmed that the preferential attachment of nanoparticles at the {111} facets was due to the lower ligand binding energy of these facets. Oriented attachment growth can control the size, shape and microstructure of nanoparticles and prepare artificial nanomaterials with anisotropic shapes, which will gain a lot of attention in the coming years.

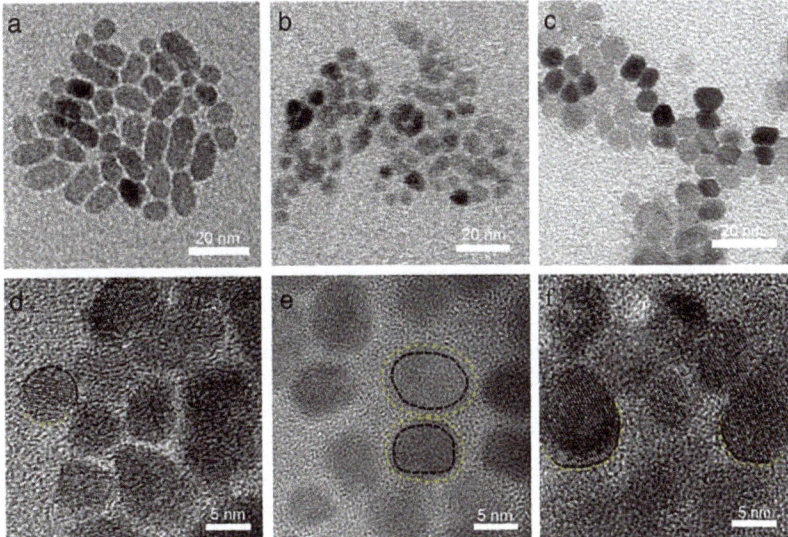

**Figure 9.** TEM images of ZnO nanoparticles at various growth stages. (**a**,**d**) Primary nucleation. (**b**,**e**) flake-like aggregates. (**c**,**f**) Well-crystallized particles. The figure is reproduced with copyright permission of ref. [77].

Random attachment growth is another way of particle attachment growth. In this case, nanoparticles with crystalline structures first form polycrystals by random attachment growth, and then achieve crystals reorientation through rotation and alignment, and finally form single crystals through internal structural reorganization or external forces. The random attachment growth theory further reveals the crystal growth mechanism and enriches the crystal growth theory. Hu [79] synthesized CdS colloidal spheres by the controllable solvothermal method and investigated their growth mechanism. The TEM results showed that small nanocrystalline nuclei were first formed, and the nanocrystalline nuclei continued to aggregate to nanoparticles, and finally these nanoparticles grew by random attachment to form CdS colloidal spheres. Liu [80] also observed the random attachment growth process during the assembly of vaterite nanocrystals. As shown in Figure 10, during the attachment growth of nanocrystals to the main crystal, the lattices between their interfaces were not matched, and the surface layer of vaterite nanoparticles with random orientations was formed around the main crystal through random attachment growth. Polycrystalline vaterite rapidly transformed into single crystals under pressure. There are still many questions in the study of random attachment growth theory. The current research mainly focuses on the experimental observation stage, and the intrinsic mechanism of random attachment growth and its difference from oriented attachment growth are the future research directions.

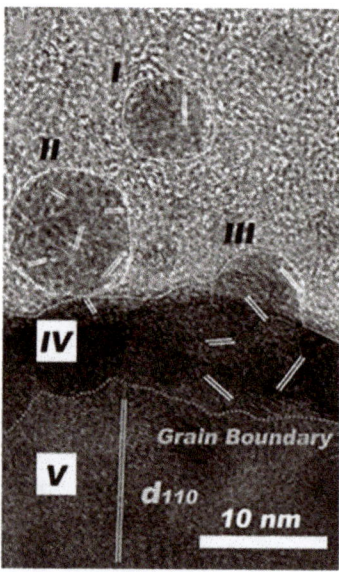

**Figure 10.** The process of random attachment of nanocrystals to the surface of vaterite. (I) Individual nanoparticle; (II) initial attachment of a nanocrystal without lattice matching; (III) later stage of attachment of a nanocrystal with the spindle vaterite surface without lattice matching; (IV) nonoriented surface layer; (V) oriented bulk. The figure is reproduced with copyright permission of ref. [80].

The major difference between particle attachment growth theory and classical growth theory is that the basic unit of crystal growth is nanoparticles with mesoscale structure, and nanoparticles grow into macroscopical crystals through oriented attachment or random attachment on the mesoscale. As an effective complement to the classical crystal growth theory, the particle attachment growth theory provides more possibilities for the functional development of novel nanomaterials.

*3.4. Mesocrystals in Mesocrystal Growth Theory*

Mesocrystals are the mesoscale intermediate between dispersed particle and single crystal with a size between 1–1000 nm [81,82]. The mesocrystal growth process is described as shown in Figure 7c: the crystal nuclei grow into primary nanocrystals through the Ostwald ripening process, and then primary nanocrystals are assembled into mesocrystals in three-dimensional directions through ordered arrangement and lattice interconnection, and finally mesocrystals evolve into a single crystal through the fusion and recrystallization process between the internal crystal planes [83]. Mesocrystal growth is also considered as a nonclassical growth process.

Mesocrystals are assembled from many primary nanocrystals, which provides more ideas for the design of nanostructures. Mesocrystal growth theory can not only be used to guide the preparation of single crystals with complex structures, but also superstructures with nanoparticles as basic units. O'Brien [84] considered that mesocrystals were in the metastable state rather than in the thermodynamically stable state and could transform to other crystal forms under certain conditions. The growth rate and equilibrium morphology of mesocrystals depend on the stability of the nanocrystals and the long-range forces on the vectors between the assembled units, and the nanocrystals must have sufficient time for ordered self-assembly rather than random binding. As shown in Figure 11, Zhang [85] obtained the CuO mesocrystals structure by oriented attachment growth of CuO nanoparticles in three-dimensional directions. The entire structure consisted of several hundred primary CuO nanoparticles whose orientations were crystallographically consistent. As the basic units that make up mesocrystals are aligned in the same direction, mesoscopic crystals tend to have similar electron diffraction patterns to single crystals. Chen [86] successfully prepared α-calcium sulfate hemihydrate spherical mesocrystals in an ethylene glycol-water system and analyzed the growth mechanism by ultramicroscopic technique. The results showed that α-calcium sulfate hemihydrate spherical mesocrystals were solid-state mesocrystals composed of nanorods oriented along the longitudinal axis, and the nanorods were formed by the continuous arrangement of irregular subunits guided by EDTA. Meanwhile, the mesocrystal growth mechanism could also be applied to the design of complex functional materials.

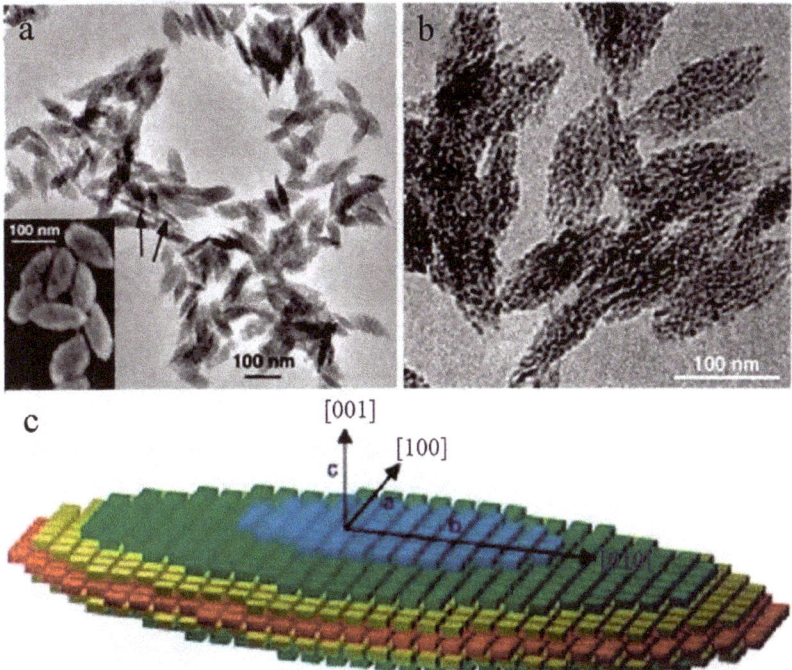

**Figure 11.** (a) An overview TEM image, (b) A HRTEM mage of CuO mesocrystals, (c) Schematic illustration of CuO mesocrystal built from aggregated nanoparticles. Arrows in (a) indicate ellipsoidal particles lying on their sides. The figure is reproduced with copyright permission of ref. [85].

Mesocrystals are widely used in the fields of photocatalysis, optoelectronics, lithium-ion batteries and biological materials due to their unique structural characteristics [87–89]. Moreover, mesocrystal growth is not limited to biomineralized systems such as sea urchins and pearl oysters but has also been observed in other reactions such as kinetic metastable systems or without crystallization additives. Tang [87] invented a method to prepare $TiO_2$ mesocrystals by controlling the hydrolysis of $TiCl_3$ in polyethylene glycol. The ultrafine $TiO_2$ nanocrystals formed by the reaction were stabilized by polyethylene glycol molecules and formed nanoparticle aggregates through oriented attachment. When $TiO_2$ nanocrystals in the solution were completely attached to the aggregates, spindle-shaped $TiO_2$ mesocrystals were formed. The synthesized $TiO_2$ mesocrystals were characterized by high crystallinity and high porosity, which enabled them to exhibit stronger visible light activity for photocatalytic NO removal. Ding [88] used a solvothermal method to prepare $Li_2FeSiO_4$ mesocrystals, a cathode material for lithium-ion batteries with excellent properties. $Li_2FeSiO_4$ mesocrystals consisted of nanoparticles along [001] zone axis, with the oriented single-crystal structure and highly exposed (001) planes. $Li_2FeSiO_4$ mesocrystals showed near-theoretical discharge capacity, excellent rate capacity, and favorable cycling stability. Sascha [89] proposed that the size of nickel hexacyanoferrate nanoparticles could be regulated by controlling the citrate concentration and total supersaturation, enabling them to orientate-attach to form nickel hexacyanoferrate colloidal mesocrystals. As shown in Figure 12, nickel hexacyanoferrate nanoparticles with cubic structures as building units were continuously assembled in an orderly manner along the same crystallographic direction to form highly catalytic active colloidal mesocrystals, which could efficiently treat caffeine in wastewater.

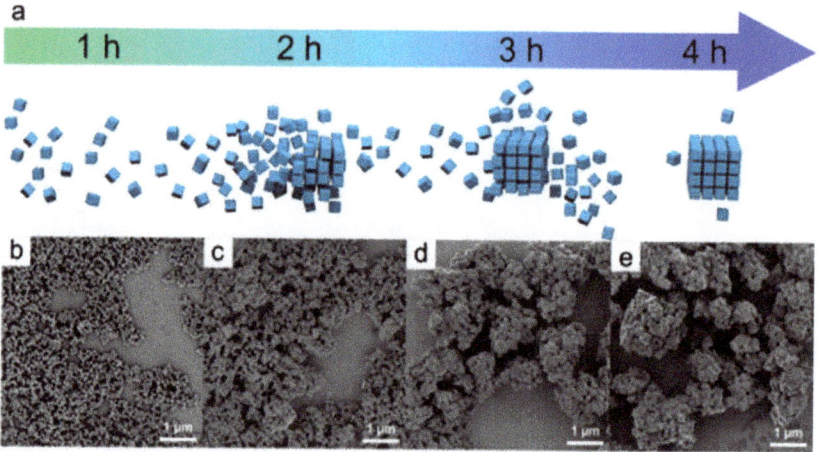

**Figure 12.** (a) Schematic illustration of the time-dependent alignment of nickel hexacyanoferrate nanoparticles. SEM image of the nickel hexacyanoferrate mesocrystals formation. (b) 1 h, (c) 2 h, (d) 3 h, (e) 4 h. The figure is reproduced with copyright permission of ref. [89].

At present, the mesocrystal growth theory has been confirmed in many systems, such as $BaSO_4$, $CaCO_3$, zeolite, $TiO_2$ [90–92]. However, the mechanism of mesocrystal formation and growth is still not fully understood. This is because mesocrystals as intermediate products have a short lifetime and are not easy to detect, and there is no ideal method for in-situ observation in solution. For example, atomic force microscopy (AFM) can only work on a certain surface. If the surface does not exist in solution, it cannot be used for the study of mesocrystals. Transmission electron microscopy can only observe the structure of larger mesocrystals, but the formation process of mesocrystals cannot be observed. Continuing in-depth research on mesocrystal growth has important theoretical and applied value for the preparation of nanostructured crystals and the development of materials with novel properties.

## 4. EMMS Model and Crystal Nucleation and Growth

As we study complex systems in science and engineering, we generally need to understand them from the macroscopic scale, and then gradually study the various mechanisms at the microscopic scale, and gradually try to establish the relationship between the two scales. However, it is very difficult to establish this correlation directly. Complex systems mostly exhibit different layers, and each layer has multi-scale structures [93]. Although the multi-scale structures at different levels are different, they all have a common property that the behavior of the system at the boundary scale of each level is relatively simple and easy to characterize and analyze. However, at the scales between them, the behavior of the system is mostly very complex, and these scales are called mesoscales. Within the mesoscale, there exists a characteristic structure, namely the mesoscale structure. Since the mesoscale structure is spatiotemporally dynamic in space and time, which makes it difficult to correlate the microscale with the macroscale, it can only be analyzed by means of averaging. Research has gradually shown that there may be a universal dominant principle at the mesoscale that the coordination of different control mechanisms in competition. For this reason, the interdisciplinary concept of mesoscience is proposed to address various issues [94].

Since the late 20th century, researchers have studied mesoscale problems in reactors [95]. Starting from the phenomenon of particle agglomeration in the gas–solid system, it is believed that the formation of mesoscale agglomeration originated from the coordination of the respective motion trends of gas and particles in the competition, thus establishing the stability condition of the mesoscale agglomeration structure and proposing the theory of energy minimization multi-scale (EMMS) [96]. The EMMS model solved the problem of quantitative simulation of heterogeneous gas–solid systems, and significantly improved the prediction performance of gas–solid two-phase computational fluid dynamics and the ability to solve practical problems. Later, the EMMS principle was extended and applied to other multiphase systems [97].

The EMMS model contains eight variables to describe the gas–solid two-phase flow system, and these eight variables are constrained by six dynamic equations [98].

Dilute-phase force balance equation: the effective load-bearing capacity of the dilute phase particles per unit volume is equal to the drag force of the dilute phase particles by the surrounding dilute phase gas.

$$\frac{3}{4} C_{Df} \frac{1-\varepsilon_f}{d_p} \rho_g U_{sf}^2 - (1-\varepsilon_f)(\rho_p - \rho_g) g = 0 \qquad (1)$$

where $C_{Df}$ is the drag coefficient of the dilute phase, $\varepsilon_f$ is the void fraction of the dilute phase, $d_p$ is the particle diameter, $\rho_g$ is the density of the gas phase, $U_{sf}$ is the apparent slip velocity of the dilute phase, and $\rho_p$ is the density of the particle.

Dense-phase force balance equation: the effective load-bearing capacity of dense phase particles per unit volume is equal to the drag force of dense phase particles by the surrounding dense phase gas and the drag force of agglomerates by the surrounding dilute phase gas.

$$\frac{3}{4} C_{Dc} \frac{1-\varepsilon_c}{d_p} \rho_g U_{sc}^2 + \frac{3}{4} C_{Di} \frac{1}{d_{cl}} \rho_g U_{si}^2 - (1-\varepsilon_c)(\rho_p - \rho_g) g = 0 \qquad (2)$$

where $C_{Dc}$ is the drag coefficient of the dense phase, $C_{Di}$ is the interphase drag coefficient, $\varepsilon_c$ is the void fraction of the dense phase, $d_{cl}$ is the diameter of the particle agglomeration, $U_{sc}$ is the apparent slip velocity of the dense phase, and $U_{si}$ is the interphase apparent slip velocity.

Interphase pressure drop balance equation: The fluid maintains the pressure drop balance of the system itself through the drag force acting on the dense two-phase particles, and the dense phase pressure drop is equal to the dilute phase pressure drop and the interphase pressure drop caused by the surrounding agglomerates.

$$C_{Df}\frac{1-\varepsilon_f}{d_p}\rho_g U_{sf}^2 + \frac{1}{1-f}C_{Di}\frac{f}{d_{cl}}\rho_g U_{si}^2 = C_{Dc}\frac{1-\varepsilon_c}{d_p}\rho_g U_{sc}^2 \tag{3}$$

where $f$ is the dense phase fraction.

Fluid continuity equation: The flow rate of the fluid flowing through the entire section is equal to the sum of the fluid flow rates in the dense and dense phases.

$$U_g = (1-f)U_f + fU_c \tag{4}$$

where $U_g$ is the superficial gas velocity of the gas phase, $U_f$ is the superficial gas velocity of the dilute phase, and $U_c$ is the superficial gas velocity of the dense phase.

Particle continuity equation: The particle phase flow through the entire section is equal to the sum of the particle mass flow in the dilute and dense phases.

$$U_p = (1-f)U_{pf} + fU_{pc} \tag{5}$$

where $U_p$ is the superficial gas velocity of the particle phase, $U_{pf}$ is the superficial gas velocity of the particles in the dilute phase, and $U_{pc}$ is the superficial gas velocity of the particles in the dense phase.

Agglomerate size equation: This correlation assumes that the agglomerate size is inversely proportional to the system input energy.

$$d_{cl} = d_p \cdot \frac{\frac{U_p g}{1-\varepsilon_{max}} - \left(U_{mf} + \frac{\varepsilon_{mf}}{1-\varepsilon_{mf}}U_p\right)g}{N_{st} \cdot \frac{\rho_p}{\rho_p - \rho_g} - \left(U_{mf} + \frac{\varepsilon_{mf}}{1-\varepsilon_{mf}}U_p\right)g} \tag{6}$$

where $U_{mf}$ is the superficial gas velocity of minimum fluidization, $\varepsilon_{mf}$ is the minimum fluidized void fraction, $\varepsilon_{max}$ is the maximum non-uniform void fraction, and $N_{st}$ is the suspended pumping energy consumption per unit mass.

Stability conditions: The stability condition is expressed by the compromise of two dominant mechanisms in competition.

$$N_{st} = \frac{W_{st}}{(1-\varepsilon_g)\rho_p} = \frac{\rho_p - \rho_g}{\rho_p} \cdot g \cdot \left(U_g - \frac{f(1-f)\left(\varepsilon_f - \varepsilon_g\right)}{1-\varepsilon_g} \cdot U_f\right) \rightarrow min \tag{7}$$

where $\varepsilon_g$ is the void fraction of the gas phase.

Due to the nonlinear nature of the equations, it remains a challenge to directly find the analytical solution.

The characteristic of the EMMS model is the description of the non-uniform structure. The multi-scale structure is firstly found in a complex system, and the complex system is regarded as being composed of internal multi-scale structures and their interactions. Then the characteristic equations of the system are constructed, the main motion modes in the system are analyzed and the equations are constructed by the conservation equations. Finally, the stability of the whole system is achieved by the mutual coordination between control mechanisms of different parts to solve the closed equation system [99]. The EMMS model has been developed into a universal EMMS principle for multiple systems, which mainly considers the influence of the mesoscale structure between the whole system and its constituent units on the system behavior, and thus builds a theoretical framework for the development of mesoscale science. The core idea is the mesoscale structure and the compromise in competition of multiple mechanisms [100]. When the system involves two dominant mechanisms, A and B, with different motion trends, the system will present a relatively uniform structure under a single dominant mechanism (A or B). When both A and B cannot dominate and must compromise with each other, a dynamic inhomogeneous structure will appear in the system [101–103]. Additionally, it has been proved that the EMMS model can be applied to the mesoscale behavior of gas–solid, liquid–solid, and other multiphase systems [104,105]. In recent years, the EMMS principle has been further extended to the mesoscale study of proteins, heterogeneous catalysis systems and crystal growth, and some progress has been made [106–108].

Although there are few practical applications of EMMS principle in the crystallization process, the intermediates in the nonclassical crystallization process are typical mesoscale structures, so the EMMS principle can also be used in the crystallization process theoretically. For example, in the crystallization of calcium carbonate, the diffusion and reaction of chemicals both dominate the structure of products. When diffusion controls the crystallization process, particles with a branched structure are produced. When crystallization is controlled by reaction, spherical particles are formed. When the compromise between diffusion and reaction dominates the process, snowflake-like particles are formed [109]. The traditional crystallization research paradigms are based on the crystallization theory. By considering the mass and heat transfer during the crystallization process, the removal of the crystallization heat and the sufficient mixing of the system are realized. However, because of the inability to maintain the consistency of the process between the crystallization theory and the complex system, the crystallization theory cannot guide the actual production process accurately. Additionally, the traditional crystallization research paradigm only focuses on the unit scale and system scale at each level and lacks the understanding of the mesoscale structure. Therefore, the theory of crystal nucleation and growth can be supplemented by the EMMS method of global distribution, local simulation, and detailed evolution. The precursor and solution system are disassembled into precursor dense phase, precursor dilute phase and interaction phase. The surface energy and volume energy of the three subsystems are calculated according to the stability conditions, so as to obtain the critical sizes of various precursors. For the nonclassical crystallization process, the existence of the precursor dense phase is essentially considered. It is possible to infer other nucleation and growth pathways and process mechanisms by considering the compromise and competition between the precursor dilute phase and the interaction phase. According to the EMMS principle, the dominant roles of ordering and disordering in the mesoscale to reach compromise in competition during nucleation should be studied. Additionally, the competition and compromise between the two dominant mechanisms of diffusion and reaction during crystal growth should also be studied [110]. The EMMS model is the mathematical modeling method at the mesoscale, which cannot be used in classical nucleation and growth processes. However, EMMS models can be applied to a variety of nonclassical nucleation and growth processes with mesoscale structures, such as two-step nucleation theory, prenucleation clusters theory, particle agglomeration theory, amorphous precursor growth theory, particle attachment growth theory and mesocrystal growth theory. The mathematical models at mesoscale have been preliminarily established in the existing nonclassical crystallization theories, but these models are still not perfect. The mathematical modeling method of EMMS at mesoscale can further modify and supplement the existing theories and models of nucleation and growth, and it provides new ideas for researchers to use the EMMS theory to study the crystallization process furtherly.

## 5. Conclusions and Outlook

An increasing number of studies have confirmed nonclassical crystallization pathways on the mesoscale. Precursors with mesoscale structures during nucleation and growth can deviate the crystallization behavior from the predictions of classical nucleation theoretical models. Therefore, it is urgent to modify and refine crystal nucleation and growth theories. In this paper, a series of nonclassical crystallization phenomena discovered in recent years and the growth paths based on experiments and simulation calculations are reviewed. The two-step nucleation theory, prenucleation clusters theory, particle agglomeration theory, amorphous precursor growth theory, particle attachment growth theory and mesocrystal growth theory involving mesoscale precursors are summarized, and the research paradigm of the EMMS model based on mesoscale science to guide the solution crystallization process is presented.

With the rapid development of PAT, computational simulation and other technologies, the research on nonclassical crystallization theory has been effectively promoted. In-situ monitoring of mesoscale crystallization processes can be achieved by in-situ liquid TEM, and the formation of various intermediate structures between microscale and macroscale, including prenucleation clusters, nanoparticles, and mesocrystals, has changed the classical crystallization point. Nonclassical crystallization is not a process of basic monomer attachment, but a process of precursor assembly and aggregation at the mesoscale. According to the idea of constructing a mathematical model based on the EMMS principle, virtual process engineering combined with mesoscale science, supercomputing and machine learning may become a brand-new crystallization process research model. On the basis of the traditional crystal research and the development of the model, virtual process engineering can combine basic disciplines with traditional chemical theory, mesoscale science, algorithm software, supercomputing and virtual reality. The physicochemical changes in the whole crystallization process can be reproduced accurately and completely [111,112]. Meanwhile, the algorithm software is used to simulate the mesoscale crystallization process. Based on supercomputing, the visualization of the whole process design and amplification are realized. Additionally, the crystallization theoretical system with multi-level, multi-scale properties and mesoscale complexity will be formed, which has broad application value in the fields of crystal engineering, crystallization separation and material sciences.

**Author Contributions:** Conceptualization, X.W. and K.L.; methodology, X.W. and K.L.; validation, X.W.; formal analysis, K.L.; investigation, X.Q.; resources, M.L.; data curation, Y.L.; writing—original draft preparation, X.W.; writing—review and editing, K.L. and M.C.; visualization, M.C. and J.O.; supervision, Y.A., W.Y., M.C., J.O. and J.G.; project administration, Y.A., W.Y., M.C., J.O. and J.G.; funding acquisition, Y.A., W.Y., M.C., J.O. and J.G. All authors have read and agreed to the published version of the manuscript.

**Funding:** This work was financially supported by the National Science Foundation of China 22108195, Key R&D Program of Zhejiang (2022C01208), the key project of State Key Laboratory of Chemical Engineering (SKL-ChE-20Z03), the Chemistry and Chemical Engineering Guangdong Laboratory (Grant No. 1912014) and the financial support of Haihe Laboratory of Sustainable Chemical Transformations, Academic and Technical Leader Training Program for Major Disciplines in Jiangxi Province (20212BCJ23001).

**Institutional Review Board Statement:** Not applicable.

**Informed Consent Statement:** Not applicable.

**Data Availability Statement:** Not applicable.

**Acknowledgments:** Kangli Li thanks the Shaoxing People's Government for a supporting of her post-doctoral research. This work was financially supported by the Institute of Shaoxing.

**Conflicts of Interest:** The authors declare no conflict of interest.

## References

1. Kato, S.; Furukawa, S.; Aoki, D.; Goseki, R.; Oikawa, K.; Tsuchiya, K.; Shimada, N.; Maruyama, A.; Numata, K.; Otsuka, H. Crystallization-induced mechanofluorescence for visualization of polymer crystallization. *Nat. Commun.* **2021**, *12*, 126. [CrossRef] [PubMed]
2. Zhang, J.; Fang, Y.; Zhao, W.; Han, R.; Wen, J.; Liu, S. Molten-Salt-Assisted $CsPbI_3$ Perovskite Crystallization for Nearly 20%-Efficiency Solar Cells. *Adv. Mater.* **2021**, *33*, 2103770. [CrossRef] [PubMed]
3. Mu, X.; Wu, F.; Tang, Y.; Wang, R.; Li, Y.; Li, K.; Li, C.; Lu, Y.; Zhou, X.; Li, Z. Boost photothermal theranostics via self-assembly-induced crystallization (SAIC). *Aggregate* **2022**, e170, 2692–4560. [CrossRef]
4. Cui, W.; He, Z.; Zhang, Y.; Fan, Q.; Feng, N. Naringenin Cocrystals Prepared by Solution Crystallization Method for Improving Bioavailability and Anti-hyperlipidemia Effects. *AAPS PharmSciTech* **2019**, *20*, 115. [CrossRef] [PubMed]
5. Wei, Z.; Zhang, Q.; Li, X. Crystallization Kinetics of α-Hemihydrate Gypsum Prepared by Hydrothermal Method in Atmospheric Salt Solution Medium. *Crystals* **2021**, *11*, 843. [CrossRef]

6. Frenkel, J. A general theory of heterophase fluctuations and pretransition phenomena. *J. Chem. Phys.* **1939**, *7*, 538–547. [CrossRef]
7. Volmer, M. Nucleus formation in supersaturated systems. *Z. Phys. Chem.* **1926**, *119*, 277–301. [CrossRef]
8. Santra, M.; Singh, R.S.; Bagchi, B. Polymorph selection during crystallization of a model colloidal fluid with a free energy landscape containing a metastable solid. *Phys. Rev. E* **2018**, *98*, 032606. [CrossRef]
9. Janković, B.; Marinović-Cincović, M.; Dramićanin, M. Kinetic study of isothermal crystallization process of $Gd_2Ti_2O_7$ pre-cursor's powder prepared through the Pechini synthetic approach. *J. Phys. Chem. Sol.* **2015**, *85*, 160–172. [CrossRef]
10. Chang, H.; Bootharaju, M.S.; Lee, S.; Kim, J.H.; Kim, B.H.; Hyeon, T. To inorganic nanoparticles via nanoclusters: Nonclassical nucleation and growth pathway. *Bull. Korean Chem. Soc.* **2021**, *42*, 1386–1399. [CrossRef]
11. Houben, L.; Weissman, H.; Wolf, S.G.; Rybtchinski, B. A mechanism of ferritin crystallization revealed by cryo-STEM tomography. *Nature* **2020**, *579*, 540–543. [CrossRef] [PubMed]
12. Wiedenbeck, E.; Kovermann, M.; Gebauer, D.; Cölfen, H. Liquid metastable precursors of ibuprofen as aqueous nucleation intermediates. *Angew. Chem. Int. Ed.* **2019**, *58*, 19103–19109. [CrossRef] [PubMed]
13. Li, J.; Kwauk, M. Exploring complex systems in chemical engineering—the multi-scale methodology. *Chem. Eng. Sci.* **2003**, *58*, 521–535. [CrossRef]
14. Huang, W.; Li, J.; Edwards, P.P. Mesoscience: Exploring the common principle at mesoscales. *Natl. Sci. Rev.* **2018**, *5*, 27–32. [CrossRef]
15. Grossmann, L.; King, B.T.; Reichlmaier, S.; Hartmann, N.; Rosen, J.; Heckl, W.M.; Björk, J.; Lackinger, M. On-surface photopoly-merization of two-dimensional polymers ordered on the mesoscale. *Nat. Chem.* **2021**, *13*, 730–736. [CrossRef]
16. Gibbs, J.W. On the equilibrium of heterogeneous substances. *Am. J. Sci.* **1878**, *3*, 441–458. [CrossRef]
17. Farkas, L. Keimbildungsgeschwindigkeit in übersättigten Dämpfen. *Z. Phys. Chem.* **1927**, *125*, 236–242. [CrossRef]
18. Becker, R.; Döring, W. Kinetische behandlung der keimbildung in übersättigten dämpfen. *Ann. Phys.* **1935**, *416*, 719–752. [CrossRef]
19. Li, C.; Liu, Z.; Goonetilleke, E.C.; Huang, X. Temperature-dependent kinetic pathways of heterogeneous ice nucleation competing between classical and non-classical nucleation. *Nat. Commun.* **2021**, *12*, 4954. [CrossRef]
20. Banner, D.J.; Firlar, E.; Rehak, P.; Phakatkar, A.H.; Foroozan, T.; Osborn, J.K.; Sorokina, L.V.; Narayanan, S.; Tahseen, T.; Baggia, Y. In Situ Liquid-Cell TEM Observation of Multiphase Classical and Nonclassical Nucleation of Calcium Oxalate. *Adv. Funct. Mater.* **2021**, *31*, 2007736. [CrossRef]
21. Katz, J.L. The critical supersaturations predicted by nucleation theory. *J. Stat. Phys.* **1970**, *2*, 137–146. [CrossRef]
22. Shore, J.D.; Perchak, D.; Shnidman, Y. Simulations of the nucleation of AgBr from solution. *J. Chem. Phys.* **2000**, *113*, 6276–6284. [CrossRef]
23. Adamski, T. Commination of crystal nucleation by a precipitation method. *Nature* **1963**, *197*, 894. [CrossRef]
24. Yau, S.-T.; Vekilov, P.G. Direct observation of nucleus structure and nucleation pathways in apoferritin crystallization. *J. Am. Chem. Soc.* **2001**, *123*, 1080–1089. [CrossRef]
25. Erdemir, D.; Lee, A.Y.; Myerson, A.S. Nucleation of crystals from solution: Classical and two-step models. *Acc. Chem. Res.* **2009**, *42*, 621–629. [CrossRef] [PubMed]
26. Davey, R.J.; Schroeder, S.L.; Ter Horst, J.H. Nucleation of organic crystals—a molecular perspective. *Angew. Chem. Int. Ed.* **2013**, *52*, 2166–2179. [CrossRef] [PubMed]
27. Chen, P.-Z.; Niu, L.-Y.; Zhang, H.; Chen, Y.-Z.; Yang, Q.-Z. Exploration of the two-step crystallization of organic micro/nano crystalline materials by fluorescence spectroscopy. *Mater. Chem. Front.* **2018**, *2*, 1323–1327. [CrossRef]
28. Fang, H.; Hagan, M.F.; Rogers, W.B. Two-step crystallization and solid–solid transitions in binary colloidal mixtures. *Proc. Natl. Acad. Sci. USA* **2020**, *117*, 27927–27933. [CrossRef] [PubMed]
29. ten Wolde, P.R.; Frenkel, D. Enhancement of protein crystal nucleation by critical density fluctuations. *Science* **1997**, *277*, 1975–1978. [CrossRef]
30. Duff, N.; Peters, B. Nucleation in a Potts lattice gas model of crystallization from solution. *J. Chem. Phys.* **2009**, *131*, 184101. [CrossRef]
31. Sun, X.; Garetz, B.A.; Myerson, A.S. Polarization switching of crystal structure in the nonphotochemical laser-induced nucleation of supersaturated aqueous l-histidine. *Cryst. Growth Des.* **2008**, *8*, 1720–1722. [CrossRef]
32. Chattopadhyay, S.; Erdemir, D.; Evans, J.M.; Ilavsky, J.; Amenitsch, H.; Segre, C.U.; Myerson, A.S. SAXS study of the nucleation of glycine crystals from a supersaturated solution. *Cryst. Growth Des.* **2005**, *5*, 523–527. [CrossRef]
33. Vekilov, P.G. The two-step mechanism of nucleation of crystals in solution. *Nanoscale* **2010**, *2*, 2346–2357. [CrossRef]
34. Harano, K.; Homma, T.; Niimi, Y.; Koshino, M.; Suenaga, K.; Leibler, L.; Nakamura, E. Heterogeneous nucleation of organic crystals mediated by single-molecule templates. *Nat. Mater.* **2012**, *11*, 877–881. [CrossRef]
35. Leunissen, M.E.; Christova, C.G.; Hynninen, A.P.; Royall, C.P.; Campbell, A.I.; Imhof, A.; Van Blaaderen, A. Ionic colloidal crystals of oppositely charged particles. *Nature* **2005**, *437*, 235–240. [CrossRef]
36. Savage, J.R.; Dinsmore, A.D. Experimental evidence for two-step nucleation in colloidal crystallization. *Phys. Rev. Lett.* **2009**, *102*, 198302. [CrossRef]
37. Pouget, E.M.; Bomans, P.H.; Goos, J.A.; Frederik, P.M.; de With, G.; Sommerdijk, N.A. The initial stages of template-controlled $CaCO_3$ formation revealed by cryo-TEM. *Science* **2009**, *323*, 1455–1458. [CrossRef]

38. Gower, L.B. Biomimetic model systems for investigating the amorphous precursor pathway and its role in biomineralization. *Chem. Rev.* **2008**, *108*, 4551–4627. [CrossRef]
39. Betts, F.; Posner, A. An X-ray radial distribution study of amorphous calcium phosphate. *Mater. Res. Bull.* **1974**, *9*, 353–360. [CrossRef]
40. Gebauer, D.; Kellermeier, M.; Gale, J.D.; Bergström, L.; Cölfen, H. Pre-nucleation clusters as solute precursors in crystallisation. *Chem. Soc. Rev.* **2014**, *43*, 2348–2371. [CrossRef]
41. Gebauer, D.; Cölfen, H. Prenucleation clusters and non-classical nucleation. *Nano Today* **2011**, *6*, 564–584. [CrossRef]
42. Kellermeier, M.; Rosenberg, R.; Moise, A.; Anders, U.; Przybylski, M.; Cölfen, H. Amino acids form prenucleation clusters: ESI-MS as a fast detection method in comparison to analytical ultracentrifugation. *Faraday Discuss.* **2012**, *159*, 23–45. [CrossRef]
43. Lu, J.; Li, Z.; Jiang, X. Polymorphism of pharmaceutical molecules: Perspectives on nucleation. *Front. Chem. Eng. Chin.* **2010**, *4*, 37–44. [CrossRef]
44. Lin, T.-J.; Chiu, C.-C. Structures and infrared spectra of calcium phosphate clusters by ab initio methods with implicit solvation models. *Phys. Chem. Chem. Phys.* **2018**, *20*, 345–356. [CrossRef]
45. Zhang, J.; Zhou, X.; Dong, C.; Sun, Y.; Yu, J. Investigation of amorphous calcium carbonate's formation under high concentration of magnesium: The prenucleation cluster pathway. *J. Cryst. Growth* **2018**, *494*, 8–16. [CrossRef]
46. Habraken, W.J.; Tao, J.; Brylka, L.J.; Friedrich, H.; Bertinetti, L.; Schenk, A.S.; Verch, A.; Dmitrovic, V.; Bomans, P.H.; Frederik, P.M. Ion-association complexes unite classical and non-classical theories for the biomimetic nucleation of calcium phosphate. *Nat. Commun.* **2013**, *4*, 1507. [CrossRef]
47. Trinh, T.T.H.; Khuu, C.Q.; Wolf, S.E.; Nguyen, A.-T. The multiple stages towards crystal formation of L-glutamic acid. *J. Cryst. Growth* **2020**, *544*, 125727. [CrossRef]
48. Wang, F.; Richards, V.N.; Shields, S.P.; Buhro, W.E. Kinetics and mechanisms of aggregative nanocrystal growth. *Chem. Mater.* **2014**, *26*, 5–21. [CrossRef]
49. De Yoreo, J.J.; Gilbert, P.U.; Sommerdijk, N.A.; Penn, R.L.; Whitelam, S.; Joester, D.; Zhang, H.; Rimer, J.D.; Navrotsky, A.; Banfield, J.F. Crystallization by particle attachment in synthetic, biogenic, and geologic environments. *Science* **2015**, *349*, 6760. [CrossRef]
50. Baumgartner, J.; Dey, A.; Bomans, P.H.; Le Coadou, C.; Fratzl, P.; Sommerdijk, N.A.; Faivre, D. Nucleation and growth of magnetite from solution. *Nat. Mater.* **2013**, *12*, 310–314. [CrossRef] [PubMed]
51. Mirabello, G.; Ianiro, A.; Bomans, P.H.; Yoda, T.; Arakaki, A.; Friedrich, H.; de With, G.; Sommerdijk, N.A. Crystallization by particle attachment is a colloidal assembly process. *Nat. Mater.* **2020**, *19*, 391–396. [CrossRef]
52. Park, J.; Elmlund, H.; Ercius, P.; Yuk, J.M.; Limmer, D.T.; Chen, Q.; Kim, K.; Han, S.H.; Weitz, D.A.; Zettl, A. 3D structure of individual nanocrystals in solution by electron microscopy. *Science* **2015**, *349*, 290–295. [CrossRef]
53. Polte, J.; Erler, R.; Thunemann, A.F.; Sokolov, S.; Ahner, T.T.; Rademann, K.; Emmerling, F.; Kraehnert, R. Nucleation and growth of gold nanoparticles studied via in situ small angle X-ray scattering at millisecond time resolution. *ACS Nano* **2010**, *4*, 1076–1082. [CrossRef]
54. Crespo, D.; Pradell, T.; Clavaguera-Mora, M.; Clavaguera, N. Microstructural evaluation of primary crystallization with diffusion-controlled grain growth. *Phys. Rev. B* **1997**, *55*, 3435. [CrossRef]
55. Brüning, K.; Schneider, K.; Roth, S.V.; Heinrich, G. Kinetics of strain-induced crystallization in natural rubber: A diffusion-controlled rate law. *Polymer* **2015**, *72*, 52–58. [CrossRef]
56. Wang, K.; Leng, Y.; Lu, X.; Ren, F.; Ge, X.; Ding, Y. Theoretical analysis of protein effects on calcium phosphate precipitation in simulated body fluid. *CrystEngComm* **2012**, *14*, 5870–5878. [CrossRef]
57. Volmer, M. Zum problem des kristallwachstums. *Z. Phys. Chem.* **1922**, *102*, 267–275. [CrossRef]
58. Burton, W.-K.; Cabrera, N.; Frank, F. The growth of crystals and the equilibrium structure of their surfaces. *Philos. Trans. R. Soc.* **1951**, *243*, 299–358.
59. Malkin, A.; Kuznetsov, Y.G.; Land, T.; DeYoreo, J.; McPherson, A. Mechanisms of growth for protein and virus crystals. *Nat. Struct. Biol.* **1995**, *2*, 956–959. [CrossRef]
60. Nozawa, J.; Uda, S.; Guo, S.; Hu, S.; Toyotama, A.; Yamanaka, J.; Okada, J.; Koizumi, H. Two-dimensional nucleation on the terrace of colloidal crystals with added polymers. *Langmuir* **2017**, *33*, 3262–3269. [CrossRef]
61. Weng, J.; Huang, Y.; Hao, D.; Ji, Y. Recent advances of pharmaceutical crystallization theories. *Chin. J. Chem. Eng.* **2020**, *28*, 935–948. [CrossRef]
62. Chernov, A. Present-day understanding of crystal growth from aqueous solutions. *Prog. Cryst. Growth Charact. Mater.* **1993**, *26*, 121–151. [CrossRef]
63. Kalt, H.; Rühle, W.; Reimann, K.; Rinker, M.; Bauser, E. Alloy-disorder-induced intervalley coupling. *Phys. Rev. B* **1991**, *43*, 12364. [CrossRef]
64. Ming, N.-B. Defect mechanisms of crystal growth and their kinetics. *J. Cryst. Growth* **1993**, *128*, 104–112. [CrossRef]
65. Rodriguez-Navarro, C.; Burgos-Cara, A.; Lorenzo, F.D.; Ruiz-Agudo, E.; Elert, K. Nonclassical crystallization of calcium hydroxide via amorphous precursors and the role of additives. *Cryst. Growth Des.* **2020**, *20*, 4418–4432. [CrossRef]
66. Yang, J.; Koo, J.; Kim, S.; Jeon, S.; Choi, B.K.; Kwon, S.; Kim, J.; Kim, B.H.; Lee, W.C.; Lee, W.B. Amorphous-phase-mediated crystallization of Ni nanocrystals revealed by high-resolution liquid-phase electron microscopy. *J. Am. Chem. Soc.* **2019**, *141*, 763–768. [CrossRef] [PubMed]

67. Uskoković, V. Disordering the disorder as the route to a higher order: Incoherent crystallization of calcium phosphate through amorphous precursors. *Cryst. Growth Des.* **2019**, *19*, 4340–4357. [CrossRef]
68. DiMasi, E.; Kwak, S.-Y.; Amos, F.F.; Olszta, M.J.; Lush, D.; Gower, L.B. Complementary control by additives of the kinetics of amorphous CaCO$_3$ mineralization at an organic interface: In-situ synchrotron X-ray observations. *Phys. Rev. Lett.* **2006**, *97*, 045503. [CrossRef]
69. Penn, R.L.; Banfield, J.F. Imperfect oriented attachment: Dislocation generation in defect-free nanocrystals. *Science* **1998**, *281*, 969–971. [CrossRef]
70. Penn, R.L.; Banfield, J.F. Morphology development and crystal growth in nanocrystalline aggregates under hydrothermal conditions: Insights from titania. *Geochim. Cosmochim. Acta* **1999**, *63*, 1549–1557. [CrossRef]
71. Tsukiyama, K.; Takasaki, M.; Oaki, Y.; Imai, H. Evolution of Co$_3$O$_4$ nanocubes through stepwise oriented attachment. *Langmuir* **2019**, *35*, 8025–8030. [CrossRef] [PubMed]
72. Gehrke, N.; Cölfen, H.; Pinna, N.; Antonietti, M.; Nassif, N. Superstructures of calcium carbonate crystals by oriented attachment. *Cryst. Growth Des.* **2005**, *5*, 1317–1319. [CrossRef]
73. Zhang, J.; Lin, Z.; Lan, Y.; Ren, G.; Chen, D.; Huang, F.; Hong, M. A multistep oriented attachment kinetics: Coarsening of ZnS nanoparticle in concentrated NaOH. *J. Am. Chem. Soc.* **2006**, *128*, 12981–12987. [CrossRef] [PubMed]
74. Kurz, W.; Rappaz, M.; Trivedi, R. Progress in modelling solidification microstructures in metals and alloys. Part II: Dendrites from 2001 to 2018. *Int. Mater. Rev.* **2021**, *66*, 30–76. [CrossRef]
75. Cho, K.-S.; Talapin, D.V.; Gaschler, W.; Murray, C.B. Designing PbSe nanowires and nanorings through oriented attachment of nanoparticles. *J. Am. Chem. Soc.* **2005**, *127*, 7140–7147. [CrossRef]
76. Pacholski, C.; Kornowski, A.; Weller, H. Self-assembly of ZnO: From nanodots to nanorods. *Angew. Chem. Int. Ed.* **2002**, *41*, 1188–1191. [CrossRef]
77. Cao, D.; Gong, S.; Shu, X.; Zhu, D.; Liang, S. Preparation of ZnO nanoparticles with high dispersibility based on oriented attachment (OA) process. *Nanoscale Res. Lett.* **2019**, *14*, 210. [CrossRef]
78. Zhu, C.; Liang, S.; Song, E.; Zhou, Y.; Wang, W.; Shan, F.; Shi, Y.; Hao, C.; Yin, K.; Zhang, T. In-situ liquid cell transmission electron microscopy investigation on oriented attachment of gold nanoparticles. *Nat. Commun.* **2018**, *9*, 421. [CrossRef]
79. Li, X.H.; Li, J.X.; Li, G.D.; Liu, D.P.; Chen, J.S. Controlled synthesis, growth mechanism, and properties of monodisperse CdS colloidal spheres. *Chem. Eur. J.* **2007**, *13*, 8754–8761. [CrossRef]
80. Liu, Z.; Pan, H.; Zhu, G.; Li, Y.; Tao, J.; Jin, B.; Tang, R. Realignment of nanocrystal aggregates into single crystals as a result of inherent surface stress. *Angew. Chem.* **2016**, *128*, 13028–13032. [CrossRef]
81. Ye, Y.; Chen, C.; Li, W.; Guo, X.; Yang, H.; Guan, H.; Xi, G. Highly sensitive W$_{18}$O$_{49}$ mesocrystal Raman scattering substrate with large-area signal uniformity. *Anal. Chem.* **2021**, *93*, 3138–3145. [CrossRef]
82. Li, L.; Liu, C.Y. Organic small molecule-assisted synthesis of high active TiO$_2$ rod-like mesocrystals. *CrystEngComm* **2010**, *12*, 2073–2078. [CrossRef]
83. Zhou, L.; O'Brien, P. Mesocrystals: A new class of solid materials. *Small* **2008**, *4*, 1566–1574. [CrossRef]
84. Zhou, L.; O'Brien, P. Mesocrystals—Properties and Applications. *J. Phys. Chem. Lett.* **2012**, *3*, 620–628. [CrossRef]
85. Zhang, Z.; Sun, H.; Shao, X.; Li, D.; Yu, H.; Han, M. Three-Dimensionally Oriented Aggregation of a Few Hundred Nanoparticles into Monocrystalline Architectures. *Adv. Mate.* **2005**, *17*, 42–47. [CrossRef]
86. Chen, Q.; Wu, L.; Zeng, Y.; Jia, C.; Lin, J.; Yates, M.Z.; Guan, B. Formation of spherical calcium sulfate mesocrystals: Orientation controlled by subunit growth. *CrystEngComm* **2019**, *21*, 5973–5979. [CrossRef]
87. Tan, B.; Zhang, X.; Li, Y.; Chen, H.; Ye, X.; Wang, Y.; Ye, J. Anatase TiO$_2$ Mesocrystals: Green Synthesis, In Situ Conversion to Porous Single Crystals, and Self-Doping Ti$^{3+}$ for Enhanced Visible Light Driven Photocatalytic Removal of NO. *Chem. Eur. J.* **2017**, *23*, 5478–5487. [CrossRef]
88. Ding, Z.; Zhang, D.; Feng, Y.; Zhang, F.; Chen, L.; Du, Y.; Ivey, D.G.; Wei, W. Tuning anisotropic ion transport in mesocrystalline lithium orthosilicate nanostructures with preferentially exposed facets. *NPG Asia Mater.* **2018**, *10*, 606–617. [CrossRef]
89. Keßler, S.; Reinalter, E.R.; Ni, B.; Cölfen, H. Rational Design of Environmentally Compatible Nickel Hexacyanoferrate Mesocrystals as Catalysts. *J. Phys. Chem. C* **2021**, *125*, 26503–26511. [CrossRef]
90. Ruiz-Agudo, C.; Ruiz-Agudo, E.; Putnis, C.V.; Putnis, A. Mechanistic principles of barite formation: From nanoparticles to micron-sized crystals. *Cryst. Growth Des.* **2015**, *15*, 3724–3733. [CrossRef]
91. Wang, T.; Cölfen, H.; Antonietti, M. Nonclassical crystallization: Mesocrystals and morphology change of CaCO$_3$ crystals in the presence of a polyelectrolyte additive. *J. Am. Chem. Soc.* **2005**, *127*, 3246–3247. [CrossRef] [PubMed]
92. Lin, F.; Ye, Z.; Kong, L.; Liu, P.; Zhang, Y.; Zhang, H.; Tang, Y. Facile Morphology and Porosity Regulation of Zeolite ZSM-5 Mesocrystals with Synergistically Enhanced Catalytic Activity and Shape Selectivity. *Nanomaterials* **2022**, *12*, 1601. [CrossRef] [PubMed]
93. Estrin, Y.; Beygelzimer, Y.; Kulagin, R.; Gumbsch, P.; Fratzl, P.; Zhu, Y.; Hahn, H. Architecturing materials at mesoscale: Some current trends. *Mater. Lett.* **2021**, *9*, 399–421. [CrossRef]
94. Guo, L.; Wu, J.; Li, J. Complexity at Mesoscales: A common challenge in developing artificial intelligence. *Engineering* **2019**, *5*, 924–929. [CrossRef]
95. Li, J. *Particle-Fluid Two-Phase Flow: The Energy-Minimization Multi-Scale Method*; Metallurgical Industry Press: Beijing, China, 1994.

96. Ge, W.; Chen, F.; Gao, J.; Gao, S.; Huang, J.; Liu, X.; Ren, Y.; Sun, Q.; Wang, L.; Wang, W. Analytical multi-scale method for multi-phase complex systems in process engineering—Bridging reductionism and holism. *Chem. Eng. Sci.* **2007**, *62*, 3346–3377. [CrossRef]
97. Li, J.; Ge, W.; Wang, W.; Yang, N.; Liu, X.; Wang, L.; He, X.; Wang, X.; Wang, J.; Kwauk, M. *From Multiscale Modeling to Meso-science*; Springer: Berlin/Heidelberg, Germany, 2013.
98. Lu, B.; Niu, Y.; Chen, F.; Ahmad, N.; Wang, W.; Li, J. Energy-minimization multiscale based mesoscale modeling and applications in gas-fluidized catalytic reactors. *Rev. Chem. Eng.* **2019**, *35*, 879–915. [CrossRef]
99. Wang, H.; Lu, Y. Mesoscale-Structure-Dependent EMMS Drag Model for an SCW Fluidized Bed: Formulation of Conservation Equations Based on Structures in Subphases. *Ind. Eng. Chem. Res.* **2021**, *60*, 18136–18153. [CrossRef]
100. Li, J.; Ge, W.; Wang, W.; Yang, N.; Huang, W. Focusing on mesoscales: From the energy-minimization multiscale model to mesoscience. *Curr. Opin. Chem. Eng.* **2016**, *13*, 10–23. [CrossRef]
101. Li, J.; Huang, W.; Chen, J.; Ge, W.; Hou, C. Mesoscience based on the EMMS principle of compromise in competition. *Chem. Eng. J.* **2018**, *333*, 327–335. [CrossRef]
102. Wang, W.; Lu, B.; Geng, J.; Li, F. Mesoscale drag modeling: A critical review. *Curr. Opin. Chem. Eng.* **2020**, *29*, 96–103. [CrossRef]
103. Geng, J.; Tian, Y.; Wang, W. Exploring a unified EMMS drag model for gas-solid fluidization. *Chem. Eng. Sci.* **2022**, *251*, 117444. [CrossRef]
104. Xu, T.; Jiang, X.; Yang, N.; Zhu, J. CFD simulation of internal-loop airlift reactor using EMMS drag model. *Particuology* **2015**, *19*, 124–132. [CrossRef]
105. Wang, S.; Zhao, Y.; Li, X.; Liu, L.; Wei, L.; Liu, Y.; Gao, J. Study of hydrodynamic characteristics of particles in liquid–solid fluidized bed with modified drag model based on EMMS. *Adv. Powder Technol.* **2014**, *25*, 1103–1110. [CrossRef]
106. Wu, N.; Ji, X.; Li, L.; Zhu, J.; Lu, X. Mesoscience in supported nano-metal catalysts based on molecular thermodynamic modeling: A mini review and perspective. *Chem. Eng. Sci.* **2021**, *229*, 116164. [CrossRef]
107. Yang, T.; Han, Y.; Li, J. Manipulating silver dendritic structures via diffusion and reaction. *Chem. Eng. Sci.* **2015**, *138*, 457–464. [CrossRef]
108. Han, M.; Xu, J.; Ren, Y.; Li, J. Simulations of flow induced structural transition of the β-switch region of glycoprotein Ibα. *Biophys. Chem.* **2016**, *209*, 9–20. [CrossRef]
109. Wang, H.; Han, Y.; Li, J. Dominant role of compromise between diffusion and reaction in the formation of snow-shaped vaterite. *Cryst. Growth Des.* **2013**, *13*, 1820–1825. [CrossRef]
110. Li, J.; Huang, W. From multiscale to mesoscience: Addressing mesoscales in mesoregimes of different levels. *Annu. Rev. Chem. Biomol. Eng.* **2018**, *9*, 41–60. [CrossRef]
111. Ge, W.; Wang, W.; Yang, N.; Li, J.; Kwauk, M.; Chen, F.; Zhou, G. Meso-scale oriented simulation towards virtual process engineering (VPE)—The EMMS paradigm. *Chem. Eng. Sci.* **2011**, *66*, 4426–4458. [CrossRef]
112. Ge, W.; Guo, L.; Liu, X.; Meng, F.; Xu, J.; Huang, W.L.; Li, J. Mesoscience-based virtual process engineering. *Comput. Chem. Eng.* **2019**, *126*, 68–82. [CrossRef]

*Review*

# Advances of Combinative Nanocrystal Preparation Technology for Improving the Insoluble Drug Solubility and Bioavailability

Qiuyan Ran [1,2], Mengwei Wang [1,2], Wenjie Kuang [1,2], Jinbo Ouyang [3], Dandan Han [1,2,*], Zhenguo Gao [1,2,*] and Junbo Gong [1,2,4]

1. State Key Laboratory of Chemical Engineering, School of Chemical Engineering and Technology, Tianjin University, Tianjin 300072, China
2. Haihe Laboratory of Sustainable Chemical Transformations, Tianjin 300192, China
3. School of Chemistry, Biology and Materials Science, East China University of Technology, Nanchang 330013, China
4. Institute of Shaoxing, Tianjin University, Shaoxing 312300, China
* Correspondence: handandan@tju.edu.cn (D.H.); zhenguogao@tju.edu.cn (Z.G.); Tel.: +86-22-27405754 (D.H.)

**Abstract:** The low solubility and bioavailability of aqueous insoluble drugs are critical challenges in the field of pharmaceuticals that need to be overcome. Nanocrystal technology, a novel pharmacological route to address the poor aqueous solubility problem of many poorly soluble drugs, has recently demonstrated great potential for industrial applications and developments. This review focuses on today's preparation technologies, containing top-down, bottom-up, and combinative technology. Among them, the highlighted combinative technology can improve the efficiency of particle size reduction and overcome the shortcomings of a single technology. Then, the characterization methods of nanocrystal production are presented in terms of particle size, morphology, structural state, and surface property. After that, we introduced performance evaluations on the stability, safety, and the *in vitro/in vivo* dissolution of drug nanocrystals. Finally, the applications and prospects of nanocrystals in drug development are presented. This review may provide some references for the further development and optimization of poorly soluble drug nanocrystals.

**Keywords:** nanocrystals; combinative technology; aqueous solubility; stability; dissolution rate

**Citation:** Ran, Q.; Wang, M.; Kuang, W.; Ouyang, J.; Han, D.; Gao, Z.; Gong, J. Advances of Combinative Nanocrystal Preparation Technology for Improving the Insoluble Drug Solubility and Bioavailability. *Crystals* **2022**, *12*, 1200. https:// doi.org/10.3390/cryst12091200

Academic Editor: Waldemar Maniukiewicz

Received: 29 July 2022
Accepted: 22 August 2022
Published: 25 August 2022

**Publisher's Note:** MDPI stays neutral with regard to jurisdictional claims in published maps and institutional affiliations.

**Copyright:** © 2022 by the authors. Licensee MDPI, Basel, Switzerland. This article is an open access article distributed under the terms and conditions of the Creative Commons Attribution (CC BY) license (https:// creativecommons.org/licenses/by/ 4.0/).

## 1. Introduction

One of the most challenging problems in pharmaceutical science is the bioavailability limitations of drugs with poor solubility. About 40% of the drugs currently on the market are struggling with poor aqueous solubility [1,2] and approximately 90% of drugs in development are classified as poorly soluble drugs [3] based on the definition of the biopharmaceutical classification system (BCS). In particular, BCS II drugs with low solubility and high permeability (Figure 1a) account for approximately 70% [4]. The Developability Classification System (DCS) for oral administered drugs was proposed based on the BCS classification system. According to the DCS, DCS II drugs can be divided into two categories, one is dissolving rate limiting DCS IIa, which is insoluble in water and organic phases, and the other is solubility limiting DCS IIb, which is usually soluble in at least some lipids (Figure 1b) [5,6]. Until now, several techniques have been proposed to solve the problems of drug insolubility, mainly involving two approaches: (i) modification of morphological properties of raw drug particles, i.e., improving the surface area to volume ratio by preparing a fine powder or promoting the porosity; and (ii) modification of some physicochemical and structural properties of insoluble active pharmaceutical ingredients (APIs), such as preparation of polymorphic forms, cocrystal, solid dispersions, etc. [7–9]. However, the second approach usually requires large screening efforts (e.g., the selection of solvents and ligands) when it comes to DCS IIa drugs. Therefore, reducing drug particle size is the best option for DCS IIa drugs [10].

Nanocrystal technology brings a new dawn for improving the solubility and bioavailability of insoluble drugs [11]. Nanocrystals (usually 1–1000 nm) are pure drug particles stabilized by suitable surfactants/polymers [2,11,12]. Nanocrystals have the following features. (i) The surface area of nanocrystals increase with decreasing particle size. According to the Noyes-Whitney equation [13], the dissolution rate of nanocrystals increase with improving the surface area. (ii) According to the Ostwald-Freundlich equation [14], downsizing the size to the nanometer range (Figure 1c) significantly enhances the solubility of a drug [15]. (iii) A mucus layer with a porous structure is present on the surface of the gastrointestinal tract. The nanograined size is small, which can rapidly permeate into the pore channels of the mucus layer and tightly adhere to them [2,16,17], so it can prolong the effective range and time of drugs in the gastrointestinal tract (Figure 1d). All of these properties contribute to enhancing the absorption and bioavailability of drugs [17]. In fact, nanocrystals were originally invented to improve the oral bioavailability of insoluble drugs [18]. With advanced research in drug nanocrystals, other advantages have also been explored, such as loading high active pharmaceutical ingredients (APIs) [4,16], improving the metabolism behavior of drugs [19], reducing toxic and side effects, promoting patient compliance [20], and so on. Consequently, the superior properties of drug nanocrystals have attracted more and more attention from pharmaceutical enterprises.

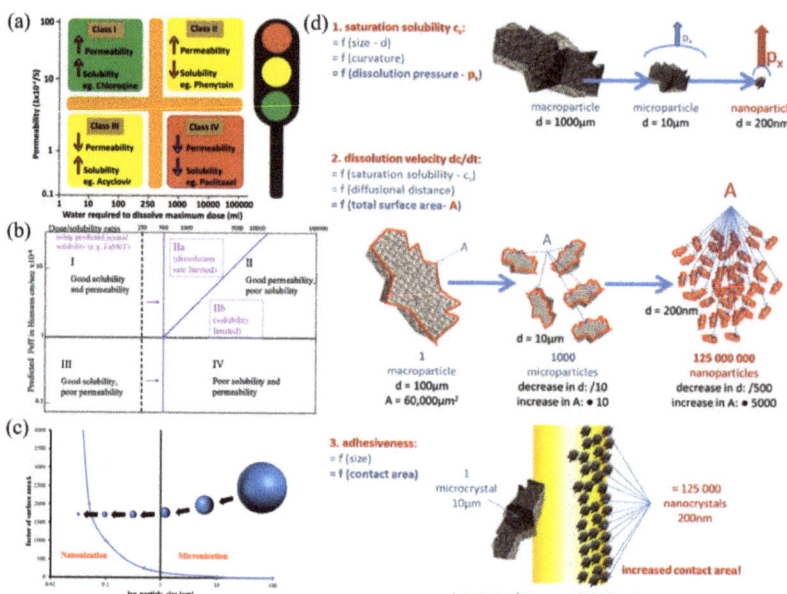

**Figure 1.** (a) description of BCS, (b) description of DCS, (c) size boundaries of nanoparticles, (d) features of nanocrystals: increased saturation solubility (upper), increased dissolution velocity (middle), and increased adhesiveness of nanomaterial, for surface: calculations were performed as cubes. Figure 1a was reprinted from ref. [4] with permission from Elsevier, Copyright® 2019. Figure 1b was reprinted from ref. [5] with permission from Elsevier, Copyright® 2010. Figure 1d was reprinted from ref. [21] with permission from Elsevier, Copyright® 2011.

In this review, the latest applications of combinative drug nanocrystal preparation technology are highlighted. Firstly, the current technologies for the fabrication of drug nanocrystals are introduced. Subsequently, the characterization and evaluation methods are summarized. Finally, applications and future prospects are briefly mentioned.

## 2. Preparation Technology

Nanocrystals are generated by reducing particle size (top-down technology), growing particles in the nanometer size range (bottom-up technology), and combining these two technologies [22].

### 2.1. Top-Down Technology

Top-down technology mainly includes wet bead milling and high pressure homogenization [23], which is easily industrialized. Drug particles decrease by mechanically generated shear and collision forces [24], accompanied by the fragmentation of crystalline species and the appearance of secondary nucleation nuclei. The formation rate of top-down technology is independent of supersaturation. Most previous reported anticancer drugs have been prepared by this technology because they do not require organic solvents and are relatively easy to scale up production [25]. In summary, top-down technology can be used for drugs that are insoluble in both the aqueous and organic phases. It processes quickly and is widely used for marketed drug nanocrystals.

#### 2.1.1. Wet Bead Milling

Wet bead milling involves crushing the drug itself into nanoparticles by high intensity mechanical force with stabilizers and water [26]. The particle size of nanocrystals is mainly relevant to the size of the milling beads [27] (usually 0.1–20 nm), the property parameters of the drug, and the setting parameters [24]. Since the temperature can be controlled in the preparation process, wet bead milling is especially suitable for preparing thermally unstable drug nanocrystals [16]. It operates easily to obtain a uniform product. However, stabilizers and wetting agents still need to be added, and several cycles are required to reach the specific particle size range. Meanwhile, the obtained product has disadvantages in the contamination caused by grinding beads [28] and poor physical storage stability due to agglomeration. Funahashi et al. [29] found that ice beads melted after the milling process, which could avoid contamination. Most of the drug nanocrystals that have been successfully translated industrially are prepared using milling, including the earliest marketed pentoxifylline capsule Verelan®PM, a fenofibrate tablet for the treatment of hypercholesterolemia Tricor®, and an anti-inflammatory drug Naprelan® [30].

#### 2.1.2. High Pressure Homogenization

High pressure homogenization (HPH) utilizes violent shearing, collision, and cavitation generated in a high pressure homogenization chamber to break down drug particles. Depending on the instrumentation and solution used, it can be divided into microfluidization, IDD-P™, Dissocubes®, and Nanopure®. Microfluidization has the 'Z' or 'Y' type chamber based on the jet stream principle. IDD-P™ uses a jet homogenizer for the homogenization of suspensions. Dissocubes® uses a piston gap homogenizer for homogenization in aqueous media. Nanopure® is suitable for the production of easily hydrolyzed drugs in reduced/non-aqueous media [31]. In general, the setting parameters of the homogenization process and the hardness of the drug mainly affect the properties of the product [32]. Through these efficient methods, the obtained product has a small particle size with narrow distribution and is not contaminated by the grinding medium. Most importantly, the method can be better combined with other methods to reduce the cycle number of homogenizations and the requirement of homogenization pressure. However, expensive equipment and demanding techniques hinder the transferability to larger scales [33]. In addition, high pressure may unintentionally lead to crystal structure changes, increase the content of amorphous states, and affect the stability of some amorphous nanosuspensions [32]. The currently marketed paliperidone palmitate intramuscular suspension Invega Sustenna, fenofibrate tablets Triglide®, and Luteolin nanocrystals [34] are prepared by the HPH technique [30].

2.1.3. Laser Ablation

Laser ablation is a new technique developed in recent years for nanocrystal preparation. During laser ablation, the solid target is irradiated and the ejected material forms nanoparticles in the surrounding liquid. Then, stirred suspensions of microparticles are broken into nanoparticles by laser-mediated fragmentation [35]. According to the laser processing time, it is divided into nanosecond, picosecond, and femtosecond laser irradiation, among which more nanoscale particles can be produced [36]. The parameters affecting the particle size include the laser intensity, scanning speed, and the properties of the suspension, etc. In this process, no organic solvent is involved, but a small fraction of the drug may undergo oxidative degradation and crystal state changes due to excessive power. This method has been successfully used to prepare paclitaxel, megestrol acetate, and curcumin nanosuspensions [37].

2.1.4. Ultrasound

Ultrasound is an efficient method to break drug particles into smaller particles through the vibration of acoustic waves. Ultrasound has been shown to enhance nucleation by creating acoustic cavitation in solution and rapidly dispersing the drug solution [38]. Because it is easily operated in the laboratory and is highly reproducible, it is also usually combined with other techniques [39]. Ultrasound-assisted precipitation of nanoparticles mainly alters the mixing process, nucleation, growth, and agglomeration [40]. The size of the nanocrystal depends on the intensity of the ultrasound treatment, the horn length, the horn immersion depth, and the cavitation depth [12].

*2.2. Bottom-Up Technology*

Bottom-up technology is mainly based on precipitation and evaporation [33]. The basic principle is to obtain drug nanocrystals from the supersaturated state of drugs and subsequently control the size distribution of the nanoparticles by appropriate methods [41]. Nucleation is especially important for the formation of small and homogeneous nanocrystals. Controlling the crystal growth is the best way to precisely control the particle size of drug particles. Many physical methods have been used to control the crystal growth, such as high gravity controlled precipitation. Compared with top-down methods, these methods provide better control of particle properties [42]. In conclusion, bottom-up technology is simple in principle and operation, but difficult to scale up due to poor reproducibility. The process may also use organic solvents.

2.2.1. Liquid Antisolvent Precipitation

Liquid antisolvent (LAS) precipitation is the preparation of nanocrystals by mixing a solution stream (organic phase) dissolved with an insoluble drug with an aqueous antisolvent. The solution–antisolvent method of nanoprecipitation is the most commonly reported. Since this method only contains nucleation and growth steps, it is simple and cost-effective [12]. The optimized nanocrystals can be prepared through two steps. However, unstable crystal particles are also recrystallized in the process, leading to the aggregation and precipitation of nanocrystals [16]. Additionally, the use of organic solvents in the preparation process leads to the problem of solvent residues, and it is unsuitable for drugs that are neither soluble in aqueous nor insoluble in non-aqueous solvents. Currently, some studies have used this method to obtain suspensions of hydrochlorothiazide [39], budesonide [43], etc.

2.2.2. Precipitation Assisted by Acid-Base Method

The carbon dioxide-assisted precipitation method using acid-base reactions usually involves dissolving the drug in a weak acid solution as the acid phase, and weak base in a solution containing stabilizer as the base phase. The acid phase is slowly added to the base phase to produce carbon dioxide, then the drug nanocrystals are precipitated by vapor effervescence [44]. This method avoids the addition of organic solvents, which is more friendly to the environment. However, it is only applicable to insoluble drugs whose solubility is related to pH and is stable to acids-bases [19]. Wang et al. [45] prepared tacrolimus nanocrystal suspensions using this method. *In vivo* pharmacokinetic results indicated that tacrolimus nanosuspensions significantly increased the oral bioavailability compared with commercial hard capsules.

2.2.3. High Gravity Controlled Precipitation

High gravity controlled precipitation (HGCP) is [46] the improvement of the precipitation method using gravity control to obtain more uniform and smaller drug nanocrystals. Reactant concentration, rotational speed, and volumetric flow rate are effective factors influencing particle size. In this process, the drug suspension in the device can be circulated for long-term mixing and reaction. However, local oversaturation of the feed stream at the turbulent edge during mixing leads to continuous nucleation, thus limiting the industrial application of this method [25]. To date, this method has been successfully used on the laboratory scale to prepare salbutamol sulfate [47] and sorafenib [48].

2.2.4. Supercritical Fluid Method

The supercritical fluid (SCF) method means drugs dissolve in a supercritical fluid (e.g., $CO_2$) and precipitate nanocrystals with the rapid vaporization of the supercritical fluid as the fluid is atomized under reduced pressure through a nozzle with a tiny aperture [11]. According to the function of supercritical fluid in the crystallization process, supercritical fluid methods include rapid expansion of supercritical solution (RESS) and supercritical antisolvent (SAS). According to the different nozzle positions, the rapid expansion of a supercritical fluid method was improved to yield the rapid expansion of a supercritical solution into a liquid solvent (RESOLV) method, in which the former places the nozzle in the air while the latter places the nozzle in aqueous solution. The state parameters of the supercritical fluid in the process, the morphology of the nozzle, and the concentration of the drug will influence the particle size of the nanocrystals [7]. This process does not require organic solvent to produce high purity products. However, the consumption of supercritical fluids is large and it is only suitable for drugs dissolved in supercritical fluids [16]. Zhang et al. [49] prepared apigenin (AP) nanocrystals using the SAS method. No substantial changes in the crystal structure were observed in the nanocrystals, but decreased particle size and smooth spherical surface were noticed. The AP nanocrystals exhibited faster dissolution profiles than the original AP in dissolution media.

2.2.5. Emulsion Polymerization Method

API is dissolved in volatile organic solvents or solvents partially mixed with aqueous as the dispersed phase, then the organic solvent is emulsified dropwise into the aqueous phase (usually including stabilizers) to form an O/W emulsion [11]. The emulsion droplet size is easy to control. After that, the emulsions are evaporated, stirred, and extracted to obtain drug nanocrystals. Factors such as emulsifier, stirring speed, evaporation rate, temperature gradient, and pH value have significant effects on product quality. Because this emulsion polymerization method requires the assistance of homogenization or ultrasound, it is suitable for laboratory operations but not for large-scale pilot production [11]. Chen [50] obtained florfenicol nanocrystals with a mean particle size of 226.1 ± 11.3 nm using the emulsification solvent volatilization method. Compared with the original powder, the solubility and dissolution of florfenicol nanocrystals were remarkably improved.

*2.3. Combinative Technology*

There are always many limitations to obtain the desired nanocrystals using a single preparation technology due to the properties of various drugs and characteristics of the instruments. When top-down technology is selectively united with bottom-up technology, forming combinative technology, the disadvantages of a single preparation technique can be overcome and the efficiency of particle size reduction can be improved. Combinative technology is divided into the Nanoedge® technology developed by Baxter [51] and the SmartCrystal® technology (Abbott/Soliqs, Ludwigshafen/Germany). Combinative technology (Figure 2) combines pre-treatment and a particle size reduction step. It can eliminate the drawbacks of instrument clogging [52] and improve the stability of nanocrystals. Combinative technology can take full advantage of various technologies and are also suitable for a wide range of insoluble drugs. However, there are limitations in terms of economics and process complexity.

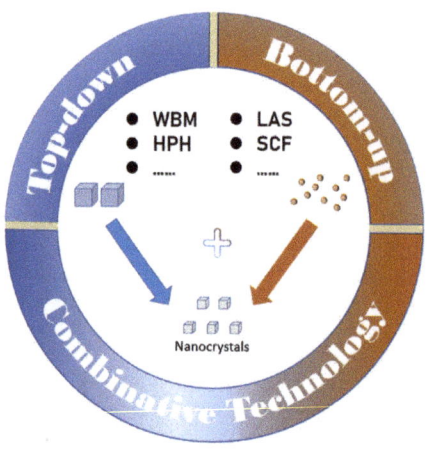

**Figure 2.** The correlation of different preparation technology.

2.3.1. Nanoedge Technology

Nanoedge technology was the first combined method for particle size reduction developed for nanodrug production. It utilizes the HPH method assisted by the precipitation method. Initial crystal particles are obtained by precipitation, reducing the high pressure homogenizer slit blockage and improving the efficiency of reducing particle size during the homogenization process [53]. Subsequently, the homogenization process from the HPH method is used to further crush the particles, preventing secondary growth and overcoming the problem of uneven particle size distribution and Oswald ripening in the precipitation method, which increases the physical stability of the nanocrystal particles. Furthermore, other alternative techniques such as ultrasound or microfluidization can be used for the high-energy process [33].

2.3.2. SmartCrystal Technology

SmartCrystal® technology mainly combines a pre-treatment step followed by a high pressure homogenization step. The pre-treatment step can be, for example, wet bead milling, spray drying, freeze drying, or precipitation, followed by HPH [54]. SmartCrystal technology is recognized as a second-generation nanocrystal preparation method. Specifically, the joint method in this technology takes the form of collection, which is a toolbox of technology optimization for the preparation of drug nanocrystals [55]. It includes H69, H42, H96, and combination technology (CT) techniques.

1. H69

H69 is similar to NanoEdge technology, combining nanoprecipitation and HPH methods. The formation of nanocrystals occurs in the high pressure homogeneous cavitation region, which contributes to ultra-small and homogeneous particle size. Li et al. [56] tried to prepare ursodeoxycholic acid nanosuspension by homogenization technology and high pressure precipitation tandem homogenization technology. The dissolution rate of ursodeoxycholic acid nanocrystalline powder was quicker than that of the raw and physical mixture powder.

2. H42 and H96

H42 and H96 are combined with spray drying, freeze drying, and HPH techniques, respectively. The solution of insoluble drug and stabilizer is pretreated by spray/freeze drying, uniformly dispersed in the stabilizer skeleton, then redispersed into the water by HPH to prepare drug nanocrystals. The combination reduces particle aggregation and improves processing efficiency. It is suitable for large-scale production. Möschwitzer et al. [57] used poloxamer 188 as a stabilizer to prepare hydrocortisone acetate powder by H42. More uniform nanosuspensions exhibiting good long-term storage stability were obtained. Yu [58] prepared meloxicam and naproxen drug nanoamorphous using the H96 technique.

3. CT

CT is a combination of top-down technology. The two most common types of wet bead milling methods used are rotor-stator and mills [59]. Taking the former as an example, artcrystals is a combined rotor-stator high-speed shear and HPH technology. Firstly, the drug suspension is pretreated by shearing in a rotor-stator high-speed shear, then the nanocrystals are homogenized at high pressure to obtain stable and homogeneous suspensions. Wadhawan et al. [60] obtained crystalline acyclovir nanocrystals with a mean particle size of 400–500 nm using high pressure homogenizer and hydroxypropyl cellulose-LF as a stabilizer followed by wet bead milling. The saturation solubility of the nanocrystals was 1.6 times higher than that of micronized acyclovir. Martena et al. [61] prepared nicergoline nanocrystals in aqueous solutions of polysorbate 80. Four different techniques, HPH, bead milling (BM), and combined techniques (HPH + BM, BM + HPH) were explored in his work. The combined technique was found to be superior, but HPH + BM produced nanocrystals with a smaller mean particle size than BM + HPH. Particle solubility increased for all nanocrystals, especially for HPH and the combination technique, which obtained nanocrystals showing a higher dissolution rate.

2.3.3. Other New Combinative Technology

1. Precipitation-lyophilization-homogenization (PLH) method

PLH is a combination of precipitation-lyophilization-homogenization method. Morakul et al. [62] obtained clarithromycin nanocrystals by this method, using poloxamer 407 and sodium dodecyl sulfate (SDS) as co-stabilizers. The obtained clarithromycin nanocrystals were cubic particles, about 400 nm, in a crystalline or partially amorphous state. It had high solubility and permeability.

2. High gravity antisolvent precipitation process (HGAP)

HGCP technology is merged with antisolvent precipitation process to form HGAP. The benefits of the HGCP are retained while the disadvantages of impurities in the product are eliminated [12]. Zhao et al. [63] prepared danazol nanocrystals with uniform size distribution by the HGAP process. The average particle size was 190 nm. The molecular state and crystalline form of Danazol nanoparticles were maintained. The nanoparticles were highly evaluated by the industry for its high recovery rate and continuous production capacity.

3. Microjet reactor technology (MRT)

MRT is similar to HPH. The drug solution is mixed in the high pressure chamber through the micro-hole of the nozzle to form a high-speed fluid sprayed into the reaction chamber, and convective shear in the reaction chamber to form turbulence. At the same time, there is cavitation, impact, and shear effect to reduce the product particle size. The influencing factors of MRT include the mixing ratio of solution and antisolvent, jet strength, stabilizer dosage, temperature, etc. This method can realize continuous large-scale production. However, the energy consumption and path clogging [37] cannot be ignored. Chen et al. [64] prepared albendazole nanocrystals by MRT with a mean particle size of $367.34 \pm 0.68$ nm under the optimal preparation process. The nanocrystals can significantly improve the dissolution performance of albendazole and facilitate the improvement of oral absorption of the drug.

4. Evaporative precipitation into aqueous solution (EPAS)

The EPAS method dissolves the API in the low-boiling-point solvent and heats above its boiling point. Thereafter, the heated solution is sprayed into heated aqueous solutions containing stabilizers [12]. Chen et al. [65] produced amorphous nanoparticle suspensions of cyclosporine A by EPAS. Due to the low crystallinity, small particle size of nanoparticles, and hydrophilic stabilizers, it has shown a high dissolution rate.

5. Antisolvent precipitation-high pressure homogenization method

Huang et al. [66] combined the antisolvent precipitation method and HPH method to prepare celecoxib nanocrystal suspensions with a particle size of 283.67 ± 20.84 nm, using polyvinylpyrrolidone K30 (PVP K30) and SDS as crystal stabilizers. The solubility of celecoxib nanocrystals was obviously higher than that of the raw celecoxib and the physical mixture. The product remained quite stable under high temperature and high moisture conditions for 10 days of storage.

6. Ultrasound probe-high pressure homogenization method

Jin et al. [67] used an ultrasound probe combined with HPH and fluidized drying process to prepare baicalin nanocrystals with an average particle size of 248 ± 6 nm and PDI 0.181 ± 0.065 by selecting mixed surfactant poloxamer 188 as a steric stabilizer and SDS as an electrostatic stabilizer. The results of pharmacokinetic experiments in rats showed that the drug bioavailability *in vivo* was significantly improved.

7. Rotary evaporation method-high pressure homogenization method

Zuo [68] prepared curcumin-artemisinin cocrystal nanomedicine by the rotary evaporation-HPH method. The particle size of nanomedicine was 234.6 nm after optimization. Curcumin-artemisinin cocrystal nanomedicine showed obvious solubility advantages and excellent stability compared with that of raw curcumin, curcumin-artemisinin cocrystals, and curcumin nanocrystals. Yu [58] also obtained quercetin drug nanoamorphous using the rotary evaporation method assisted by the HPH method.

8. Melt quench-high pressure homogenization method

Yu [58] also used the combined melt quench-high pressure homogenization technique to prepare nanoamorphous indomethacin. The particle size of the prepared suspension was 245 nm. The solubility of the nanosuspensions was significantly enhanced. However, the stability of the nanoamorphous was poor. The particle size started to increase significantly within 7 days and even reached 890 nm after 30 days due to the presence of water and the occurrence of recrystallization.

9. Antisolvent precipitation-ultrasound method

Zhang et al. [69] obtained fenofibrate nanocrystals using the ultrasound probe-precipitation method. However, one of the disadvantages of ultrasonic probes is that they can leave metal particles and thus are not suitable for industrial production. Liu et al. [70] used alpha tocopherol succinate as an auxiliary stabilizer in the organic phase to prepare carvedilol nanosuspensions by this method. The mean particle size of the nanoparticles was 212 nm and it was stable at 25 °C for 1 week. The dissolution rate of the nanosuspension was significantly increased. *In vivo* tests indicated that the nanosuspensions showed approximately two-fold increase in each index compared with commercial tablets. Additionally, the method is fast, inexpensive, and easy to control. Paclitaxel nanocrystals [71], zaleplon nanocrystals [72], and nintedanib nanocrystals [73] are also prepared using this method.

## 3. Characterization and Evaluation

After obtaining nanocrystal products, characterization and performance evaluation are crucial to the further application of the products. On the one hand, the purpose of characterization is to understand the properties of the sample such as size, morphology, and crystalline form. Based on this, the performance of the product can be quantitatively controlled on different quality characteristics, and then the quality controllability of the product can be achieved [19]. On the other hand, evaluation is to obtain the nanocrystal performance index (stability, cytotoxicity, dissolution rate) to explore more effective drug delivery strategies.

*3.1. Characterization*

3.1.1. Particle Size and Distribution

The particle size range of drug nanocrystals has a certain size requirement, which is generally less than 1 μm. Therefore, it is very important for drug nanocrystals to precisely control the particle size of the drug to obtain narrow and uniform particle size distribution. The particle size not only affects the drug loading and release behavior of API, but also is closely related to pharmacokinetics, biodistribution, and even the delivery mechanism of drug nanocrystals. To summarize, particle size and distribution are considered to be the most important index parameters of nanocrystals, which is one of the key elements in the development and control quality of nanocrystals [41].

The particle size and distribution can be evaluated with the help of offline or online evaluation tools. The measurement is usually performed by dynamic light scattering (DLS), also known as photon correlation spectroscopy (PCS). It observes Brownian motion using laser irradiation of particles and relates Brownian motion to particle size by analyzing the light intensity fluctuations of the scattered light. The measurement result is the hydrodynamic particle size, and the particle size distribution is generally expressed using PDI. In general, a PDI value of 0.1–0.25 represents a narrow particle size distribution and also indicates that the nanocrystal system is stable [74]. DLS can obtain accurate and statistical particle size distributions, but it is necessary to make nanocrystals into well-dispersed suspensions. From the variation of Z-average and PDI values, small increases in drug nanocrystal size can also be assessed by DLS. Therefore, DLS is considered effective for measuring the particle size of submicron and nanoparticles.

Laser Diffraction (LD) analyzers are designed based on the phenomenon of light diffraction, which occurs when light passes through particles and the angle of the diffracted light is inversely proportional to the size of the particles. Typical characterization parameters of LD are 50%, 90%, and 99% of the diameter, expressed as D50, D90, and D99, respectively (i.e., D50 means that 50% of the particle volume is below a given size). The measurement range of LD and PCS varies, where LD is generally used to measure particle sizes larger than 0.05 μm, while PCS usually gives the practical size in the range from 3 nm to 3 μm [75]. Furthermore, it is important to note that the data of particle size acquired by LD and PCS for nanosuspensions are different because the data from LD is volume-based, while PCS gives the average particle size based on light intensity-weighted particle size [2].

Small-angle X-ray scattering (SAXS) uses the X-ray small-angle scattering effect to measure the particle size distribution of nanocrystals. This is a simple method with high accuracy, but it is also relatively expensive. Typically, it can measure the size distribution of particles in the range of 1–300 nm.

3.1.2. Morphological Characterization

Both the size and morphological state of nanoparticles can influence the structural properties of nanocrystals. Scanning electron microscopy (SEM), transmission electron microscopy (TEM), and atomic force microscopy (AFM) are frequently applied to characterize the morphological appearance of drug nanocrystals. It should be mentioned that some people use electron microscopy images supplemented with statistics to obtain the mean value of the particle size of nanocrystals, which is the actual particle size. In contrast, the previous section obtained by DLS or LD is only the hydrodynamic size of the nanoparticles.

SEM uses narrow-focused beams of high energy electrons to scan the sample for the purpose of characterizing the microscopic morphology of the substance. The imaging stereo effect is good, and the resolution can reach 1 nm. Therefore, it is now widely used to observe nanomaterials. In particular, when formulated nanosuspensions are converted into dry powders, SEM analysis is critical to monitor changes in particle shape and size. SEM images of different nanocrystals obtained in the literature are as follows (Figure 3).

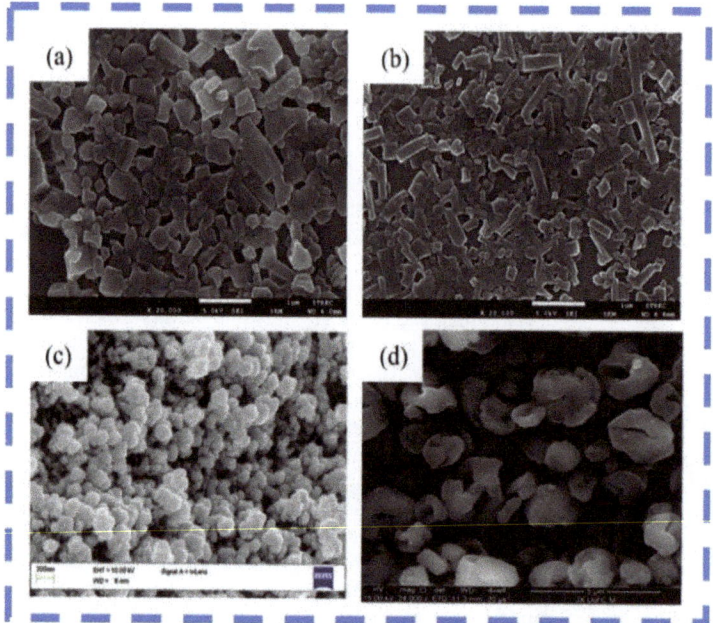

**Figure 3.** (**a,b**) SEM micrographs of clarithromycin nanocrystals obtained by media milling method; (**c**) SEM micrographs of carvedilol nanosuspensions using the antisolvent precipitation-ultrasonication method; (**d**) Transform Breviscapine (BVC) nanocrystals into BVC nanocrystals embedded microparticles (BVC-NEP) via spray-drying, adding matrix formers Maltodextrin. Figure 3a,b were reprinted from ref. [76] with permission from Elsevier, Copyright® 2018. Figure 3c was reprinted from ref. [70] with permission from Springer Nature, Copyright® 2012. Figure 3d was reprinted from ref. [77] with permission from Elsevier, Copyright® 2021.

TEM uses an electron beam as the light source to project accelerated and aggregated electron beams onto the very thin sample. In order for the electron beam to penetrate, the thickness of the sample should be less than 100 nm. TEM requires an appropriate concentration of the wet sample. The current resolution is up to 0.2 nm. TEM images of different nanocrystals obtained in the literature are as follows (Figure 4).

AFM is a new generation of scanning probe microscopy. It is capable of imaging in any environment (including liquids), and the low force of the needle tip on the sample surface prevents damage to the sample. It does not require the sample to be electrically conductive. The sample can be directly observed at the nanoscale without special treatment. A three-dimensional image of the sample surface can be obtained by collecting feedback signals from the force applied to the sample by the probe [80]. AFM can also provide information on the shape and structure of nanocrystals that cannot be accessed by other methods [81]. Overall, AFM has become an important tool for conducting real-time observations at the nanoscale.

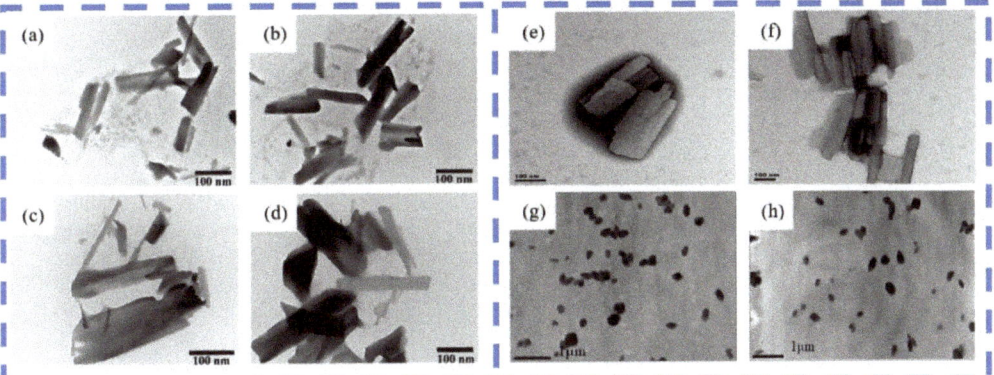

**Figure 4.** Breviscapine (BVC) nanocrystals modified by d-Tocopherol acid polyethylene glycol 1000 succinate (TPGS) at different concentrations. (**a**) BVC-NC/@5%TPGS, (**b**) BVC-NC/@10%TPGS, (**c**) BVC-NC/@20%TPGS, (**d**) BVC-NC/@30%TPGS; (**e,f**) 10-Hydroxycamptothecin nanocrystals; (**g,h**) TEM of the resveratrol nanocrystals. Figure 4a–d were reprinted from ref. [77] with permission from Elsevier, Copyright® 2021. Figure 4e,f were reprinted from ref. [78] with permission from Elsevier, Copyright® 2021. Figure 4g,h were reprinted from ref. [79] with permission from Elsevier, Copyright® 2022.

3.1.3. Structural Characterization

The crystalline form is also a quality factor of nanocrystal that needs to be noted. During the process of nanosizing, the crystalline state of drug particles may change, such as from crystalline state to amorphous state, which has a significant difference in physical properties, such as density, hardness, solubility, and stability. Therefore, the structure characterizations of nanocrystal need to be carefully considered.

Differential scanning calorimetry (DSC) is a common method for thermal analysis of nanocrystals. DSC measures the crystallinity of drug nanoparticles by detecting the glass transition temperature, melting point, and associated enthalpy. DSC also can be used to determine the interactions between excipients and drugs in nanocrystals as well as numerous thermodynamic and kinetic parameters [2]. The method has a wide temperature range and high resolution so that it has an indestructible place in structural characterization. In addition, thermal analysis can be investigated by thermogravimetric measurements or differential thermal analysis (DTA). Thermogravimetric analysis is a technique to study the thermal stability, heat flow, and structural deformation of particles in the inert environment [82]. DTA can be used for phase changes and other thermal processes, such as the determination of melting points [83]. In combination, thermogravimetric and differential thermal analysis (TG-DTA) can characterize multiple thermal properties of a nanocrystal sample at the same time. It is valuable for the investigation of thermal stability as well as for the determination of volatile content and other compositional analyses [2]. X-ray powder diffraction (XPRD) analysis based on constructive interference between monochromatic X-rays and samples, is another tool for characterizing the crystallinity of nanocrystals [52]. It can be used to determine the structural parameters and crystal structure of some crystalline substances, as well as amorphous substances.

Fourier transform infrared spectroscopy (FTIR) is a method of measuring interferograms and performing Fourier transform on the interferograms to measure infrared spectra with high resolution. FTIR can study the interaction between drugs and excipients in nanocrystal formulations at the molecular level. When the two components interact, FTIR spectroscopy shows the interaction by changing the vibrational frequency of the molecules [84].

### 3.1.4. Surface Property

The particle surface charge situation is closely related to the stability of drug nanocrystals [85], which is generally reflected by the zeta potential (ZP) values. Currently, the measurement methods include electrophoresis, electroosmosis, flow potential, and ultrasound methods, among which laser Doppler electrophoresis is commonly used for measurement. The higher the ZP values, the greater the electrostatic repulsion among the particles, and the more stable the system. Usually, when the particles have sufficient ZP values, it will provide effective charge repulsion to prevent particle aggregation. In general, an absolute ZP value above 30 mV provides great stability and about 20 mV provides only short-term stability. However, for larger molecular weights of APIs, the stabilizer acts mainly through steric stabilization. In this case, a ZP value of only 20 mV or even less can provide sufficient stabilization [86].

## 3.2. Performance Evaluation
### 3.2.1. Stability

The stability of nanoparticles is essential for the generation of nanosized effects. The stability of nanocrystals includes chemical stability (degradation and spoilage) and physical stability (sedimentation, agglomeration, crystal growth, and crystalline state) [87]. The large specific surface area of nanocrystals results in high free energy, which may result in physical instability, such as aggregation or agglomeration [88]. Therefore, surfactants/polymers are necessary to stabilize the nanocrystals. The choices of stabilizers can also affect the performances of drug nanocrystals *in vivo* and further formulations.

Stabilizers can mainly be divided into ionic and nonionic stabilizers (Table 1). One mechanism by which stabilizers work is to reduce the surface tension at the interphase interface, where electrostatic repulsion from ionic surfactants stabilizes the nanosuspension. The other mechanism is steric stabilization, where polymers cover the particles to prevent particle-to-particle aggregation [89]. The basic principles that must be followed in the selection of stabilizers are that the addition dose must be acceptable and safe to humans. For instance, stabilizers will not cause any allergic or immune response [90]. The current screening of surfactants/stabilizers is primarily based on reported references and experience. Experience indicates that binary or ternary mixtures of electrostatic stabilizers typically generate enhanced stabilization [2]. The stability is generally divided into long-term and short-term stability, which can be measured by the size, ZP values, and morphology of the nanocrystals [91].

**Table 1.** The types and usage of commonly used stabilizers.

| Types | Names | Usage |
|---|---|---|
| Ionic surfactants | Sodium dodecyl sulfate | Slightly toxic, electrostatic stabilizer |
| non-ionic surfactants | Tween | Not for the preparation of solid powder nanoparticles |
| | Vitamin E TPGS | Available mainly in oral dosage form |
| | Soluplus® | Available |
| amphiphilic copolymers | Poloxamers | Non-toxic material, or oral, parenteral and topical preparations |
| | Hydroxypropyl cellulose (HPC), Hydroxyethyl cellulose (HEC), Methyl cellulose (MC) | HPC and MC: oral and topical applications, HEC: ophthalmic and topical drug delivery |
| | Hydroxypropyl Methyl Cellulose (HPMC) | Non-toxic, non-irritant, available for oral and ocular |
| | Polyvinylpyrrolidone (PVP) | Widely used, but not for parenteral injection |
| | Polyvinylalcohol (PVA) | Widely used |
| Others | Lecithin | Available for oral and parenteral |

Guo et al. [92] prepared nitrendipine nanocrystals (NTD-NCs) using a media milling method. Rectangular nanocrystals with 1.25% ($w/v$) HPMC-E5 and 0.4% ($w/v$) SDS as stabilizers were obtained with a particle size of 256.5 ± 6.6 nm. The size of the nanocrystals was monitored for 30 days (at 4 °C, 25 °C, and 40 °C). As shown in Figure 5, there was no change in monitoring results after 30 days of storing at 25 ± 2 °C ($p > 0.05$). When the product was stored at 4 ± 2 °C, it was noticed that there was a slight increase in particle size within 20 days. It is speculated to be due to recrystallization of free drug molecules from the existing larger crystal surfaces. When the product was stored at 4 ± 2 °C, a significant increase in particle size occurred. It is speculated that the Ostwald ripening phenomenon occurred. The particle sizes of the freshly prepared nanocrystals were monitored at 4 °C, 25 °C, and 40 °C for 30 days. As shown in Figure 5, the particle size remained constant after storing at 25 ± 2 °C for 30 days ($p > 0.05$). When stored at 4 ± 2 °C, a slight increase of particle size was observed from 268 nm to 300 nm for 20 days. This may be due to the recrystallization of free drug molecules on the surface of existing larger crystals. A significant increase in particle size was found at 40 ± 2 °C, where the Ostwald ripening phenomenon may have occurred. This indicates that the nanocrystals are unstable at higher temperatures. Therefore, appropriate temperature conditions are critical for the storage of NTD-NCs.

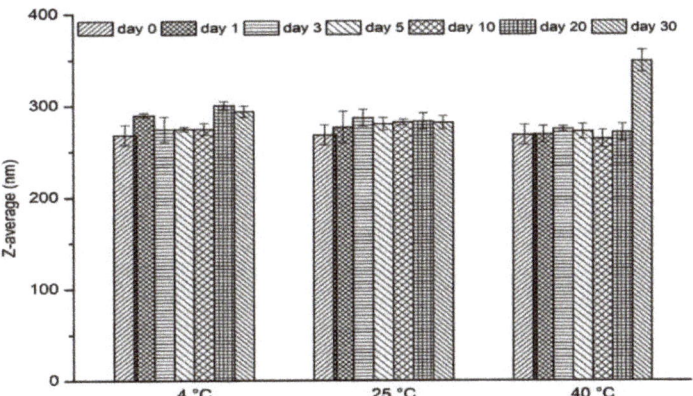

**Figure 5.** Particle size monitoring data of NTD nanocrystals stored at different temperature conditions for 30 days. Figure 5 was reprinted from ref. [92] with permission from Elsevier, Copyright® 2019.

Al Shaal et al. [93] prepared apigenin nanocrystals and assessed the long-term physical stability. The nanocrystals were stored under different conditions (refrigerated, room temperature, and 40 °C) for 6 months. According to PCS and LD data, neither significant increase in particle size nor change in particle size distribution can be found. No clearly visible aggregation was seen for the formulations on the day of production and after 6 months of storage at room temperature.

Luo et al. [94] prepared silymarin nanocrystal micro pills with PVP K30 and SDS as stabilizers. The produced nanocrystal micro pills were placed in a weighing open bottle for 10 days under high temperature/humidity/light conditions. The samples were taken on days 0, 5, and 10, respectively. The results showed that no significant changes occurred but the samples should be stored in a dry environment due to slight moisture absorption under high humidity conditions.

3.2.2. Cytotoxicity

Safety is the primary concern for pharmaceuticals, so toxicity assessment is a priority for new drug registration [90]. Cytotoxicity assays are assessed by measuring the number or growth of cells before and after exposure to the sample [95]. It can be divided into leachate test, direct contact test, and indirect contact test. Cell viability is often used as an important evaluation indicator. Cytotoxicity experiments can provide the prerequisites for *in vivo* animal experiments, so it plays a vital role in the evaluation of drugs. Additionally, embryo toxicity assessment is more rapid than whole organism methods (<96 h).

Sheng et al. [96] prepared oridonin nanocrystals (ORI-NCs) and studied the *in vitro* cytotoxicity of ORI-NCs on Madin-Darby canine kidney (MDCK) cells. Both ORI and ORI-NCs exhibited an effect on inhibiting MDCK cell proliferation. The decrease in cell viability with increasing concentration indicates that ORI-NCs and ORI drugs exert concentration-dependent effects on MDCK cells. With increasing concentration, the cell viability of ORI-NCs decreased more significantly than that of ORI.

Choi [97] explored the cytotoxicity characteristics of cilostazol nanocrystals (CLT-NCs) in the Caco-2 cells. CLT-NC was successfully prepared by the ultrasonic probe method with a mean size of 500–600 nm. Cytotoxicity of CLT-NC and CLT was assessed in Caco-2 cells. The experimental results demonstrated that all CLT concentrations tested (0.02–20.0 μg/mL) did not exhibit toxicity (Figure 6a). Therefore, CLT-NC is safe for cells at the related concentrations.

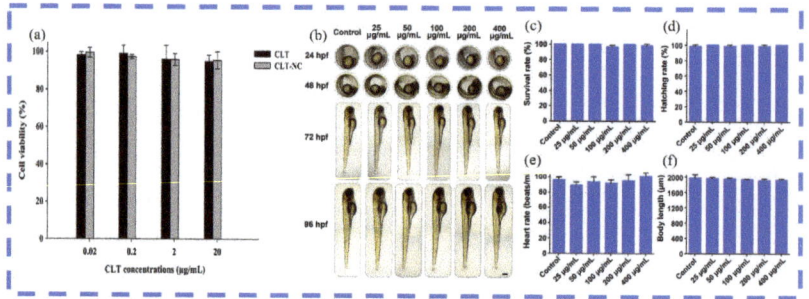

**Figure 6.** (**a**) Cell viability was assessed in Caco-2 cells after induction of CLT and CLT-NC; (**b**) Zebrafish embryo morphology at the indicated hpf following GB-NCs treatment (Scale bar stands for 250 μm); The survival rate (**c**), hatching rate (**d**), heart rate (**e**) and body length (**f**) of zebrafish larvae at 96 hpf in the presence of GB-NCs. Figure 6a was reprinted from ref. [97] with permission from Springer Nature, Copyright® 2020. Figure 6b–f were reprinted from ref. [98] with permission from Elsevier, Copyright® 2020.

Liu et al. [98] obtained ginkgolide B nanocrystals (GB-NCs), a potent anti-parkinsonism compound. The toxicity assays were done in zebrafish embryos. The zebrafish embryos showed no morphological changes following GB-NCs treatment (Figure 6b), and no differences in hatching, heart rate, body length, or survival occurred at 96 h post-fertilization (hpf) in GB-NCs relative to control groups (Figure 6c–f). These data confirmed that the GB-NCs are non-toxic to zebrafish.

3.2.3. *In Vitro* Dissolution

Dissolution or release of drugs is an important quality attribute of nanomedicines, which may have significant effects on drug absorption, *in vivo* safety and efficacy. *In vitro* dissolution or release can reflect the *in vivo* behavior of nanomedicines to some extent [41]. The flow cell method is often used to study the dissolution or release of drugs *in vitro*. A constant-flow pump is used to pump the release medium at the desired temperature into contact with the sample at the lower end of the flow cell at a suitable flow rate, and the release medium is filtered through the upper end of the flow cell to measure the concentration of the drug in the release medium at set time intervals [99]. Parameters such as dissolution rate and saturation solubility were determined to help predict the *in vivo* performance of drug nanocrystals. To a large extent, the dissolution rate depends on the size and surface area of the drug particles. The most important evaluation indicator is the dissolution curve, which can distinguish differences in product bioequivalence.

Zhu et al. [73] fabricated rod-shaped Nintedanib nanocrystals (BIBF-NCs) by the antisolvent precipitation-ultrasonic method with sodium carboxymethylcellulose as a stabilizer. The particle size was 325.30 ± 1.03 nm and the zeta potential was 32.70 ± 1.24 mV. Then, the *in vitro* dissolution of BIBF powder and BIBF-NCs in pH 6.8 PBS during 120 min were explored. Compared with the dissolution rate of BIBF powder at 10 min ((6.10 ± 1.55)%, $p < 0.05$), the BIBF-NCs was (73.74 ± 5.33)%. At 120 min both still exhibited large dissolution rate differences (Figure 7a). This phenomenon is attributed to smaller particle size reducing the diffusion layer thickness and increasing the surface area available for dissolution. Meanwhile, BIBF-NCs (rod-shape) have a larger surface area, resulting in higher dissolution rates. Additionally, the cumulative release of the drug in the gastrointestinal tract was simulated. The results showed that the cumulative release of BIBF-NCs was significantly higher than that of BIBF crude powder (Figure 7b).

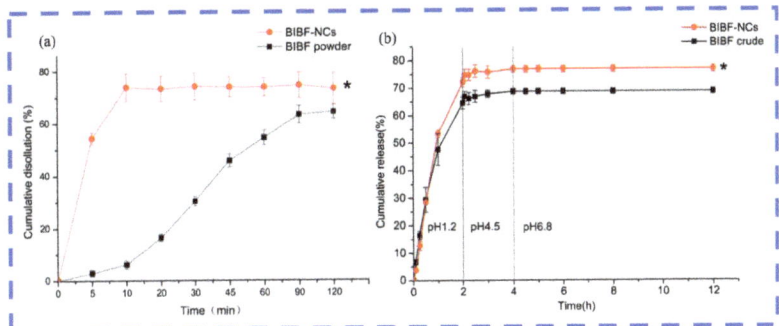

**Figure 7.** (**a**) Dissolution curves of BIBF powder and BIBF-NCs in pH 6.8 PBS (containing 0.5% ($w/v$) Tween 80); (**b**) BIBF release profile of the BIBF powder and BIBF-NCs in gastric and intestinal simulated liquid media. Data represent the mean ± SD, n = 3. (* $p < 0.05$, compared to BIBF crude). Figure 7a,b were reprinted from ref. [73] with permission from Elsevier, Copyright® 2022.

### 3.2.4. *In Vivo* Dissolution

In order to observe the functional, metabolic, and morphological changes caused by drug nanocrystals, it is often necessary to deliver the drugs to animals. There are various ways of drug delivery with an emphasis on oral and injection. After the intervention with the drug, the corresponding tissues are taken for index testing. For example, the drug concentration in blood and certain organs is measured. Parameters such as dissolution rate, bioavailability, or relative bioavailability are calculated to study the *in vivo* effects of nanocrystals. Other qualitative and quantitative means such as detection of lesion sites can also be used. Further *in vivo* pharmacokinetic and pharmacodynamic investigations can be performed to calculate the key parameters of peak concentration, time to peak, and area under the curve (AUC).

Soroushnia et al. [100] prepared midazolam nanosuspensions with a particle size of 197 ± 7 nm and zeta potential of 31 ± 4 mV using an ultrasound technique. Then, *in vivo* tests were carried out in rabbits. Plasma concentration-time curves (Figure 8) and pharmacokinetic parameters (Table 2) were obtained for the midazolam nanosuspensions. In the *in vivo* evaluation, a higher Cmax (111.90% higher), a higher AUC0-t (275.08%), and a shorter Tmax (15 min) were observed for the midazolam nanosuspension, indicating that the midazolam nanosuspension is more readily absorbed.

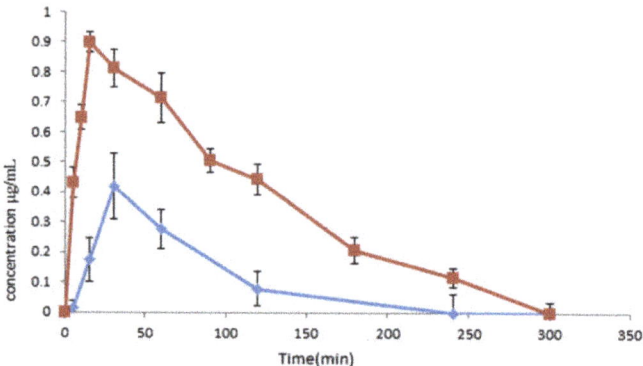

**Figure 8.** The mean plasma concentration-time curve of the midazolam coarse suspension (♦) and the midazolam nanosuspension (■), (n = 3). Figure 8 was reprinted from ref. [100] with permission from Springer Nature, Copyright® 2021.

**Table 2.** Main pharmacokinetic parameters of midazolam coarse suspension and nanosuspension in rabbits after buccal administration (n = 5). Table 2 was reprinted from ref. [100] with permission from Springer Nature, Copyright® 2021.

| Parameters | Midazolam Coarse Suspension | Midazolam Nanosuspension |
|---|---|---|
| $C_{max}$ (ng/mL) | 420 ± 12 | 890 ± 14 |
| $T_{max}$ (min) | 30 ± 3 | 15 ± 2 |
| AUC (ng/mL·h) | 28.9 ± 2.6 | 108.4 ± 8.2 |

## 4. Applications

Nanocrystal formulations can be administered by various routes (Figure 9), where oral routes of drug delivery account for more than 60% of current drug nanocrystal delivery routes [101,102]. Nanocrystal-based formulations (Table 3) are widely used to treat cancer, inflammatory, cardiovascular, depression, and other diseases [35].

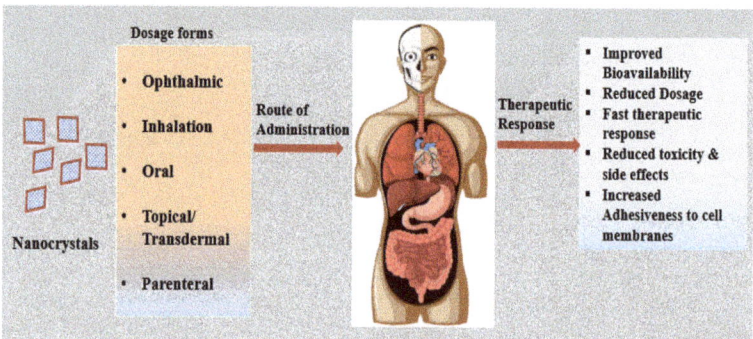

**Figure 9.** A schematic diagram of the nanocrystal drug delivery route and therapeutic response. Figure 9 was reprinted from ref. [102] with permission from Springer Nature, Copyright® 2020.

The oral route is considered to be one of the most appropriate, safe, and preferred routes [103]. During oral administration, dissolution of the nanocrystals begins in the intestine, where the drug particles are slightly absorbed by the cells. In contrast, cellular interactions and drug uptake play a key role in nanocrystals administered via parenteral routes. It can enhance the bioavailability, reduce toxicity, and release in specific targeting sites, such as the brain, lung, liver, kidney, or colon [4]. For pulmonary administration, the nebulized nanosuspension had a significantly higher inhalable fraction and also showed stronger mucus adhesion. Ocularly administered nanodropable dosage forms also have high drug loads and long-lasting drug effects [35]. In conclusion, the application and combination of various routes of drug administration will provide the basis for nanocrystals to play an increasingly important role in disease treatment.

**Table 3.** Examples of existing commercial products for drug nanocrystals.

| Brand Name (Chemical Name) | Preparation Technology | Route of Delivery | Uses | Problems with Old Treatment | Inference |
| --- | --- | --- | --- | --- | --- |
| Verelan PM * (Verapamil) | Milling | Oral | Anti-arrhythmic agent | Use organic acid and addition of enteric materials | Increase oral bioavailability |
| Rapamune * (Rapamycin) | Milling | Oral | Immuno-suppressive agent | Poor solubility | Require low dissolution medium |
| Neprelan * (Naproxen sodium) | Milling | Oral | Anti-inflammatory agent | Fed condition effects | Increase rate of absorption |
| Invega * Sustenna (Paliperidone palmitate) | HPH | Parenteral | Anti-depressant agent | Low solubility and dissolution rate | Increase solubility |

Note: * only represents the brand name of the drug.

## 5. Conclusions and Perspective

In this work, we reviewed the preparation method for drug nanocrystals, containing top-down (e.g., WBM, HPH), bottom-up (e.g., LAS, SCF), and combinative (e.g., Nanoedge, SmartCrystal) technology. After that, characterization methods and application scope of nanocrystals including DLS, SEM, XPRD, etc. were discussed. Next, evaluation paths and examples on stability, cytotoxicity, and *in vitro/in vivo* dissolution rate were introduced, which can reflect the practical performance of drug nanocrystals. Lastly, the applicability of drug nanocrystals was broadly demonstrated by various routes (e.g., oral, parenteral).

Although great progress has been made in nanocrystal technology, there are still many difficulties to be faced. For drug nanocrystal preparation technology, it is necessary to continuously improve the productive efficiency and stability of nanocrystals. For this purpose, the continuous strengthening of basic research on nanocrystal preparation, such as the optimization of nanomedicine formulations and process parameters is required. Furthermore, on industrial production of nanocrystals, the transferability from laboratory scale to industrial scale should be seriously considered.

What follows is the characterization and evaluation of nanocrystal products including the properties of the nanoparticles themselves and *in vivo/in vitro* behaviors. Until now, the release of nanocrystals is not yet fully understood. More detailed preclinical *in vivo* experiments are needed to explain the mechanism of transportation and absorption of nanocrystals *in vivo*, both for efficacy and safety. Additionally, more robust and in-depth *in vitro/in vivo* correlation evaluations are also required.

In addition, for specific applications area of nanocrystals, the high drug-carrying capacity of nanocrystals makes them a hot spot for the study of targeted insoluble drug formulations. However, further research is still needed on targeted drug nanocrystal delivery formulations, rates, and mechanisms.

Overall, nanocrystal technology has the potential for industrial applications and the ability to take on insolubility problems. This makes it a prospering technology for optimizing the bioavailability of insoluble drugs. Therefore, continuous efforts are needed to introduce drug nanocrystals into the pharmaceutical market.

**Author Contributions:** Investigation, Q.R., M.W. and J.O.; writing—original draft preparation, Q.R.; writing—review, D.H. and Z.G.; writing—modifications, Q.R., M.W. and W.K.; supervision, J.O. and J.G. All authors have read and agreed to the published version of the manuscript.

**Funding:** This work was financially supported by Shandong Provincial Key R&D Program (Major Key Technology Project) 2021CXGC01051 and Academic and technical leader training program for major disciplines in Jiangxi Province (20212BCJ23001).

**Institutional Review Board Statement:** Not applicable.

**Informed Consent Statement:** Not applicable.

**Acknowledgments:** This work was financially supported by Shandong Provincial Key R&D Program (Major Key Technology Project) 2021CXGC01051 and Academic and technical leader training program for major disciplines in Jiangxi Province (20212BCJ23001). The financial support of Haihe Laboratory of Sustainable Chemical Transformations and the Institute of Shaoxing, Tianjin University.

**Conflicts of Interest:** The authors declare no conflict of interest.

## References

1. Loftsson, T.; Brewster, M.E. Pharmaceutical applications of cyclodextrins: Basic science and product development. *J. Pharm. Pharmacol.* **2010**, *62*, 1607–1621. [CrossRef] [PubMed]
2. Chogale, M.M.; Ghodake, V.N.; Patravale, V.B. Performance parameters and characterizations of nanocrystals: A brief review. *Pharmaceutics* **2016**, *8*, 26. [CrossRef] [PubMed]
3. Price, D.J.; Nair, A.; Kuentz, M.; Dressman, J.; Saal, C. Calculation of drug-polymer mixing enthalpy as a new screening method of precipitation inhibitors for supersaturating pharmaceutical formulations. *Eur. J. Pharm. Sci.* **2019**, *132*, 142–156. [CrossRef] [PubMed]
4. Mohammad, I.S.; Hu, H.; Yin, L.; He, W. Drug nanocrystals: Fabrication methods and promising therapeutic applications. *Int. J. Pharm.* **2019**, *562*, 187–202. [CrossRef] [PubMed]
5. Butler, J.M.; Dressman, J.B. The Developability Classification System: Application of Biopharmaceutics Concepts to Formulation Development. *J. Pharm. Sci.* **2010**, *99*, 4940–4954. [CrossRef]
6. Al-Kassas, R.; Bansal, M.; Shaw, J. Nanosizing techniques for improving bioavailability of drugs. *J. Control. Release* **2017**, *260*, 202–212. [CrossRef]
7. Padrela, L.; Rodrigues, M.A.; Duarte, A.; Dias, A.M.A.; Braga, M.E.M.; de Sousa, H.C. Supercritical carbon dioxide-based technologies for the production of drug nanoparticles/nanocrystals—A comprehensive review. *Adv. Drug Deliv. Rev.* **2018**, *131*, 22–78. [CrossRef]
8. Kalepu, S.; Nekkanti, V. Insoluble drug delivery strategies: Review of recent advances and business prospects. *Acta Pharm. Sin. B* **2015**, *5*, 442–453. [CrossRef]
9. Kumar, S.; Bhargava, D.; Thakkar, A.; Arora, S. Drug carrier systems for solubility enhancement of BCS class II drugs: A critical review. *Crit. Rev. Ther. Drug Carr. Syst.* **2013**, *30*, 217–256. [CrossRef]
10. Peltonen, L.; Hirvonen, J. Drug nanocrystals—Versatile option for formulation of poorly soluble materials. *Int. J. Pharm.* **2018**, *537*, 73–83. [CrossRef]
11. Zheng, A.; Shi, J. Research progress in nanocrystal drugs. *J. Int. Pharm. Res.* **2012**, *39*, 177–183.
12. Sinha, B.; Müller, R.H.; Möschwitzer, J.P. Bottom-up approaches for preparing drug nanocrystals: Formulations and factors affecting particle size. *Int. J. Pharm.* **2013**, *453*, 126–141. [CrossRef]
13. Kesisoglou, F.; Panmai, S.; Wu, Y. Nanosizing–oral formulation development and biopharmaceutical evaluation. *Adv. Drug Deliv. Rev.* **2007**, *59*, 631–644. [CrossRef]
14. Müller, R.H.; Peters, K. Nanosuspensions for the formulation of poorly soluble drugs: I. Preparation by a size-reduction technique. *Int. J. Pharm.* **1998**, *160*, 229–237. [CrossRef]
15. Yu, Q.; Wu, X.; Zhu, Q.; Wu, W.; Chen, Z.; Li, Y.; Lu, Y. Enhanced transdermal delivery of meloxicam by nanocrystals: Preparation, *in vitro* and *in vivo* evaluation. *Asian J. Pharm.* **2018**, *13*, 518–526. [CrossRef] [PubMed]
16. Yue, P.; Liu, Y.; Xie, J.; Chen, Y.; Yang, M. Review and prospect on preparation technology of drug nanocrystals in the past thirty years. *Acta Pharm. Sin.* **2018**, *53*, 529–537.
17. Gao, L.; Liu, G.; Ma, J.; Wang, X.; Zhou, L.; Li, X.; Wang, F. Application of Drug Nanocrystal Technologies on Oral Drug Delivery of Poorly Soluble Drugs. *Pharm. Res.* **2013**, *30*, 307–324. [CrossRef]
18. Lu, Y.; Qi, J.; Dong, X.; Zhao, W.; Wu, W. The *in vivo* fate of nanocrystals. *Drug Discov. Today* **2017**, *22*, 744–750. [CrossRef]
19. Mou, D.; Liao, Y.; Zhou, X.; Wan, J.; Yang, X. Advances in Nanocrystal Medicine. *Her. Med.* **2020**, *39*, 1257–1261.
20. Zhou, Y.; Du, J.; Wang, L.; Wang, Y. Nanocrystals Technology for Improving Bioavailability of Poorly Soluble Drugs: A Mini-Review. *J. Nanosci. Nanotechnol.* **2017**, *17*, 18–28. [CrossRef]
21. Müller, R.H.; Gohla, S.; Keck, C.M. State of the art of nanocrystals—Special features, production, nanotoxicology aspects and intracellular delivery. *Eur. J. Pharm. Biopharm.* **2011**, *78*, 1–9. [CrossRef] [PubMed]

22. Gao, L.; Zhang, D.; Chen, M. Drug nanocrystals for the formulation of poorly soluble drugs and its application as a potential drug delivery system. *J. Nanopart. Res.* **2008**, *10*, 845–862. [CrossRef]
23. Lu, Y.; Li, Y.; Wu, W. Injected nanocrystals for targeted drug delivery. *Acta Pharm. Sin. B* **2016**, *6*, 106–113. [CrossRef]
24. Bitterlich, A.; Laabs, C.; Krautstrunk, I.; Dengler, M.; Juhnke, M.; Grandeury, A.; Bunjes, H.; Kwade, A. Process parameter dependent growth phenomena of naproxen nanosuspension manufactured by wet media milling. *Eur. J. Pharm. Biopharm.* **2015**, *92*, 171–179. [CrossRef] [PubMed]
25. Chen, Z.; Wu, W.; Lu, Y. What is the future for nanocrystal-based drug-delivery systems? *Ther. Deliv.* **2020**, *11*, 225–229. [CrossRef] [PubMed]
26. Raval, A.; Patel, M.M. Preparation and Characterization of Nanoparticles for Solubility and Dissolution Rate Enhancement of Meloxicam. *Int. Res. J. Pharm.* **2011**, *1*, 42–49.
27. Liu, T.; Müller, R.H.; Möschwitzer, J.P. Effect of drug physico-chemical properties on the efficiency of top-down process and characterization of nanosuspension. *Expert Opin. Drug Deliv.* **2015**, *12*, 1741–1754. [CrossRef]
28. Jarvis, M.; Krishnan, V.; Mitragotri, S. Nanocrystals: A perspective on translational research and clinical studies. *Bioeng. Transl. Med.* **2019**, *4*, 5–16. [CrossRef]
29. Funahashi, I.; Kondo, K.; Ito, Y.; Yamada, M.; Niwa, T. Novel contamination-free wet milling technique using ice beads for poorly water-soluble compounds. *Int. J. Pharm.* **2019**, *563*, 413–425. [CrossRef]
30. Liu, J.; Xu, Y.; Li, M.; Qian, H. Research progress of nanomedicine. *Pharm. Clin. Res.* **2020**, *28*, 51–55.
31. Shegokar, R.; Müller, R.H. Nanocrystals: Industrially feasible multifunctional formulation technology for poorly soluble actives. *Int. J. Pharm.* **2010**, *399*, 129–139. [CrossRef] [PubMed]
32. Keck, C.M.; Müller, R.H. Drug nanocrystals of poorly soluble drugs produced by high pressure homogenisation. *Eur. J. Pharm. Biopharm.* **2006**, *62*, 3–16. [CrossRef]
33. Raghava Srivalli, K.M.; Mishra, B. Drug nanocrystals: A way toward scale-up. *Saudi Pharm. J.* **2016**, *24*, 386–404. [CrossRef] [PubMed]
34. Liu, J.; Sun, Y.; Cheng, M.; Liu, Q.; Liu, W.; Gao, C.; Feng, J.; Jin, Y.; Tu, L. Improving Oral Bioavailability of Luteolin Nanocrystals by Surface Modification of Sodium Dodecyl Sulfate. *AAPS PharmSciTech* **2021**, *22*, 133. [CrossRef]
35. Joshi, K.; Chandra, A.; Jain, K.; Talegaonkar, S. Nanocrystalization: An Emerging Technology to Enhance the Bioavailability of Poorly Soluble Drugs. *Pharm. Nanotechnol.* **2019**, *7*, 259–278. [CrossRef] [PubMed]
36. Sugiyama, T.; Asahi, T.; Masuhara, H. Formation of 10 nm-sized Oxo(phtalocyaninato)vanadium(IV) Particles by Femtosecond Laser Ablation in Water. *Chem. Lett.* **2004**, *33*, 724–725. [CrossRef]
37. Tian, Y.; Peng, Y.; Zhang, Z.; Zhang, H.; Gao, X. Research progress on preparation technology of nanocrystal drugs. *Acta Pharm. Sin.* **2021**, *56*, 1902–1910.
38. Guo, Z.; Zhang, M.; Li, H.; Wang, J.; Kougoulos, E. Effect of ultrasound on anti-solvent crystallization process. *J. Cryst. Growth* **2005**, *273*, 555–563. [CrossRef]
39. Chen, J. The Pharmacokinetics Evaluation and Preparation of Hydrochlorothiazide and Valsartan Nanocrystals. Master's Thesis, Guangdong Pharmaceutical University, Guangzhou, China, 2016.
40. Thorat, A.A.; Dalvi, S.V. Liquid antisolvent precipitation and stabilization of nanoparticles of poorly water soluble drugs in aqueous suspensions: Recent developments and future perspective. *Chem. Eng. J.* **2012**, *181–182*, 1–34. [CrossRef]
41. Technical Guidelines for Nanomedicine Quality Control Research (for Trial Implementation). Available online: https://www.nmpa.gov.cn/ (accessed on 18 March 2021).
42. Chow, A.H.L.; Tong, H.H.Y.; Chattopadhyay, P.; Shekunov, B.Y. Particle Engineering for Pulmonary Drug Delivery. *Pharm. Res.* **2007**, *24*, 411–437. [CrossRef]
43. Ding, Y.; Kang, B.; Wang, J. Preparation and *in vitro* evaluation of budesonide inhalation nanosuspension. *Chin. J. Pharm.* **2017**, *48*, 1131–1137.
44. Han, X.; Wang, M.; Ma, Z.; Xue, P.; Wang, Y. A new approach to produce drug nanosuspensions CO2-assisted effervescence to produce drug nanosuspensions. *Colloids Surf. B. Biointerfaces* **2016**, *143*, 107–110. [CrossRef]
45. Wang, Y.; Han, X.; Wang, J.; Wang, Y. Preparation, characterization and *in vivo* evaluation of amorphous tacrolimus nanosuspensions produced using CO2-assisted in situ nanoamorphization method. *Int. J. Pharm.* **2016**, *505*, 35–41. [CrossRef] [PubMed]
46. Chen, J.F.; Wang, Y.H.; Guo, F.; Wang, X.M.; Zheng, C. Synthesis of Nanoparticles with Novel Technology: High-Gravity Reactive Precipitation. *Ind. Eng. Chem. Res.* **2000**, *39*, 948–954. [CrossRef]
47. Chiou, H.; Li, L.; Hu, T.; Chan, H.-K.; Chen, J.-F.; Yun, J. Production of salbutamol sulfate for inhalation by high-gravity controlled antisolvent precipitation. *Int. J. Pharm.* **2007**, *331*, 93–98. [CrossRef]
48. Yin, Y.; Deng, H.; Wu, K.; He, B.; Dai, W.; Zhang, H.; Le, Y.; Wang, X.; Zhang, Q. A multiaspect study on transcytosis mechanism of sorafenib nanogranules engineered by high-gravity antisolvent precipitation. *J. Control. Release* **2020**, *323*, 600–612. [CrossRef] [PubMed]
49. Zhang, J.; Huang, Y.; Liu, D.; Gao, Y.; Qian, S. Preparation of apigenin nanocrystals using supercritical antisolvent process for dissolution and bioavailability enhancement. *Eur. J. Pharm. Sci.* **2013**, *48*, 740–747. [CrossRef] [PubMed]
50. Chen, H. Development and Bioavailability Evaluation in Hens of Florfenicol Nanocrystals. Master's Thesis, Northwest A&F University, Xianyang, China, 2021.

51. Kipp, J.E.; Wong, J.C.T.; Doty, M.J.; Rebbeck, C.L. Microprecipitation Method for Preparing Submicron Suspensions. 2001. Available online: https://patents.google.com/patent/US7037528B2/en (accessed on 18 March 2021).
52. Pardhi, V.P.; Verma, T.; Flora, S.J.S.; Chandasana, H.; Shukla, R. Nanocrystals: An Overview of Fabrication, Characterization and Therapeutic Applications in Drug Delivery. *Curr. Pharm. Des.* **2018**, *24*, 5129–5146. [CrossRef]
53. Möschwitzer, J.P. Drug nanocrystals in the commercial pharmaceutical development process. *Int. J. Pharm.* **2013**, *453*, 142–156. [CrossRef]
54. Romero, G.B.; Chen, R.; Keck, C.M.; Müller, R.H. Industrial concentrates of dermal hesperidin smartCrystals®—Production, characterization & long-term stability. *Int. J. Pharm.* **2015**, *482*, 54–60.
55. Gholap, A.; Borude, S.; Mahajan, A.; Amol, M.; Gholap, D. Smart Crystals Technology: A Review. *Pharmacologyonline* **2011**, *3*, 238–243.
56. Li, Y.; Wang, Y.; Yue, P.F.; Hu, P.Y.; Wu, Z.F.; Yang, M.; Yuan, H.L. A novel high-pressure precipitation tandem homogenization technology for drug nanocrystals production—A case study with ursodeoxycholic acid. *Pharm. Dev. Technol.* **2014**, *19*, 662–670. [CrossRef] [PubMed]
57. Möschwitzer, J.; Müller, R.H. New method for the effective production of ultrafine drug nanocrystals. *J. Nanosci. Nanotechnol.* **2006**, *6*, 3145–3153. [CrossRef] [PubMed]
58. Yu, X. Preparation and Formulation of Novel Drug Nanoparticles Based on High Efficient Solubilization. Master's Thesis, Qingdao University of Science & Technology, Qingdao, China, 2021.
59. Malamatari, M.; Taylor, K.M.G.; Malamataris, S.; Douroumis, D.; Kachrimanis, K. Pharmaceutical nanocrystals: Production by wet milling and applications. *Drug Discov. Today* **2018**, *23*, 534–547. [CrossRef]
60. Wadhawan, J.; Parmar, P.K.; Bansal, A.K. Nanocrystals for improved topical delivery of medium soluble drug: A case study of acyclovir. *J. Drug Deliv. Sci. Technol.* **2021**, *65*, 102662. [CrossRef]
61. Martena, V.; Shegokar, R.; Di Martino, P.; Müller, R.H. Effect of four different size reduction methods on the particle size, solubility enhancement and physical stability of nicergoline nanocrystals. *Drug Dev. Ind. Pharm.* **2014**, *40*, 1199–1205. [CrossRef]
62. Morakul, B.; Suksiriworapong, J.; Chomnawang, M.T.; Langguth, P.; Junyaprasert, V.B. Dissolution enhancement and *in vitro* performance of clarithromycin nanocrystals produced by precipitation–lyophilization–homogenization method. *Eur. J. Pharm. Biopharm.* **2014**, *88*, 886–896. [CrossRef]
63. Zhao, H.; Wang, J.; Zhang, H.; Shen, Z.; Yun, J.; Chen, J. Facile Preparation of Danazol Nanoparticles by High-Gravity Anti-solvent Precipitation (HGAP) Method. *Chin. J. Chem. Eng.* **2009**, *17*, 318–323. [CrossRef]
64. Chen, G.; Zhang, Y.; Tian, C.; Ran, X.; Zhu, L.; Chen, B.; Zhao, J. Preparation and characterization of albendazole nanocrystals. *Chin. J. Hosp. Pharm.* **2020**, *40*, 260–264.
65. Chen, X.; Young, T.J.; Sarkari, M.; Williams, R.O.; Johnston, K.P. Preparation of cyclosporine A nanoparticles by evaporative precipitation into aqueous solution. *Int. J. Pharm.* **2002**, *242*, 3–14. [CrossRef]
66. Huang, T.; Lin, Q.; Qian, Y.; Xu, X.; Zhou, J. Preparation and Characteristics of Celecoxib Nanocrystals. *Chin. J. Pharm.* **2015**, *46*, 358–363.
67. Jin, S.; Yuan, H.; Jin, S.; Lv, Q.; Bai, J.; Han, J. Preparation of baicalin nanocrystal pellets and preliminary study on its pharmacokinetics. *China J. Chin. Mater. Med.* **2013**, *38*, 1156–1159.
68. Zuo, X. Preparation of New Nnanopharmaceutical Combination Method and Its Preparation. Master's Thesis, Qingdao University of Science & Technology, Qingdao, China, 2019.
69. Zhang, H.; Meng, Y.; Wang, X.; Dai, W.; Wang, X.; Zhang, Q. Pharmaceutical and pharmacokinetic characteristics of different types of fenofibrate nanocrystals prepared by different bottom-up approaches. *Drug Deliv.* **2014**, *21*, 588–594. [CrossRef] [PubMed]
70. Liu, D.; Xu, H.; Tian, B.; Yuan, K.; Pan, H.; Ma, S.; Yang, X.; Pan, W. Fabrication of carvedilol nanosuspensions through the antisolvent precipitation-ultrasonication method for the improvement of dissolution rate and oral bioavailability. *AAPS PharmSciTech* **2012**, *13*, 295–304. [CrossRef] [PubMed]
71. Lu, Y.; Wang, Z.-h.; Li, T.; McNally, H.; Park, K.; Sturek, M. Development and evaluation of transferrin-stabilized paclitaxel nanocrystal formulation. *J. Control. Release* **2014**, *176*, 76–85. [CrossRef]
72. Latif, R.; Makar, R.R.; Hosni, E.A.; El Gazayerly, O.N. The potential of intranasal delivery of nanocrystals in powder form on the improvement of zaleplon performance: In-vitro, in-vivo assessment. *Drug Dev. Ind. Pharm.* **2021**, *47*, 268–279. [CrossRef]
73. Zhu, Y.; Fu, Y.; Zhang, A.; Wang, X.; Zhao, Z.; Zhang, Y.; Yin, T.; Gou, J.; Wang, Y.; He, H.; et al. Rod-shaped nintedanib nanocrystals improved oral bioavailability through multiple intestinal absorption pathways. *Eur. J. Pharm. Sci.* **2022**, *168*, 106047. [CrossRef]
74. Shah, S.M.H.; Ullah, F.; Khan, S.; Shah, S.M.M.; Isreb, M. Fabrication and Evaluation of Smart Nanocrystals of Artemisinin for Antimalarial and Antibacterial Efficacy. *Afr. J. Tradit. Complement. Altern. Med.* **2017**, *14*, 251–262. [CrossRef]
75. Pu, X.; Sun, J.; Li, M.; He, Z. Formulation of Nanosuspensions as a New Approach for the Delivery of Poorly Soluble Drugs. *Curr. Nanosci.* **2009**, *5*, 417–427. [CrossRef]
76. Soisuwan, S.; Teeranachaideekul, V.; Wongrakpanich, A.; Langguth, P.; Junyaprasert, V.B. In vitro performances and cellular uptake of clarithromycin nanocrystals produced by media milling technique. *Powder Technol.* **2018**, *338*, 471–480. [CrossRef]
77. Chen, Y.; Gui, Y.; Luo, Y.; Liu, Y.; Tu, L.; Ma, Y.; Yue, P.; Yang, M. Design and evaluation of inhalable nanocrystals embedded microparticles with enhanced redispersibility and bioavailability for breviscapine. *Powder Technol.* **2021**, *377*, 128–138. [CrossRef]

78. Wang, Y.; Xuan, J.; Zhao, G.; Wang, D.; Ying, N.; Zhuang, J. Improving stability and oral bioavailability of hydroxycamptothecin via nanocrystals in microparticles (NCs/MPs) technology. *Int. J. Pharm.* **2021**, *604*, 120729. [CrossRef] [PubMed]
79. Ančić, D.; Oršolić, N.; Odeh, D.; Tomašević, N.; Pepić, I.; Ramić, S. Resveratrol and its nanocrystals: A promising approach for cancer therapy? *Toxicol. Appl. Pharmacol.* **2022**, *435*, 115851. [CrossRef] [PubMed]
80. Yau, S.T.; Thomas, B.R.; Vekilov, P.G. Molecular mechanisms of crystallization and defect formation. *Phys. Rev. Lett.* **2000**, *85*, 353–356. [CrossRef]
81. Du, J.; Li, X.; Zhao, H.; Zhou, Y.; Wang, L.; Tian, S.; Wang, Y. Nanosuspensions of poorly water-soluble drugs prepared by bottom-up technologies. *Int. J. Pharm.* **2015**, *495*, 738–749. [CrossRef] [PubMed]
82. Parhi, B.; Bharatiya, D.; Swain, S.K. Application of quercetin flavonoid based hybrid nanocomposites: A review. *Saudi Pharm. J.* **2020**, *28*, 1719–1732. [CrossRef]
83. Tiede, K.; Boxall, A.B.A.; Tear, S.P.; Lewis, J.; David, H.; Hassellöv, M. Detection and characterization of engineered nanoparticles in food and the environment. *Food Addit. Contam. Part A* **2008**, *25*, 795–821. [CrossRef]
84. Doyle, W.M. Principles and Applications of Fourier Transform Infra-Red (FTIR) Process Analysis. 2017. Available online: https://www.semanticscholar.org/paper/Principles-and-Applications-of-Fourier-Transform-(-Doyle/8b9108726fe76043badeecd1c75ed6e72352b8a1 (accessed on 18 March 2021).
85. Gao, Q. Preparation and Characterization of ITZ Capsules Based on Nanocrystal Technology. Master's Thesis, Hebei University of Science and Technology, Shijiazhuang, China, 2015.
86. Honary, S.; Zahir, F. Effect of Zeta Potential on the Properties of Nano-Drug Delivery Systems—A Review (Part 2). *Trop. J. Pharm. Res.* **2013**, *12*, 265–273.
87. Agarwal, V.; Kaushik, N.; Sharma, P.K. Nanocrystal Approaches for Poorly Soluble Drugs and their Role in Development of Marketed Formulation. *Drug Deliv. Lett.* **2021**, *11*, 275–294. [CrossRef]
88. Blom, K.; Senkowski, W.; Jarvius, M.; Berglund, M.; Rubin, J.; Lenhammar, L.; Parrow, V.; Andersson, C.; Loskog, A.; Fryknäs, M.; et al. The anticancer effect of mebendazole may be due to M1 monocyte/macrophage activation via ERK1/2 and TLR8-dependent inflammasome activation. *Immunopharmacol. Immunotoxicol.* **2017**, *39*, 199–210. [CrossRef]
89. Parmar, P.K.; Wadhawan, J.; Bansal, A.K. Pharmaceutical nanocrystals: A promising approach for improved topical drug delivery. *Drug Discov. Today* **2021**, *26*, 2329–2349. [CrossRef] [PubMed]
90. Gao, L.; Liu, G.; Ma, J.; Wang, X.; Zhou, L.; Li, X. Drug nanocrystals: In vivo performances. *J. Control. Release* **2012**, *160*, 418–430. [CrossRef] [PubMed]
91. Li, Y.-c.; Dong, L.; Jia, A.; Chang, X.-m.; Xue, H. Preparation of solid lipid nanoparticles loaded with traditional Chinese medicine by high-pressure homogenization. *J. South. Med. Univ.* **2006**, *26*, 541–544.
92. Guo, M.; Dong, Y.; Wang, Y.; Ma, M.; He, Z.; Fu, Q. Fabrication, characterization, stability and *in vitro* evaluation of nitrendipine nanocrystals by media milling. *Powder Technol.* **2019**, *358*, 20–28. [CrossRef]
93. Al Shaal, L.; Mueller, R.H.; Shegokar, R. smartCrystal combination technology—Scale up from lab to pilot scale and long term stability. *Pharmazie* **2010**, *65*, 877–884.
94. Luo, K.; Li, X.; Luo, J.; Yang, L.; Lin, H.; Mou, Q. Characterization of Silymarin Nanocrystal Pellets and Investigation of Its Stability. *Chin. J. Exp. Tradit. Med. Formulae* **2017**, *23*, 7–11.
95. Assad, M.; Jackson, N. Biocompatibility Evaluation of Orthopedic Biomaterials and Medical Devices: A Review of Safety and Efficacy Models. In *Encyclopedia of Biomedical Engineering*; Narayan, R., Ed.; Elsevier: Oxford, UK, 2019; pp. 281–309.
96. Sheng, H.; Zhang, Y.; Nai, J.; Wang, S.; Dai, M.; Lin, G.; Zhu, L.; Zhang, Q. Preparation of oridonin nanocrystals and study of their endocytosis and transcytosis behaviours on MDCK polarized epithelial cells. *Pharm. Biol.* **2020**, *58*, 518–527. [CrossRef]
97. Choi, J.-S. Design of Cilostazol Nanocrystals for Improved Solubility. *J. Pharm. Innov.* **2020**, *15*, 416–423. [CrossRef]
98. Liu, Y.; Liu, W.; Xiong, S.; Luo, J.; Li, Y.; Zhao, Y.; Wang, Q.; Zhang, Z.; Chen, X.; Chen, T. Highly stabilized nanocrystals delivering Ginkgolide B in protecting against the Parkinson's disease. *Int. J. Pharm.* **2020**, *577*, 119053. [CrossRef] [PubMed]
99. Xie, Y.; Yue, P.; Dan, J.; Xu, J.; Zheng, Q.; Yang, M. Research Progress of in Vitro Release Evaluation Methods for Nano Preparation. *Chin. Pharm. J.* **2016**, *51*, 861–866.
100. Soroushnia, A.; Ganji, F.; Vasheghani-Farahani, E.; Mobedi, H. Preparation, optimization, and evaluation of midazolam nanosuspension: Enhanced bioavailability for buccal administration. *Prog. Biomater.* **2021**, *10*, 19–28. [CrossRef] [PubMed]
101. Chen, M.L.; John, M.; Lee, S.L.; Tyner, K.M. Development Considerations for Nanocrystal Drug Products. *AAPS J.* **2017**, *19*, 642–651. [CrossRef] [PubMed]
102. Jahangir, M.A.; Imam, S.S.; Muheem, A.; Chettupalli, A.; Al-Abbasi, F.A.; Nadeem, M.S.; Kazmi, I.; Afzal, M.; Alshehri, S. Nanocrystals: Characterization Overview, Applications in Drug Delivery, and Their Toxicity Concerns. *J. Pharm. Innov.* **2022**, *17*, 237–248. [CrossRef]
103. Shojaei, A.H. Buccal mucosa as a route for systemic drug delivery: A review. *J. Pharm. Pharm. Sci.* **1998**, *1*, 15–30. [PubMed]

Article

# Measurement and Correlation of the Solubility of Florfenicol in Four Binary Solvent Mixtures from $T$ = (278.15 to 318.15) K

Xinyuan Zhang [1], Pingping Cui [1], Qiuxiang Yin [1,2] and Ling Zhou [1,*]

[1] State Key Laboratory of Chemical Engineering, School of Chemical Engineering and Technology, Tianjin University, Tianjin 300072, China
[2] Collaborative Innovation Center of Chemical Science and Engineering of Tianjin, Tianjin 300072, China
* Correspondence: zhouling@tju.edu.cn

**Abstract:** Florfenicol is an excellent antibiotic and is widely used in animal bacterial diseases. However, its poor water solubility leads to various problems, such as poor absorption and bioavailability. The development of nanocrystals is one of the most useful methods for solubilizing florfenicol, which often requires solubility data of florfenicol in different mixed solvents. In this work, the solubility of florfenicol was determined by the gravimetric method in methanol + water, ethanol + water, 1-propanol + water, and isopropanol + water binary solvents at temperatures from 278.15 to 318.15 K. In these four mixed solvents, the solubility of florfenicol increased with the increase in temperature. The solubility of florfenicol in methanol + water mixed solvent increases with the decrease in water ratio, while the solubility of florfenicol in ethanol + water, 1-propanol + water, or isopropanol + water mixed solvents increased first and then decreased with the decrease in water ratio, indicating a cosolvency phenomenon. The modified Apelblat model, CNIBS/R-K model, Jouyban–Acree model, and NRTL model were used to correlate the solubility data of florfenicol in four binary solvents. RMSD values indicated that the calculated values are in good agreement with the experimental solubility data for all four models, among which the CNIBS/R-K model provides the best correlation.

**Keywords:** florfenicol; solubility; binary solvent system; cosolvency; correlation model

## 1. Introduction

Florfenicol ($C_{12}H_{14}Cl_2FNO_4S$, Figure 1) is broad-spectrum chloramphenicol antibiotic for bacterial diseases in animals [1]. Florfenicol has the advantages of good antibacterial effect, high safety, low toxicity and side effects, and low possibility for the development of drug resistance [2]. However, the solubility of florfenicol in water is only 0.9 mg/mL at 25 °C under atmospheric pressure. Its poor water solubility leads to problems such as poor absorption in animals and poor drug bioavailability [3], which limit the clinical application of florfenicol and the diversity of pharmaceutical preparations to a certain extent [4]. Therefore, it is of great importance to improve its bioavailability by solubilization.

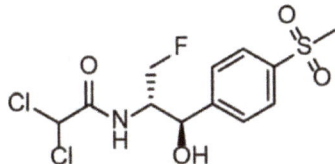

**Figure 1.** Chemical structure of florfenicol.

Many studies have focused on the solubilization of drugs, and various solubilization methods have been developed such as chemical modification [5], cyclodextrin inclusion [6], and application of nanocrystals [7]. Among these methods, the development of nanocrystals is preferable for the solubilization of florfenicol due to the advantages of avoiding the involvement of unexpected chemicals or impurities, which can produce florfenicol products with higher drug purity, and less toxicity and side effects [8]. Antisolvent crystallization is an effective method to prepare nanocrystals. Meanwhile, florfenicol exhibited good solubility in alcohol solvents such as methanol, ethanol, 1-propanol, and isopropanol, which makes them potential solvents for the antisolvent crystallization of florfenicol. The determination of the solubility of florfenicol in the binary solvent mixtures is essential for the design of the antisolvent crystallization process and the production of nanocrystals. However, the fundamental data in these mixed solvents are rarely reported in the literature [9–13].

Many thermodynamic models have been commonly used to correlate solubility data and check the accuracy of the determined data. These models can describe the solid–liquid phase equilibrium relationship in the solution such as the dependence of solubility data on temperature and cosolvent composition, which allow the models to have practical application value in the field of engineering calculation and engineering design. In this work, the solubility of florfenicol in methanol + water, ethanol + water, 1-propanol + water, and isopropanol + water binary solvents was determined using the gravimetric method at temperatures ranging from 278.15 to 318.15 K under atmospheric pressure. To extend the applicability of the experimental solubility, the solubility data were then correlated by the modified Apelblat model, the CNIBS/R-K model, the Jouyban–Acree model, and the NRTL model, respectively.

## 2. Materials and Methods

### 2.1. Materials

Florfenicol was offered by Ruipu Bio-Pharmacy Co., Ltd. (Tianjin, China) with a mass fraction purity higher than 99.5%. The organic solvents including methanol, ethanol, 1-propanol, and isopropanol used in this work were offered by Jiangtian Chemical Co., Ltd. (Tianjin, China) and the mass fraction purities of all selected solvents are higher than 99.5%. Deionized water was supplied by Yuanli Chemical Co., Ltd. (Tianjin, China).

### 2.2. Characterization of Florfenicol

The crystal form of florfenicol samples was identified by Powder X-ray diffraction (PXRD, Rigaku, Japan, D/MAX 2500). The region of scanning angle was from 2 to 45° with a scanning rate of 5°/min. All the X-ray diffraction measurements were carried out at room temperature and atmospheric pressure. The melting properties of florfenicol were measured using a differential scanning calorimeter (DSC, Mettler Toledo, Zurich, Switzerland). The measurement was under the protection of nitrogen at a heating rate of 2 K/min.

### 2.3. Solubility Measurement

The solubility of florfenicol in methanol + water, ethanol + water, 1-propanol + water, and isopropanol + water was determined using a gravimetric method [14] in this work. The mass fraction of alcohol in four kinds of binary solvents varies from $\omega_A = 0$ to 1, with an interval of 0.1. The experimental procedures are described as follows: firstly, 10 mL of binary solvents was added into a 20 mL glass vial and placed into a big jacket vessel in which the temperature was controlled by a high-precision constant temperature water bath (XOYS-2009, accuracy: ± 0.1 K). After the temperature of the mixture solvents remained stable, an excess amount of florfenicol solid was added to the glass vial. The mixture in the sealed vial was continuously stirred using magnetic stirring for 12 h to achieve solid–liquid equilibrium. The duration of 12 h is determined by a preliminary experiment. Then, the stirring was stopped and the suspension was left for another 1 h to assure the undissolved solids settle down completely. The supernatant was then drawn out using a syringe fitted with a Millipore filter (0.45 μm), which was precooled/preheated to the measurement temperature. After that, the supernatant was transferred into a pre-weighed glass beaker rapidly. Then the beaker with supernatant inside was weighed again using an analytical balance (AL204-C, Metter Toledo, Zurich, Switzerland) with an accuracy of ±0.0001 g. Finally, the beaker was dried in a vacuum oven (type DZF-2BC, Tianjin Taisite Instrument Co., Ltd., Tianjin, China) at 313.15K and weighed periodically until the weight did not change. All the experiments were repeated at least three times to reduce accidental errors and the average value of three measurements was used to calculate the mole ratio fraction solubility of florfenicol ($x_F$) according to Equation (1).

$$x_F = \frac{m_F/M_F}{m_F/M_F + m_w/M_w + m_A/M_A} \tag{1}$$

where $m_F$, $m_w$, and $m_A$ represent the mass of florfenicol, water, and alcohol solvents (methanol, ethanol, 1-propanol, or isopropanol), respectively. $M_F$, $M_w$, and $M_A$ represent the molar mass of florfenicol, water, and alcohol solvents (methanol, ethanol, 1-propanol, or isopropanol), respectively.

After the solubility measurement experiments, the undissolved florfenicol solid in the equilibrium saturated solution was filtered and tested by PXRD.

## 3. Theoretical Basis

A variety of thermodynamic models have been proposed to correlate solubility data. These models can be used to check the accuracy of the determined data and describe the solid–liquid phase equilibrium relationship in the solution. In this work, the experimental solubility data were correlated by the modified Apelblat model, CNIBS/R-K model, the Jouyban–Acree model, and the NRTL model.

### 3.1. Modified Apelblat Model

The modified Apelblat model was proposed by Apelblat et al. [15,16] and can be used to correlate the solid–liquid equilibrium solubility of the solute in pure solvents and binary solvents. This semi-empirical model is applied to describe the relationship between mole fraction solubility and temperature. The equation is expressed as follows:

$$\ln x_F = A + \frac{B}{T} + C \ln T \tag{2}$$

where $x_F$ is the mole fraction solubility of the solute. $T$ is the absolute temperature. $A$, $B$, and $C$ are model parameters.

### 3.2. CNIBS/R-K Model

The CNIBS/R-K model was proposed by Acree et al. [17,18] and is used to study the solid–liquid equilibrium solubility of the solutes in binary solvents. This equation describes the relationship between mole fraction solubility of solute and solvent composition. The equation is defined as Equation (3):

$$\ln x_F = x_a \ln X_a + x_b \ln X_b + x_a x_b \sum_{i=0}^{N} S_i (x_a - x_b)^i \tag{3}$$

where $X_a$ and $X_b$ refer to the saturated mole solubility of the solute in a pure solvent $a$ and $b$ at the same temperature, respectively. $x_a$ and $x_b$ are the initial mole fraction composition of solvent $a$ and $b$ in binary solvent mixtures in the absence of solute. The $S_i$ is the model parameter and $N$ refers to the amount of solvent. For the binary solvent system, the value of $N$ is 2 and $x_a = 1 - x_b$. Therefore, Equation (3) can be simplified to Equation (4):

$$\ln x_F = B_0 + B_1 x_a + B_2 x_a^2 + B_3 x_a^3 + B_4 x_a^4 \tag{4}$$

where $B_0$, $B_1$, $B_2$, $B_3$, and $B_4$ are model parameters.

### 3.3. The Jouyban–Acree Model

The Jouyban–Acree model is a more general model that can be used to describe the effects of both solvent composition and temperature on the solid–liquid equilibrium solubility of solute [19]. Based on the CNIBS/R-K model, the temperature parameter $T$ is introduced as the second variable. The equation is shown in Equation (5):

$$\ln x_F = x_a \ln X_a + x_b \ln X_b + x_a x_b \sum_{i=0}^{N} \frac{J_i(x_a - x_b)^i}{T} \quad (5)$$

where $J_i$ is a model parameter. Other symbols have the same meanings as those in the CNIBS/R-K model. By applying the Apelblat equation [20], the Jouyban–Acree model can be transformed into Equation (6):

$$\ln x_F = A_0 + \frac{A_1}{T} + A_2 \ln T + A_3 x_a + \frac{A_4 x_a}{T} + \frac{A_5 x_a^2}{T} + \frac{A_6 x_a^3}{T} + \frac{A_7 x_a^4}{T} + A_8 x_a \ln T \quad (6)$$

where $A_0 \sim A_8$ are model parameters.

### 3.4. NRTL Model

The NRTL model was proposed by Renon et al. and can be used to calculate the activity coefficients $\gamma_i$ of non-polar or polar miscible systems [21]. The model is expressed as follows:

$$\ln \gamma_i = \frac{(x_j G_{ji} + x_k G_{kj})(x_j G_{ji} \tau_{ji} + x_k G_{ki} \tau_{ki})}{(x_i + x_j G_{ji} + x_k G_{ki})^2} + \frac{[\tau_{ij} G_{ij} x_j^2 + G_{ij} G_{kj} x_j x_k (\tau_{ij} - \tau_{kj})]}{(x_j + x_i G_{ij} + x_k G_{kj})^2} + \frac{[\tau_{jk} G_{jk} x_k^2 + G_{ik} G_{jk} x_j x_k (\tau_{ik} - \tau_{jk})]}{(x_k + x_i G_{ik} + x_j G_{jk})^2} \quad (7)$$

with

$$G_{ij} = \exp(-\alpha_{ij} \tau_{ij}) \quad (8)$$

$$\tau_{ij} = \frac{\Delta g_{ij}}{RT} \quad (9)$$

where $i \neq j$, $\Delta g_{ij}$ is the interaction parameter that is related to the Gibbs energy that is listed in Table 8, $G_{ij}$ and $\tau_{ij}$ are model parameters. $\alpha_{ij}$ is a random parameter.

The calculated solubility corrected by the activity coefficient is shown as Equation (10):

$$\ln x_i = \frac{\Delta_{fus} H}{R} \left( \frac{1}{T_m} - \frac{1}{T} \right) - \ln \gamma_i \quad (10)$$

where $\Delta_{fus} H$ and $T_m$ are the melting enthalpy and melting point of florfenicol.

## 4. Results and Discussion

### 4.1. Solid-State Characterization

The PXRD patterns of florfenicol (Figure 2) in different solvents show that the characteristic peaks of florfenicol in the equilibrium saturated solution remained consistent with that of the raw material. This indicates that florfenicol did not undergo a phase transition during the solubility experiment. The PXRD pattern of the florfenicol in this work was consistent with the form I data in the literature [9]. The melting temperature ($T_m$) and the enthalpy of fusion ($\Delta_{fus} H$) of florfenicol were calculated from the DSC result (Figure 3). The melting temperature ($T_m$) of florfenicol is 426.21 K and the enthalpy of fusion ($\Delta_{fus} H$) of florfenicol is 34.13 kJ mol$^{-1}$. These two results are consistent with the values from other literature [12].

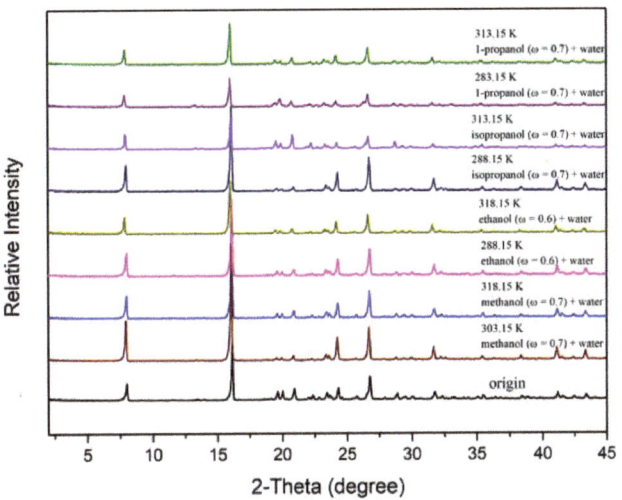

**Figure 2.** PXRD patterns of florfenicol.

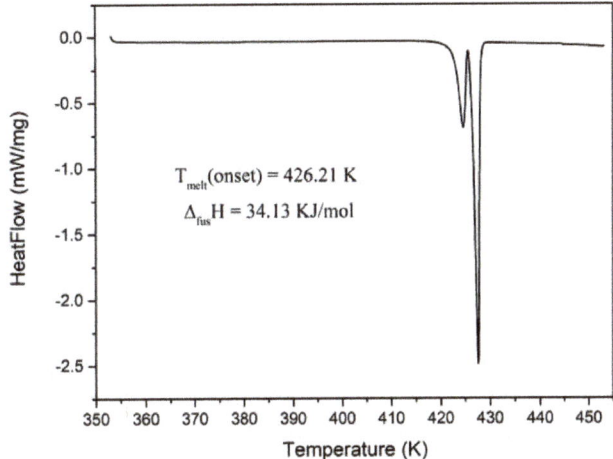

**Figure 3.** Thermal analysis spectrum (DSC) of florfenicol.

*4.2. Solubility Data*

The experimental and calculated molar ratio solubility of florfenicol in binary solvents of methanol + water, ethanol + water, 1-propanol + water, and isopropanol + water is listed in Tables 1–4, and are plotted in Figures 4–7.

The experimental results show that at constant solvent composition, the solubility of florfenicol in all four binary solvent mixtures increases with increasing temperature. Moreover, at the same temperature, the proportion of water in the binary solvent has a great influence on the solubility of florfenicol. The solubility of florfenicol increases first and then decreases with the decrease in the ratio of water in ethanol + water, 1-propanol + water, and isopropanol + water binary solvent, indicating a cosolvency phenomenon occurs in these three binary solvents. The peak position of the maximum solubility of florfenicol slightly shifted from 0.9 to 0.7 with the increase in the experimental temperature. While the solubility of florfenicol in methanol + water mixed solvent increases with the decrease in water ratio and there was no obvious cosolvency phenomenon. Furthermore, at the same temperature and mass fraction of water, the solubility of florfenicol in the tested solvent systems follows the order: (ethanol + water) > (1-propanol + water) > (isopropanol + water), which is the same as the order of the polarity of alcohol. Considering that florfenicol is a polar molecule, the effect of binary solvent mixtures on the solubility can be explained by the 'Similar Dissolution Rule'.

**Table 1.** Experimental and calculated molar ratio solubility of florfenicol in binary solvent mixtures of methanol + water from $T$ = 278.15 to 318.15 K.

| $\omega_A$ | $10^2 \cdot x_F^{exp}$ | $10^2 \cdot x_F^{cal,Apel}$ | $10^2 \cdot x_F^{cal,RK}$ | $10^2 \cdot x_F^{cal,JA}$ | $10^2 \cdot x_F^{cal,NRTL}$ |
|---|---|---|---|---|---|
| | | $T$ = 278.15 K | | | |
| 0.0 | 0.0019 | 0.0019 | 0.0020 | 0.0011 | 0.0012 |
| 0.1 | 0.0033 | 0.0038 | 0.0034 | 0.0038 | 0.0023 |
| 0.2 | 0.0059 | 0.0069 | 0.0060 | 0.0066 | 0.0048 |
| 0.3 | 0.0114 | 0.0132 | 0.0112 | 0.0118 | 0.0102 |
| 0.4 | 0.0211 | 0.0248 | 0.0211 | 0.0220 | 0.0207 |
| 0.5 | 0.0399 | 0.0445 | 0.0400 | 0.0410 | 0.0403 |
| 0.6 | 0.0738 | 0.0764 | 0.0738 | 0.0751 | 0.0742 |
| 0.7 | 0.1294 | 0.1253 | 0.1294 | 0.1309 | 0.1281 |
| 0.8 | 0.2096 | 0.1968 | 0.2096 | 0.2120 | 0.2056 |
| 0.9 | 0.3151 | 0.3010 | 0.3151 | 0.3213 | 0.3049 |
| 1.0 | 0.4815 | 0.4654 | 0.4815 | 0.5052 | 0.4176 |
| | | $T$ = 283.15 K | | | |
| 0.0 | 0.0022 | 0.0023 | 0.0021 | 0.0018 | 0.0018 |
| 0.1 | 0.0041 | 0.0041 | 0.0040 | 0.0032 | 0.0032 |
| 0.2 | 0.0075 | 0.0074 | 0.0077 | 0.0060 | 0.0066 |
| 0.3 | 0.0147 | 0.0140 | 0.0148 | 0.0115 | 0.0135 |
| 0.4 | 0.0282 | 0.0271 | 0.0282 | 0.0228 | 0.0269 |
| 0.5 | 0.0530 | 0.0509 | 0.0526 | 0.0448 | 0.0517 |
| 0.6 | 0.0940 | 0.0920 | 0.0943 | 0.0842 | 0.0939 |
| 0.7 | 0.1597 | 0.1570 | 0.1597 | 0.1450 | 0.1604 |
| 0.8 | 0.2523 | 0.2490 | 0.2522 | 0.2202 | 0.2553 |
| 0.9 | 0.3724 | 0.3697 | 0.3724 | 0.2961 | 0.3756 |
| 1.0 | 0.5408 | 0.5464 | 0.5408 | 0.4018 | 0.5098 |
| | | $T$ = 288.15 K | | | |
| 0.0 | 0.0028 | 0.0027 | 0.0027 | 0.0028 | 0.0025 |
| 0.1 | 0.0051 | 0.0047 | 0.0049 | 0.0049 | 0.0044 |
| 0.2 | 0.0090 | 0.0084 | 0.0090 | 0.0091 | 0.0088 |
| 0.3 | 0.0166 | 0.0159 | 0.0170 | 0.0174 | 0.0177 |
| 0.4 | 0.0325 | 0.0312 | 0.0324 | 0.0340 | 0.0347 |
| 0.5 | 0.0614 | 0.0606 | 0.0614 | 0.0662 | 0.0657 |
| 0.6 | 0.1122 | 0.1131 | 0.1123 | 0.1232 | 0.1180 |
| 0.7 | 0.1935 | 0.1978 | 0.1933 | 0.2105 | 0.2001 |
| 0.8 | 0.3044 | 0.3143 | 0.3046 | 0.3181 | 0.3161 |
| 0.9 | 0.4400 | 0.4540 | 0.4400 | 0.4267 | 0.4616 |
| 1.0 | 0.6309 | 0.6453 | 0.6309 | 0.5777 | 0.6229 |
| | | $T$ = 293.15 K | | | |
| 0.0 | 0.0035 | 0.0034 | 0.0032 | 0.0041 | 0.0034 |
| 0.1 | 0.0060 | 0.0057 | 0.0058 | 0.0071 | 0.0060 |
| 0.2 | 0.0110 | 0.0101 | 0.0108 | 0.0129 | 0.0117 |
| 0.3 | 0.0200 | 0.0191 | 0.0207 | 0.0246 | 0.0230 |
| 0.4 | 0.0403 | 0.0380 | 0.0400 | 0.0477 | 0.0446 |
| 0.5 | 0.0763 | 0.0748 | 0.0766 | 0.0920 | 0.0832 |
| 0.6 | 0.1416 | 0.1420 | 0.1412 | 0.1698 | 0.1482 |
| 0.7 | 0.2431 | 0.2508 | 0.2433 | 0.2884 | 0.2494 |
| 0.8 | 0.3809 | 0.3959 | 0.3809 | 0.4343 | 0.3917 |
| 0.9 | 0.5418 | 0.5573 | 0.5418 | 0.5818 | 0.5685 |
| 1.0 | 0.7594 | 0.7663 | 0.7594 | 0.7871 | 0.7621 |

**Table 1.** *Cont.*

| $\omega_A$ | $10^2 \cdot x_F^{exp}$ | $10^2 \cdot x_F^{cal,Apel}$ | $10^2 \cdot x_F^{cal,RK}$ | $10^2 \cdot x_F^{cal,JA}$ | $10^2 \cdot x_F^{cal,NRTL}$ |
|---|---|---|---|---|---|
| \multicolumn{6}{c}{$T = 298.15$ K} | | | | | |
| 0.0 | 0.0043 | 0.0043 | 0.0038 | 0.0056 | 0.0047 |
| 0.1 | 0.0074 | 0.0072 | 0.0069 | 0.0096 | 0.0081 |
| 0.2 | 0.0130 | 0.0129 | 0.0131 | 0.0174 | 0.0154 |
| 0.3 | 0.0251 | 0.0244 | 0.0255 | 0.0328 | 0.0298 |
| 0.4 | 0.0503 | 0.0484 | 0.0505 | 0.0631 | 0.0569 |
| 0.5 | 0.0993 | 0.0956 | 0.0991 | 0.1208 | 0.1051 |
| 0.6 | 0.1866 | 0.1817 | 0.1865 | 0.2215 | 0.1860 |
| 0.7 | 0.3246 | 0.3194 | 0.3245 | 0.3743 | 0.3120 |
| 0.8 | 0.5036 | 0.4975 | 0.5037 | 0.5621 | 0.4882 |
| 0.9 | 0.6936 | 0.6838 | 0.6935 | 0.7531 | 0.7030 |
| 1.0 | 0.9242 | 0.9145 | 0.9242 | 1.0196 | 0.9328 |
| \multicolumn{6}{c}{$T = 303.15$ K} | | | | | |
| 0.0 | 0.0056 | 0.0056 | 0.0055 | 0.0072 | 0.0063 |
| 0.1 | 0.0096 | 0.0095 | 0.0096 | 0.0123 | 0.0107 |
| 0.2 | 0.0179 | 0.0173 | 0.0177 | 0.0222 | 0.0201 |
| 0.3 | 0.0337 | 0.0327 | 0.0337 | 0.0415 | 0.0384 |
| 0.4 | 0.0655 | 0.0644 | 0.0656 | 0.0792 | 0.0722 |
| 0.5 | 0.1273 | 0.1262 | 0.1272 | 0.1504 | 0.1321 |
| 0.6 | 0.2366 | 0.2366 | 0.2369 | 0.2742 | 0.2323 |
| 0.7 | 0.4078 | 0.4087 | 0.4073 | 0.4615 | 0.3882 |
| 0.8 | 0.6215 | 0.6239 | 0.6218 | 0.6921 | 0.6041 |
| 0.9 | 0.8370 | 0.8387 | 0.8369 | 0.9286 | 0.8632 |
| 1.0 | 1.0917 | 1.0963 | 1.0917 | 1.2598 | 1.1367 |
| \multicolumn{6}{c}{$T = 308.15$ K} | | | | | |
| 0.0 | 0.0072 | 0.0075 | 0.0073 | 0.0088 | 0.0085 |
| 0.1 | 0.0128 | 0.0131 | 0.0128 | 0.0150 | 0.0142 |
| 0.2 | 0.0237 | 0.0243 | 0.0237 | 0.0268 | 0.0261 |
| 0.3 | 0.0451 | 0.0461 | 0.0452 | 0.0498 | 0.0491 |
| 0.4 | 0.0877 | 0.0895 | 0.0877 | 0.0943 | 0.0914 |
| 0.5 | 0.1689 | 0.1716 | 0.1685 | 0.1781 | 0.1660 |
| 0.6 | 0.3091 | 0.3131 | 0.3095 | 0.3232 | 0.2907 |
| 0.7 | 0.5225 | 0.5252 | 0.5224 | 0.5424 | 0.4841 |
| 0.8 | 0.7803 | 0.7806 | 0.7803 | 0.8133 | 0.7498 |
| 0.9 | 1.0273 | 1.0280 | 1.0273 | 1.0938 | 1.0627 |
| 1.0 | 1.3203 | 1.3197 | 1.3203 | 1.4890 | 1.3893 |
| \multicolumn{6}{c}{$T = 313.15$ K} | | | | | |
| 0.0 | 0.0104 | 0.0102 | 0.0102 | 0.0102 | 0.0113 |
| 0.1 | 0.0185 | 0.0188 | 0.0185 | 0.0173 | 0.0186 |
| 0.2 | 0.0347 | 0.0356 | 0.0346 | 0.0308 | 0.0337 |
| 0.3 | 0.0663 | 0.0679 | 0.0661 | 0.0569 | 0.0628 |
| 0.4 | 0.1260 | 0.1292 | 0.1264 | 0.1072 | 0.1158 |
| 0.5 | 0.2366 | 0.2400 | 0.2367 | 0.2013 | 0.2094 |
| 0.6 | 0.4209 | 0.4206 | 0.4205 | 0.3639 | 0.3659 |
| 0.7 | 0.6845 | 0.6773 | 0.6847 | 0.6096 | 0.6073 |
| 0.8 | 0.9904 | 0.9747 | 0.9904 | 0.9147 | 0.9346 |
| 0.9 | 1.2745 | 1.2593 | 1.2745 | 1.2344 | 1.3127 |
| 1.0 | 1.6084 | 1.5947 | 1.6084 | 1.6881 | 1.7014 |

**Table 1.** Cont.

| $\omega_A$ | $10^2 \cdot x_F^{exp}$ | $10^2 \cdot x_F^{cal,Apel}$ | $10^2 \cdot x_F^{cal,RK}$ | $10^2 \cdot x_F^{cal,JA}$ | $10^2 \cdot x_F^{cal,NRTL}$ |
|---|---|---|---|---|---|
| \multicolumn{6}{c}{T = 318.15 K} |
| 0.0 | 0.0142 | 0.0142 | 0.0142 | 0.0113 | 0.0149 |
| 0.1 | 0.0281 | 0.0279 | 0.0280 | 0.0191 | 0.0242 |
| 0.2 | 0.0549 | 0.0544 | 0.0547 | 0.0338 | 0.0435 |
| 0.3 | 0.1053 | 0.1044 | 0.1052 | 0.0621 | 0.0803 |
| 0.4 | 0.1953 | 0.1935 | 0.1956 | 0.1165 | 0.1477 |
| 0.5 | 0.3465 | 0.3444 | 0.3465 | 0.2178 | 0.2664 |
| 0.6 | 0.5738 | 0.5731 | 0.5738 | 0.3925 | 0.4630 |
| 0.7 | 0.8731 | 0.8765 | 0.8728 | 0.6568 | 0.7612 |
| 0.8 | 1.2054 | 1.2144 | 1.2057 | 0.9871 | 1.1579 |
| 0.9 | 1.5324 | 1.5415 | 1.5323 | 1.3382 | 1.6139 |
| 1.0 | 1.9252 | 1.9337 | 1.9252 | 1.8405 | 2.0787 |

[a] $\omega_A$ represents the mass fraction of alcohols (methanol, ethanol, 1-propanol, or isopropanol) in binary solvent mixtures; $x_F^{exp}$ is the experimental mole fraction solubility of florfenicol in the binary solvents; $x_F^{cal,Apel}$, $x_F^{cal,RK}$, $x_F^{cal,JA}$, and $x_F^{cal,NRTL}$ are the mole fraction solubility calculated by Equations (2), (4), (6), and (10), respectively. [b] The standard uncertainty of temperature is $u_c(T) = 0.1$ K. The relative standard uncertainty of pressure is $u_r(P) = 0.05$. The relative standard uncertainty of binary solvent composition and solubility measurement is $u_r(\omega_A) = 0.002$ and $u_r(x_F) = 0.05$.

**Table 2.** Experimental and calculated molar ratio solubility of florfenicol in binary solvent mixtures of ethanol + water from T = 278.15 to 318.15 K.

| $\omega_A$ | $10^2 \cdot x_F^{exp}$ | $10^2 \cdot x_F^{cal,Apel}$ | $10^2 \cdot x_F^{cal,RK}$ | $10^2 \cdot x_F^{cal,JA}$ | $10^2 \cdot x_F^{cal,NRTL}$ |
|---|---|---|---|---|---|
| \multicolumn{6}{c}{T = 278.15 K} |
| 0.0 | 0.0019 | 0.0019 | 0.0019 | 0.0009 | 0.0013 |
| 0.1 | 0.0037 | 0.0036 | 0.0036 | 0.0042 | 0.0032 |
| 0.2 | 0.0072 | 0.0070 | 0.0071 | 0.0080 | 0.0072 |
| 0.3 | 0.0139 | 0.0133 | 0.0139 | 0.0152 | 0.0151 |
| 0.4 | 0.0271 | 0.0255 | 0.0271 | 0.0286 | 0.0300 |
| 0.5 | 0.0517 | 0.0493 | 0.0518 | 0.0528 | 0.0563 |
| 0.6 | 0.0942 | 0.0916 | 0.0942 | 0.0929 | 0.0987 |
| 0.7 | 0.1572 | 0.1566 | 0.1570 | 0.1514 | 0.1575 |
| 0.8 | 0.2294 | 0.2321 | 0.2295 | 0.2205 | 0.2190 |
| 0.9 | 0.2828 | 0.2863 | 0.2828 | 0.2822 | 0.2536 |
| 1.0 | 0.2991 | 0.2986 | 0.2991 | 0.3410 | 0.2396 |
| \multicolumn{6}{c}{T = 283.15 K} |
| 0.0 | 0.0022 | 0.0023 | 0.0021 | 0.0016 | 0.0018 |
| 0.1 | 0.0043 | 0.0044 | 0.0043 | 0.0034 | 0.0043 |
| 0.2 | 0.0088 | 0.0088 | 0.0088 | 0.0073 | 0.0094 |
| 0.3 | 0.0178 | 0.0173 | 0.0178 | 0.0155 | 0.0195 |
| 0.4 | 0.0356 | 0.0339 | 0.0357 | 0.0324 | 0.0382 |
| 0.5 | 0.0688 | 0.0656 | 0.0689 | 0.0645 | 0.0713 |
| 0.6 | 0.1250 | 0.1201 | 0.1247 | 0.1177 | 0.1249 |
| 0.7 | 0.2036 | 0.1990 | 0.2038 | 0.1876 | 0.1989 |
| 0.8 | 0.2878 | 0.2839 | 0.2877 | 0.2462 | 0.2747 |
| 0.9 | 0.3383 | 0.3353 | 0.3383 | 0.2554 | 0.3142 |
| 1.0 | 0.3351 | 0.3343 | 0.3351 | 0.2241 | 0.2931 |

**Table 2.** *Cont.*

| $\omega_A$ | $10^2 \cdot x_F^{exp}$ | $10^2 \cdot x_F^{cal,Apel}$ | $10^2 \cdot x_F^{cal,RK}$ | $10^2 \cdot x_F^{cal,JA}$ | $10^2 \cdot x_F^{cal,NRTL}$ |
|---|---|---|---|---|---|
| | | $T = 288.15$ K | | | |
| 0.0 | 0.0028 | 0.0027 | 0.0027 | 0.0026 | 0.0025 |
| 0.1 | 0.0055 | 0.0055 | 0.0054 | 0.0055 | 0.0057 |
| 0.2 | 0.0113 | 0.0113 | 0.0109 | 0.0116 | 0.0123 |
| 0.3 | 0.0223 | 0.0228 | 0.0222 | 0.0244 | 0.0249 |
| 0.4 | 0.0441 | 0.0455 | 0.0444 | 0.0503 | 0.0483 |
| 0.5 | 0.0854 | 0.0879 | 0.0853 | 0.0988 | 0.0896 |
| 0.6 | 0.1531 | 0.1585 | 0.1534 | 0.1781 | 0.1564 |
| 0.7 | 0.2488 | 0.2555 | 0.2483 | 0.2809 | 0.2490 |
| 0.8 | 0.3462 | 0.3521 | 0.3465 | 0.3660 | 0.3423 |
| 0.9 | 0.4005 | 0.3999 | 0.4004 | 0.3781 | 0.3887 |
| 1.0 | 0.3804 | 0.3827 | 0.3804 | 0.3310 | 0.3589 |
| | | $T = 293.15$ K | | | |
| 0.0 | 0.0035 | 0.0034 | 0.0035 | 0.0040 | 0.0034 |
| 0.1 | 0.0072 | 0.0071 | 0.0070 | 0.0082 | 0.0076 |
| 0.2 | 0.0144 | 0.0148 | 0.0143 | 0.0172 | 0.0160 |
| 0.3 | 0.0291 | 0.0305 | 0.0290 | 0.0358 | 0.0319 |
| 0.4 | 0.0582 | 0.0615 | 0.0581 | 0.0729 | 0.0615 |
| 0.5 | 0.1113 | 0.1186 | 0.1118 | 0.1415 | 0.1137 |
| 0.6 | 0.2009 | 0.2106 | 0.2005 | 0.2525 | 0.1998 |
| 0.7 | 0.3206 | 0.3311 | 0.3207 | 0.3947 | 0.3183 |
| 0.8 | 0.4345 | 0.4420 | 0.4346 | 0.5114 | 0.4339 |
| 0.9 | 0.4780 | 0.4851 | 0.4780 | 0.5274 | 0.4827 |
| 1.0 | 0.4395 | 0.4473 | 0.4395 | 0.4619 | 0.4403 |
| | | $T = 298.15$ K | | | |
| 0.0 | 0.0043 | 0.0043 | 0.0044 | 0.0057 | 0.0046 |
| 0.1 | 0.0091 | 0.0092 | 0.0092 | 0.0116 | 0.0101 |
| 0.2 | 0.0191 | 0.0197 | 0.0191 | 0.0240 | 0.0208 |
| 0.3 | 0.0396 | 0.0412 | 0.0396 | 0.0493 | 0.0411 |
| 0.4 | 0.0802 | 0.0838 | 0.0803 | 0.0992 | 0.0791 |
| 0.5 | 0.1557 | 0.1609 | 0.1554 | 0.1903 | 0.1478 |
| 0.6 | 0.2773 | 0.2816 | 0.2773 | 0.3365 | 0.2623 |
| 0.7 | 0.4356 | 0.4326 | 0.4358 | 0.5224 | 0.4199 |
| 0.8 | 0.5724 | 0.5612 | 0.5722 | 0.6745 | 0.5651 |
| 0.9 | 0.6041 | 0.5976 | 0.6042 | 0.6958 | 0.6120 |
| 1.0 | 0.5410 | 0.5330 | 0.5410 | 0.6116 | 0.5476 |
| | | $T = 303.15$ K | | | |
| 0.0 | 0.0056 | 0.0056 | 0.0056 | 0.0076 | 0.0062 |
| 0.1 | 0.0125 | 0.0122 | 0.0125 | 0.0154 | 0.0133 |
| 0.2 | 0.0277 | 0.0266 | 0.0277 | 0.0314 | 0.0271 |
| 0.3 | 0.0598 | 0.0563 | 0.0596 | 0.0637 | 0.0538 |
| 0.4 | 0.1221 | 0.1149 | 0.1225 | 0.1269 | 0.1049 |
| 0.5 | 0.2338 | 0.2193 | 0.2334 | 0.2414 | 0.2001 |
| 0.6 | 0.3994 | 0.3787 | 0.3995 | 0.4233 | 0.3585 |
| 0.7 | 0.5917 | 0.5696 | 0.5917 | 0.6539 | 0.5662 |
| 0.8 | 0.7342 | 0.7201 | 0.7341 | 0.8426 | 0.7382 |
| 0.9 | 0.7548 | 0.7470 | 0.7548 | 0.8714 | 0.7785 |
| 1.0 | 0.6579 | 0.6465 | 0.6579 | 0.7706 | 0.6807 |

**Table 2.** *Cont.*

| $\omega_A$ | $10^2 \cdot x_F^{exp}$ | $10^2 \cdot x_F^{cal,Apel}$ | $10^2 \cdot x_F^{cal,RK}$ | $10^2 \cdot x_F^{cal,JA}$ | $10^2 \cdot x_F^{cal,NRTL}$ |
|---|---|---|---|---|---|
| | | $T = 308.15$ K | | | |
| 0.0 | 0.0072 | 0.0075 | 0.0073 | 0.0095 | 0.0083 |
| 0.1 | 0.0163 | 0.0166 | 0.0163 | 0.0191 | 0.0174 |
| 0.2 | 0.0359 | 0.0363 | 0.0360 | 0.0387 | 0.0350 |
| 0.3 | 0.0770 | 0.0776 | 0.0772 | 0.0778 | 0.0692 |
| 0.4 | 0.1579 | 0.1583 | 0.1578 | 0.1536 | 0.1358 |
| 0.5 | 0.2987 | 0.3003 | 0.2987 | 0.2896 | 0.2614 |
| 0.6 | 0.5078 | 0.5118 | 0.5075 | 0.5047 | 0.4737 |
| 0.7 | 0.7470 | 0.7551 | 0.7475 | 0.7765 | 0.7492 |
| 0.8 | 0.9252 | 0.9329 | 0.9249 | 1.0004 | 0.9681 |
| 0.9 | 0.9465 | 0.9463 | 0.9466 | 1.0392 | 0.9999 |
| 1.0 | 0.7872 | 0.7974 | 0.7872 | 0.9268 | 0.8442 |
| | | $T = 313.15$ K | | | |
| 0.0 | 0.0104 | 0.0102 | 0.0104 | 0.0114 | 0.0110 |
| 0.1 | 0.0230 | 0.0229 | 0.0229 | 0.0225 | 0.0228 |
| 0.2 | 0.0503 | 0.0503 | 0.0503 | 0.0452 | 0.0457 |
| 0.3 | 0.1070 | 0.1079 | 0.1070 | 0.0900 | 0.0908 |
| 0.4 | 0.2173 | 0.2194 | 0.2172 | 0.1762 | 0.1813 |
| 0.5 | 0.4078 | 0.4128 | 0.4081 | 0.3300 | 0.3575 |
| 0.6 | 0.6875 | 0.6950 | 0.6872 | 0.5720 | 0.6612 |
| 0.7 | 0.9975 | 1.0077 | 0.9977 | 0.8778 | 1.0460 |
| 0.8 | 1.2073 | 1.2196 | 1.2072 | 1.1325 | 1.3160 |
| 0.9 | 1.2005 | 1.2139 | 1.2005 | 1.1837 | 1.3043 |
| 1.0 | 0.9957 | 0.9989 | 0.9957 | 1.0672 | 1.0691 |
| | | $T = 318.15$ K | | | |
| 0.0 | 0.0142 | 0.0142 | 0.0143 | 0.0128 | 0.0147 |
| 0.1 | 0.0321 | 0.0321 | 0.0320 | 0.0252 | 0.0299 |
| 0.2 | 0.0705 | 0.0705 | 0.0707 | 0.0501 | 0.0601 |
| 0.3 | 0.1517 | 0.1514 | 0.1511 | 0.0990 | 0.1215 |
| 0.4 | 0.3061 | 0.3055 | 0.3063 | 0.1924 | 0.2499 |
| 0.5 | 0.5713 | 0.5697 | 0.5717 | 0.3581 | 0.5117 |
| 0.6 | 0.9506 | 0.9477 | 0.9505 | 0.6182 | 0.9689 |
| 0.7 | 1.3575 | 1.3526 | 1.3573 | 0.9475 | 1.5270 |
| 0.8 | 1.6141 | 1.6075 | 1.6144 | 1.2258 | 1.8612 |
| 0.9 | 1.5814 | 1.5753 | 1.5813 | 1.2914 | 1.7629 |
| 1.0 | 1.2723 | 1.2695 | 1.2723 | 1.1795 | 1.3705 |

[a] $\omega_A$ represents the mass fraction of alcohols (methanol, ethanol, 1-propanol, or isopropanol) in binary solvent mixtures; $x_F^{exp}$ is the experimental mole fraction solubility of florfenicol in the binary solvents; $x_F^{cal, Apel}$, $x_F^{cal, RK}$, $x_F^{cal, JA}$, and $x_F^{cal, NRTL}$ are the mole fraction solubility calculated by Equations (2), (4), (6) and (10), respectively.
[b] The standard uncertainty of temperature is $u_c(T) = 0.1$ K. The relative standard uncertainty of pressure is $u_r(P) = 0.05$. The relative standard uncertainty of binary solvent composition and solubility measurement is $u_r(\omega_A) = 0.002$ and $u_r(x_F) = 0.05$.

**Table 3.** Experimental and calculated molar ratio solubility of florfenicol in binary solvent mixtures of 1-propanol + water from $T$ = 278.15 to 318.15 K.

| $\omega_A$ | $10^2 \cdot x_F^{exp}$ | $10^2 \cdot x_F^{cal,Apel}$ | $10^2 \cdot x_F^{cal,RK}$ | $10^2 \cdot x_F^{cal,JA}$ | $10^2 \cdot x_F^{cal,NRTL}$ |
|---|---|---|---|---|---|
| \multicolumn{6}{c}{$T$ = 278.15 K} |
| 0.0 | 0.0019 | 0.0019 | 0.0019 | 0.0008 | 0.0010 |
| 0.1 | 0.0034 | 0.0035 | 0.0034 | 0.0040 | 0.0027 |
| 0.2 | 0.0062 | 0.0064 | 0.0062 | 0.0072 | 0.0057 |
| 0.3 | 0.0113 | 0.0115 | 0.0113 | 0.0127 | 0.0115 |
| 0.4 | 0.0202 | 0.0203 | 0.0202 | 0.0218 | 0.0217 |
| 0.5 | 0.0347 | 0.0343 | 0.0347 | 0.0356 | 0.0379 |
| 0.6 | 0.0553 | 0.0553 | 0.0553 | 0.0535 | 0.0601 |
| 0.7 | 0.0783 | 0.0784 | 0.0783 | 0.0720 | 0.0831 |
| 0.8 | 0.0946 | 0.0947 | 0.0946 | 0.0865 | 0.0967 |
| 0.9 | 0.0960 | 0.0957 | 0.0960 | 0.0969 | 0.0922 |
| 1.0 | 0.0770 | 0.0762 | 0.0770 | 0.0888 | 0.0712 |
| \multicolumn{6}{c}{$T$ = 283.15 K} |
| 0.0 | 0.0022 | 0.0023 | 0.0022 | 0.0015 | 0.0015 |
| 0.1 | 0.0041 | 0.0043 | 0.0041 | 0.0032 | 0.0037 |
| 0.2 | 0.0079 | 0.0081 | 0.0079 | 0.0067 | 0.0078 |
| 0.3 | 0.0149 | 0.0151 | 0.0149 | 0.0137 | 0.0154 |
| 0.4 | 0.0274 | 0.0276 | 0.0274 | 0.0266 | 0.0288 |
| 0.5 | 0.0475 | 0.0473 | 0.0475 | 0.0475 | 0.0499 |
| 0.6 | 0.0751 | 0.0757 | 0.0751 | 0.0747 | 0.0788 |
| 0.7 | 0.1042 | 0.1054 | 0.1042 | 0.0991 | 0.1086 |
| 0.8 | 0.1239 | 0.1252 | 0.1239 | 0.1087 | 0.1258 |
| 0.9 | 0.1275 | 0.1272 | 0.1275 | 0.1026 | 0.1201 |
| 1.0 | 0.0961 | 0.0960 | 0.0961 | 0.0671 | 0.0926 |
| \multicolumn{6}{c}{$T$ = 288.15 K} |
| 0.0 | 0.0028 | 0.0027 | 0.0028 | 0.0025 | 0.0022 |
| 0.1 | 0.0054 | 0.0054 | 0.0054 | 0.0053 | 0.0052 |
| 0.2 | 0.0105 | 0.0105 | 0.0105 | 0.0109 | 0.0107 |
| 0.3 | 0.0202 | 0.0203 | 0.0202 | 0.0220 | 0.0207 |
| 0.4 | 0.0376 | 0.0377 | 0.0376 | 0.0423 | 0.0383 |
| 0.5 | 0.0655 | 0.0654 | 0.0655 | 0.0748 | 0.0661 |
| 0.6 | 0.1033 | 0.1036 | 0.1033 | 0.1170 | 0.1041 |
| 0.7 | 0.1417 | 0.1416 | 0.1417 | 0.1545 | 0.1425 |
| 0.8 | 0.1652 | 0.1649 | 0.1652 | 0.1696 | 0.1642 |
| 0.9 | 0.1666 | 0.1676 | 0.1666 | 0.1606 | 0.1561 |
| 1.0 | 0.1209 | 0.1211 | 0.1209 | 0.1063 | 0.1198 |
| \multicolumn{6}{c}{$T$ = 293.15 K} |
| 0.0 | 0.0035 | 0.0034 | 0.0035 | 0.0039 | 0.0031 |
| 0.1 | 0.0070 | 0.0068 | 0.0070 | 0.0081 | 0.0072 |
| 0.2 | 0.0140 | 0.0139 | 0.0140 | 0.0164 | 0.0145 |
| 0.3 | 0.0274 | 0.0275 | 0.0274 | 0.0328 | 0.0278 |
| 0.4 | 0.0518 | 0.0520 | 0.0518 | 0.0625 | 0.0510 |
| 0.5 | 0.0906 | 0.0907 | 0.0906 | 0.1097 | 0.0880 |
| 0.6 | 0.1422 | 0.1420 | 0.1422 | 0.1704 | 0.1384 |
| 0.7 | 0.1920 | 0.1899 | 0.1920 | 0.2245 | 0.1883 |
| 0.8 | 0.2195 | 0.2163 | 0.2195 | 0.2466 | 0.2149 |
| 0.9 | 0.2191 | 0.2192 | 0.2191 | 0.2346 | 0.2030 |
| 1.0 | 0.1519 | 0.1529 | 0.1519 | 0.1572 | 0.1544 |

**Table 3.** Cont.

| $\omega_A$ | $10^2 \cdot x_F^{exp}$ | $10^2 \cdot x_F^{cal,Apel}$ | $10^2 \cdot x_F^{cal,RK}$ | $10^2 \cdot x_F^{cal,JA}$ | $10^2 \cdot x_F^{cal,NRTL}$ |
|---|---|---|---|---|---|
| \multicolumn{6}{c}{$T = 298.15$ K} | | | | | |
| 0.0 | 0.0043 | 0.0043 | 0.0044 | 0.0057 | 0.0044 |
| 0.1 | 0.0091 | 0.0089 | 0.0091 | 0.0115 | 0.0098 |
| 0.2 | 0.0190 | 0.0186 | 0.0190 | 0.0232 | 0.0196 |
| 0.3 | 0.0382 | 0.0377 | 0.0382 | 0.0457 | 0.0373 |
| 0.4 | 0.0729 | 0.0722 | 0.0729 | 0.0863 | 0.0688 |
| 0.5 | 0.1272 | 0.1260 | 0.1272 | 0.1502 | 0.1189 |
| 0.6 | 0.1952 | 0.1946 | 0.1952 | 0.2321 | 0.1859 |
| 0.7 | 0.2545 | 0.2543 | 0.2545 | 0.3052 | 0.2496 |
| 0.8 | 0.2826 | 0.2828 | 0.2826 | 0.3359 | 0.2805 |
| 0.9 | 0.2846 | 0.2846 | 0.2846 | 0.3212 | 0.2638 |
| 1.0 | 0.1930 | 0.1932 | 0.1930 | 0.2181 | 0.1985 |
| \multicolumn{6}{c}{$T = 303.15$ K} | | | | | |
| 0.0 | 0.0056 | 0.0056 | 0.0055 | 0.0076 | 0.0062 |
| 0.1 | 0.0119 | 0.0119 | 0.0119 | 0.0153 | 0.0134 |
| 0.2 | 0.0255 | 0.0254 | 0.0255 | 0.0306 | 0.0264 |
| 0.3 | 0.0523 | 0.0522 | 0.0523 | 0.0598 | 0.0505 |
| 0.4 | 0.1009 | 0.1007 | 0.1009 | 0.1118 | 0.0932 |
| 0.5 | 0.1755 | 0.1755 | 0.1755 | 0.1932 | 0.1620 |
| 0.6 | 0.2653 | 0.2668 | 0.2653 | 0.2970 | 0.2528 |
| 0.7 | 0.3370 | 0.3402 | 0.3370 | 0.3899 | 0.3338 |
| 0.8 | 0.3649 | 0.3684 | 0.3649 | 0.4301 | 0.3680 |
| 0.9 | 0.3678 | 0.3670 | 0.3678 | 0.4138 | 0.3433 |
| 1.0 | 0.2444 | 0.2443 | 0.2444 | 0.2849 | 0.2544 |
| \multicolumn{6}{c}{$T = 308.15$ K} | | | | | |
| 0.0 | 0.0072 | 0.0075 | 0.0072 | 0.0097 | 0.0086 |
| 0.1 | 0.0159 | 0.0163 | 0.0159 | 0.0192 | 0.0183 |
| 0.2 | 0.0345 | 0.0352 | 0.0345 | 0.0380 | 0.0359 |
| 0.3 | 0.0720 | 0.0731 | 0.0720 | 0.0735 | 0.0685 |
| 0.4 | 0.1402 | 0.1412 | 0.1402 | 0.1364 | 0.1282 |
| 0.5 | 0.2446 | 0.2448 | 0.2446 | 0.2341 | 0.2262 |
| 0.6 | 0.3670 | 0.3657 | 0.3670 | 0.3583 | 0.3525 |
| 0.7 | 0.4574 | 0.4544 | 0.4574 | 0.4699 | 0.4573 |
| 0.8 | 0.4808 | 0.4784 | 0.4808 | 0.5198 | 0.4894 |
| 0.9 | 0.4701 | 0.4703 | 0.4701 | 0.5035 | 0.4469 |
| 1.0 | 0.3095 | 0.3092 | 0.3095 | 0.3520 | 0.3254 |
| \multicolumn{6}{c}{$T = 313.15$ K} | | | | | |
| 0.0 | 0.0104 | 0.0102 | 0.0104 | 0.0116 | 0.0120 |
| 0.1 | 0.0230 | 0.0226 | 0.0230 | 0.0227 | 0.0252 |
| 0.2 | 0.0499 | 0.0494 | 0.0499 | 0.0445 | 0.0494 |
| 0.3 | 0.1035 | 0.1032 | 0.1035 | 0.0854 | 0.0957 |
| 0.4 | 0.1995 | 0.1990 | 0.1995 | 0.1573 | 0.1821 |
| 0.5 | 0.3419 | 0.3419 | 0.3419 | 0.2682 | 0.3239 |
| 0.6 | 0.5001 | 0.5013 | 0.5001 | 0.4091 | 0.5001 |
| 0.7 | 0.6047 | 0.6060 | 0.6047 | 0.5361 | 0.6294 |
| 0.8 | 0.6181 | 0.6190 | 0.6181 | 0.5951 | 0.6514 |
| 0.9 | 0.5984 | 0.5987 | 0.5984 | 0.5808 | 0.5835 |
| 1.0 | 0.3922 | 0.3915 | 0.3922 | 0.4125 | 0.4159 |

Table 3. Cont.

| $\omega_A$ | $10^2 \cdot x_F^{exp}$ | $10^2 \cdot x_F^{cal,Apel}$ | $10^2 \cdot x_F^{cal,RK}$ | $10^2 \cdot x_F^{cal,JA}$ | $10^2 \cdot x_F^{cal,NRTL}$ |
|---|---|---|---|---|---|
| | | $T = 318.15$ K | | | |
| 0.0 | 0.0142 | 0.0142 | 0.0143 | 0.0131 | 0.0167 |
| 0.1 | 0.0320 | 0.0320 | 0.0320 | 0.0255 | 0.0347 |
| 0.2 | 0.0701 | 0.0702 | 0.0701 | 0.0495 | 0.0686 |
| 0.3 | 0.1464 | 0.1469 | 0.1464 | 0.0942 | 0.1359 |
| 0.4 | 0.2818 | 0.2817 | 0.2818 | 0.1720 | 0.2653 |
| 0.5 | 0.4785 | 0.4780 | 0.4785 | 0.2917 | 0.4793 |
| 0.6 | 0.6868 | 0.6868 | 0.6868 | 0.4434 | 0.7325 |
| 0.7 | 0.8071 | 0.8068 | 0.8071 | 0.5811 | 0.8876 |
| 0.8 | 0.7988 | 0.7985 | 0.7988 | 0.6475 | 0.8778 |
| 0.9 | 0.7579 | 0.7578 | 0.7579 | 0.6370 | 0.7647 |
| 1.0 | 0.4953 | 0.4958 | 0.4953 | 0.4600 | 0.5314 |

[a] $\omega_A$ represents the mass fraction of alcohols (methanol, ethanol, 1-propanol, or isopropanol) in binary solvent mixtures; $x_F^{exp}$ is the experimental mole fraction solubility of florfenicol in the binary solvents; $x_F^{cal,Apel}$, $x_F^{cal,RK}$, $x_F^{cal,JA}$, and $x_F^{cal,NRTL}$ are the mole fraction solubility calculated by Equations (2), (4), (6), and (10), respectively. [b] The standard uncertainty of temperature is $u_c(T) = 0.1$ K. The relative standard uncertainty of pressure is $u_r(P) = 0.05$. The relative standard uncertainty of binary solvent composition and solubility measurement is $u_r(\omega_A) = 0.002$ and $u_r(x_F) = 0.05$.

Table 4. Experimental and calculated molar ratio solubility of florfenicol in binary solvent mixtures of isopropanol + water from $T = 278.15$ to $318.15$ K.

| $\omega_A$ | $10^2 \cdot x_F^{exp}$ | $10^2 \cdot x_F^{cal,Apel}$ | $10^2 \cdot x_F^{cal,RK}$ | $10^2 \cdot x_F^{cal,JA}$ | $10^2 \cdot x_F^{cal,NRTL}$ |
|---|---|---|---|---|---|
| | | $T = 278.15$ K | | | |
| 0.0 | 0.0019 | 0.0019 | 0.0020 | 0.0009 | 0.0011 |
| 0.1 | 0.0033 | 0.0033 | 0.0033 | 0.0037 | 0.0025 |
| 0.2 | 0.0057 | 0.0057 | 0.0057 | 0.0063 | 0.0051 |
| 0.3 | 0.0098 | 0.0099 | 0.0098 | 0.0108 | 0.0098 |
| 0.4 | 0.0166 | 0.0166 | 0.0166 | 0.0179 | 0.0177 |
| 0.5 | 0.0275 | 0.0274 | 0.0273 | 0.0284 | 0.0297 |
| 0.6 | 0.0421 | 0.0423 | 0.0425 | 0.0421 | 0.0456 |
| 0.7 | 0.0599 | 0.0597 | 0.0596 | 0.0565 | 0.0624 |
| 0.8 | 0.0724 | 0.0726 | 0.0725 | 0.0681 | 0.0742 |
| 0.9 | 0.0750 | 0.0751 | 0.0750 | 0.0765 | 0.0748 |
| 1.0 | 0.0707 | 0.0700 | 0.0707 | 0.0785 | 0.0629 |
| | | $T = 283.15$ K | | | |
| 0.0 | 0.0022 | 0.0023 | 0.0022 | 0.0015 | 0.0016 |
| 0.1 | 0.0039 | 0.0040 | 0.0039 | 0.0030 | 0.0035 |
| 0.2 | 0.0071 | 0.0072 | 0.0071 | 0.0059 | 0.0071 |
| 0.3 | 0.0127 | 0.0129 | 0.0126 | 0.0115 | 0.0133 |
| 0.4 | 0.0222 | 0.0223 | 0.0222 | 0.0213 | 0.0237 |
| 0.5 | 0.0370 | 0.0375 | 0.0372 | 0.0369 | 0.0395 |
| 0.6 | 0.0581 | 0.0579 | 0.0579 | 0.0572 | 0.0604 |
| 0.7 | 0.0801 | 0.0805 | 0.0802 | 0.0760 | 0.0823 |
| 0.8 | 0.0963 | 0.0964 | 0.0963 | 0.0842 | 0.0975 |
| 0.9 | 0.1006 | 0.1000 | 0.1006 | 0.0805 | 0.0982 |
| 1.0 | 0.0882 | 0.0882 | 0.0882 | 0.0620 | 0.0821 |

**Table 4.** *Cont.*

| $\omega_A$ | $10^2 \cdot x_F^{exp}$ | $10^2 \cdot x_F^{cal,Apel}$ | $10^2 \cdot x_F^{cal,RK}$ | $10^2 \cdot x_F^{cal,JA}$ | $10^2 \cdot x_F^{cal,NRTL}$ |
|---|---|---|---|---|---|
| \multicolumn{6}{c}{$T$ = 288.15 K} | | | | | |
| 0.0 | 0.0028 | 0.0027 | 0.0028 | 0.0026 | 0.0023 |
| 0.1 | 0.0051 | 0.0050 | 0.0051 | 0.0050 | 0.0049 |
| 0.2 | 0.0094 | 0.0093 | 0.0094 | 0.0096 | 0.0096 |
| 0.3 | 0.0173 | 0.0171 | 0.0171 | 0.0184 | 0.0179 |
| 0.4 | 0.0301 | 0.0304 | 0.0304 | 0.0339 | 0.0316 |
| 0.5 | 0.0514 | 0.0514 | 0.0513 | 0.0582 | 0.0525 |
| 0.6 | 0.0794 | 0.0794 | 0.0794 | 0.0897 | 0.0801 |
| 0.7 | 0.1086 | 0.1085 | 0.1085 | 0.1187 | 0.1088 |
| 0.8 | 0.1275 | 0.1276 | 0.1276 | 0.1316 | 0.1281 |
| 0.9 | 0.1314 | 0.1321 | 0.1314 | 0.1262 | 0.1284 |
| 1.0 | 0.1110 | 0.1112 | 0.1110 | 0.0982 | 0.1066 |
| \multicolumn{6}{c}{$T$ = 293.15 K} | | | | | |
| 0.0 | 0.0035 | 0.0034 | 0.0036 | 0.0039 | 0.0032 |
| 0.1 | 0.0066 | 0.0065 | 0.0067 | 0.0076 | 0.0067 |
| 0.2 | 0.0126 | 0.0123 | 0.0126 | 0.0145 | 0.0130 |
| 0.3 | 0.0235 | 0.0230 | 0.0235 | 0.0275 | 0.0240 |
| 0.4 | 0.0421 | 0.0418 | 0.0422 | 0.0501 | 0.0423 |
| 0.5 | 0.0715 | 0.0709 | 0.0712 | 0.0853 | 0.0701 |
| 0.6 | 0.1089 | 0.1089 | 0.1091 | 0.1307 | 0.1066 |
| 0.7 | 0.1461 | 0.1461 | 0.1462 | 0.1726 | 0.1441 |
| 0.8 | 0.1687 | 0.1682 | 0.1686 | 0.1916 | 0.1683 |
| 0.9 | 0.1728 | 0.1729 | 0.1728 | 0.1847 | 0.1676 |
| 1.0 | 0.1395 | 0.1404 | 0.1395 | 0.1451 | 0.1378 |
| \multicolumn{6}{c}{$T$ = 298.15 K} | | | | | |
| 0.0 | 0.0043 | 0.0043 | 0.0043 | 0.0057 | 0.0045 |
| 0.1 | 0.0084 | 0.0085 | 0.0084 | 0.0107 | 0.0090 |
| 0.2 | 0.0165 | 0.0164 | 0.0165 | 0.0204 | 0.0176 |
| 0.3 | 0.0316 | 0.0314 | 0.0316 | 0.0383 | 0.0322 |
| 0.4 | 0.0582 | 0.0577 | 0.0580 | 0.0692 | 0.0566 |
| 0.5 | 0.0981 | 0.0981 | 0.0986 | 0.1170 | 0.0939 |
| 0.6 | 0.1500 | 0.1494 | 0.1497 | 0.1783 | 0.1430 |
| 0.7 | 0.1962 | 0.1963 | 0.1962 | 0.2349 | 0.1916 |
| 0.8 | 0.2203 | 0.2208 | 0.2204 | 0.2613 | 0.2210 |
| 0.9 | 0.2248 | 0.2247 | 0.2248 | 0.2532 | 0.2184 |
| 1.0 | 0.1772 | 0.1774 | 0.1772 | 0.2012 | 0.1775 |
| \multicolumn{6}{c}{$T$ = 303.15 K} | | | | | |
| 0.0 | 0.0056 | 0.0056 | 0.0057 | 0.0076 | 0.0063 |
| 0.1 | 0.0112 | 0.0113 | 0.0113 | 0.0143 | 0.0121 |
| 0.2 | 0.0223 | 0.0224 | 0.0224 | 0.0269 | 0.0236 |
| 0.3 | 0.0432 | 0.0433 | 0.0434 | 0.0500 | 0.0432 |
| 0.4 | 0.0801 | 0.0803 | 0.0801 | 0.0897 | 0.0762 |
| 0.5 | 0.1365 | 0.1363 | 0.1359 | 0.1505 | 0.1273 |
| 0.6 | 0.2041 | 0.2051 | 0.2046 | 0.2284 | 0.1932 |
| 0.7 | 0.2638 | 0.2634 | 0.2637 | 0.3005 | 0.2569 |
| 0.8 | 0.2898 | 0.2887 | 0.2897 | 0.3350 | 0.2917 |
| 0.9 | 0.2899 | 0.2898 | 0.2899 | 0.3266 | 0.2843 |
| 1.0 | 0.2244 | 0.2244 | 0.2244 | 0.2626 | 0.2280 |

**Table 4.** *Cont.*

| $\omega_A$ | $10^2 \cdot x_F^{exp}$ | $10^2 \cdot x_F^{cal,Apel}$ | $10^2 \cdot x_F^{cal,RK}$ | $10^2 \cdot x_F^{cal,JA}$ | $10^2 \cdot x_F^{cal,NRTL}$ |
|---|---|---|---|---|---|
| | | $T = 308.15$ K | | | |
| 0.0 | 0.0072 | 0.0075 | 0.0071 | 0.0096 | 0.0086 |
| 0.1 | 0.0151 | 0.0153 | 0.0146 | 0.0179 | 0.0162 |
| 0.2 | 0.0302 | 0.0309 | 0.0300 | 0.0334 | 0.0317 |
| 0.3 | 0.0592 | 0.0604 | 0.0596 | 0.0615 | 0.0582 |
| 0.4 | 0.1119 | 0.1125 | 0.1116 | 0.1094 | 0.1039 |
| 0.5 | 0.1895 | 0.1899 | 0.1898 | 0.1825 | 0.1748 |
| 0.6 | 0.2818 | 0.2816 | 0.2818 | 0.2758 | 0.2659 |
| 0.7 | 0.3533 | 0.3529 | 0.3530 | 0.3625 | 0.3477 |
| 0.8 | 0.3753 | 0.3760 | 0.3756 | 0.4053 | 0.3855 |
| 0.9 | 0.3715 | 0.3710 | 0.3714 | 0.3978 | 0.3703 |
| 1.0 | 0.2842 | 0.2839 | 0.2842 | 0.3239 | 0.2920 |
| | | $T = 313.15$ K | | | |
| 0.0 | 0.0104 | 0.0102 | 0.0104 | 0.0115 | 0.0120 |
| 0.1 | 0.0215 | 0.0212 | 0.0215 | 0.0212 | 0.0218 |
| 0.2 | 0.0439 | 0.0432 | 0.0438 | 0.0391 | 0.0432 |
| 0.3 | 0.0864 | 0.0853 | 0.0861 | 0.0714 | 0.0803 |
| 0.4 | 0.1588 | 0.1583 | 0.1591 | 0.1263 | 0.1447 |
| 0.5 | 0.2655 | 0.2654 | 0.2658 | 0.2094 | 0.2457 |
| 0.6 | 0.3869 | 0.3865 | 0.3864 | 0.3153 | 0.3725 |
| 0.7 | 0.4713 | 0.4721 | 0.4716 | 0.4142 | 0.4763 |
| 0.8 | 0.4877 | 0.4880 | 0.4876 | 0.4645 | 0.5134 |
| 0.9 | 0.4711 | 0.4717 | 0.4711 | 0.4592 | 0.4821 |
| 1.0 | 0.3601 | 0.3595 | 0.3601 | 0.3790 | 0.3737 |
| | | $T = 318.15$ K | | | |
| 0.0 | 0.0142 | 0.0142 | 0.0142 | 0.0129 | 0.0164 |
| 0.1 | 0.0296 | 0.0297 | 0.0297 | 0.0237 | 0.0292 |
| 0.2 | 0.0611 | 0.0613 | 0.0612 | 0.0435 | 0.0589 |
| 0.3 | 0.1212 | 0.1215 | 0.1211 | 0.0787 | 0.1115 |
| 0.4 | 0.2239 | 0.2240 | 0.2239 | 0.1382 | 0.2055 |
| 0.5 | 0.3717 | 0.3717 | 0.3714 | 0.2280 | 0.3545 |
| 0.6 | 0.5303 | 0.5305 | 0.5307 | 0.3422 | 0.5336 |
| 0.7 | 0.6308 | 0.6305 | 0.6305 | 0.4495 | 0.6646 |
| 0.8 | 0.6312 | 0.6310 | 0.6313 | 0.5059 | 0.6889 |
| 0.9 | 0.5961 | 0.5959 | 0.5961 | 0.5039 | 0.6295 |
| 1.0 | 0.4548 | 0.4552 | 0.4548 | 0.4218 | 0.4777 |

[a] $\omega_A$ represents the mass fraction of alcohols (methanol, ethanol, 1-propanol, or isopropanol) in binary solvent mixtures; $x_F^{exp}$ is the experimental mole fraction solubility of florfenicol in the binary solvents; $x_F^{cal, Apel}$, $x_F^{cal, RK}$, $x_F^{cal, JA}$, and $x_F^{cal, NRTL}$ are the mole fraction solubility calculated by Equations (2), (4), (6), and (10), respectively.
[b] The standard uncertainty of temperature is $u_c(T) = 0.1$ K. The relative standard uncertainty of pressure is $u_r(P) = 0.05$. The relative standard uncertainty of binary solvent composition and solubility measurement is $u_r(\omega_A) = 0.002$ and $u_r(x_F) = 0.05$.

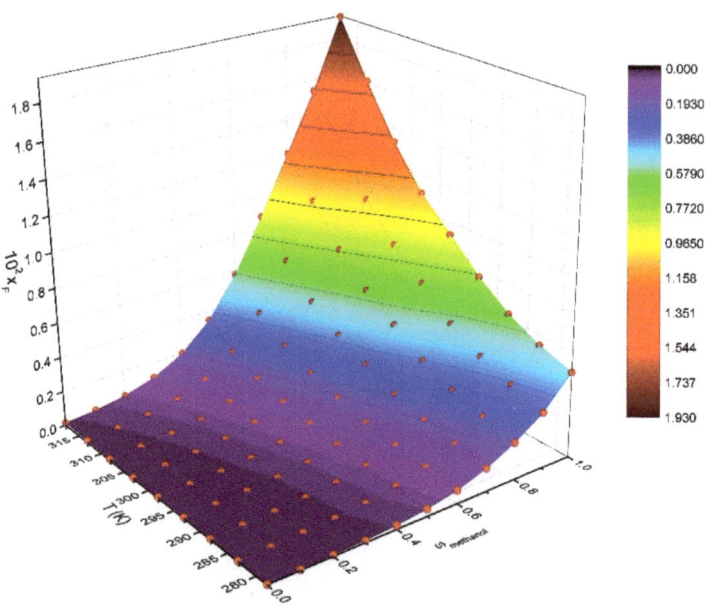

**Figure 4.** Molar ratio solubility data of florfenicol in methanol + water binary solvents at $T = 278.15$ K to 318.15 K.

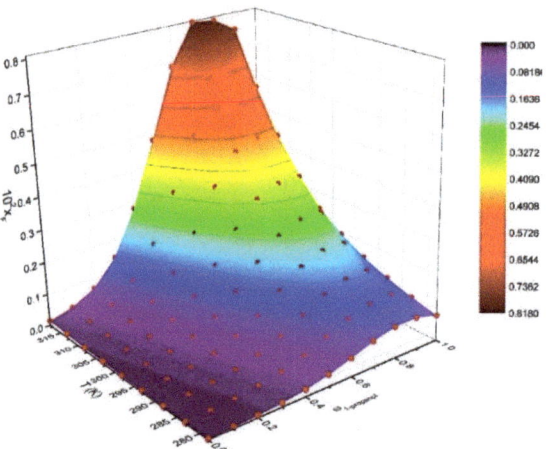

**Figure 5.** Molar ratio solubility data of florfenicol in ethanol + water binary solvents at $T = 278.15$ K to 318.15 K.

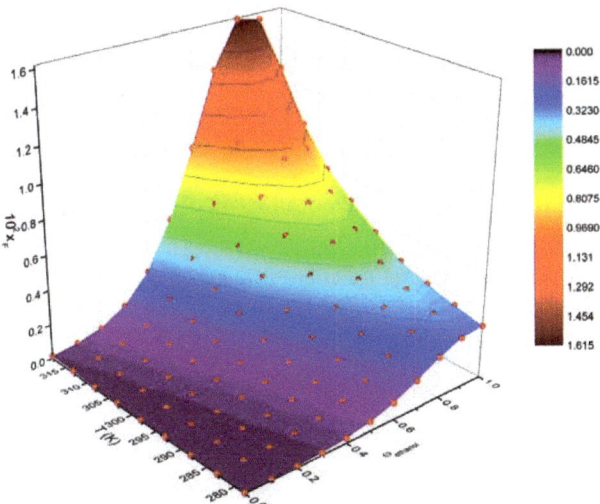

**Figure 6.** Molar ratio solubility data of florfenicol in 1-propanol + water binary solvents at $T$ = 278.15 K to 318.15 K.

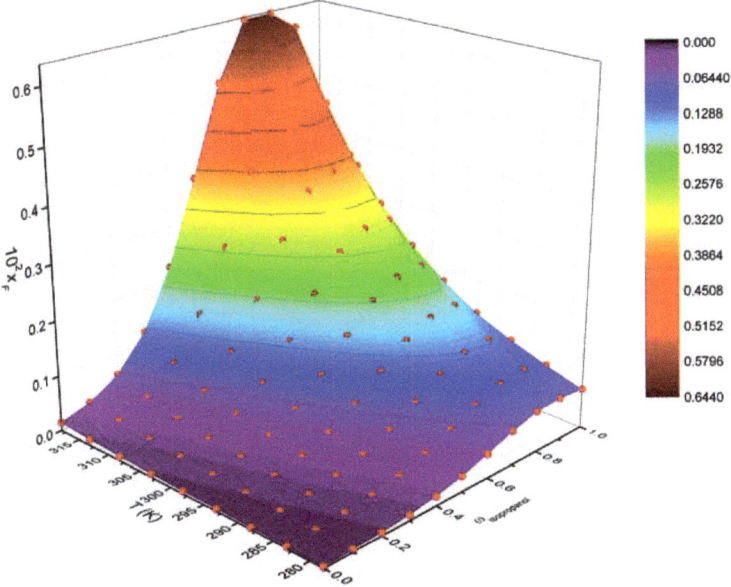

**Figure 7.** Molar ratio solubility data of florfenicol in isopropanol + water binary solvents at $T$ = 278.15 K to 318.15 K.

*4.3. Data Correlation*

The experimental solubility data in this work were correlated by the modified Apelblat model, the CNIBS/R-K model, the Jouyban–Acree model, and the NRTL model. Root-mean-square deviations (RMSD) were used to evaluate the accuracy and applicability of these models. It is defined as follows:

$$\text{RMSD} = \sqrt{\frac{1}{N}\sum_{j=1}^{N}\left(x_j^{\text{cal}} - x_j^{\text{exp}}\right)^2} \tag{11}$$

where $N$ stands for the total number of experiments, $x_j^{\text{exp}}$ refers to the experimental mole fraction solubility, and $x_j^{\text{cal}}$ refers to the calculated mole fraction solubility of florfenicol.

The model parameters and RMSD values are listed in Tables 5–8. The RMSD values obtained by the modified Apelblat model, the Jouyban–Acree model, and the NRTL model are less than 0.001. The RMSD values obtained by the CNIBS/R-K model are less than 0.00001, indicating that the calculated solubility data of the CNIBS/R-K model is in best agreement with the experimental data.

**Table 5.** Model parameters of modified Apelblat model for molar ratio solubility of florfenicol in binary solvents.

| $\omega_A$ | $A$ | $10^{-3}\,B$ | $C$ | $10^5$ RMSD |
|---|---|---|---|---|
| | | Methanol + water | | |
| 0.0 | −808.13 | 31,693.03 | 121.41 | 0.13 |
| 0.1 | −1247.43 | 51,196.87 | 187.13 | 0.29 |
| 0.2 | −1405.06 | 58,101.88 | 210.83 | 0.66 |
| 0.3 | −1448.94 | 60,069.52 | 217.49 | 1.09 |
| 0.4 | −1265.91 | 52,002.49 | 190.23 | 2.21 |
| 0.5 | −964.28 | 38,656.35 | 145.27 | 2.72 |
| 0.6 | −612.52 | 23,123.44 | 92.78 | 2.42 |
| 0.7 | −294.48 | 9157.60 | 45.29 | 4.71 |
| 0.8 | −121.01 | 1711.19 | 19.30 | 9.82 |
| 0.9 | −144.46 | 3117.79 | 22.64 | 10.82 |
| 1.0 | −245.25 | 7999.35 | 37.51 | 10.12 |
| | | Ethanol + water | | |
| 0.0 | −808.13 | 31,693.03 | 121.41 | 0.13 |
| 0.1 | −692.87 | 26,257.05 | 104.52 | 0.16 |
| 0.2 | −574.18 | 20,756.06 | 87.06 | 0.48 |
| 0.3 | −484.37 | 16,577.92 | 73.89 | 1.46 |
| 0.4 | −389.78 | 12,315.84 | 59.92 | 3.15 |
| 0.5 | −352.72 | 10,766.72 | 54.44 | 6.18 |
| 0.6 | −356.69 | 11,178.37 | 54.99 | 8.71 |
| 0.7 | −388.18 | 12,934.37 | 59.56 | 9.84 |
| 0.8 | −435.46 | 15,461.08 | 66.42 | 8.77 |
| 0.9 | −509.61 | 19,190.21 | 77.25 | 6.60 |
| 1.0 | −568.06 | 22,266.08 | 85.68 | 6.53 |

**Table 5.** Cont.

| $\omega_A$ | $A$ | $10^{-3} B$ | $C$ | $10^5$ RMSD |
|---|---|---|---|---|
| \multicolumn{5}{c}{1-Propanol + water} | | | | |
| 0.0 | −808.13 | 31,693.03 | 121.41 | 0.13 |
| 0.1 | −730.72 | 27,893.79 | 110.19 | 0.18 |
| 0.2 | −611.35 | 22,249.17 | 92.70 | 0.26 |
| 0.3 | −495.99 | 16,869.26 | 75.74 | 0.46 |
| 0.4 | −391.10 | 12,093.60 | 60.25 | 0.53 |
| 0.5 | −297.79 | 7967.05 | 46.40 | 0.45 |
| 0.6 | −243.35 | 5785.18 | 38.21 | 0.71 |
| 0.7 | −187.66 | 3677.54 | 29.72 | 1.77 |
| 0.8 | −119.49 | 1030.92 | 19.34 | 1.85 |
| 0.9 | −33.04 | −2688.41 | 6.35 | 0.47 |
| 1.0 | −198.02 | 4989.83 | 30.72 | 0.52 |
| \multicolumn{5}{c}{Isopropanol + water} | | | | |
| 0.0 | −803.53 | 31,693.18 | 121.42 | 12.57 |
| 0.1 | −670.31 | 25,192.77 | 101.17 | 0.16 |
| 0.2 | −614.35 | 22,428.78 | 93.09 | 0.35 |
| 0.3 | −540.78 | 18,938.95 | 82.35 | 0.60 |
| 0.4 | −406.90 | 12,831.21 | 62.55 | 0.35 |
| 0.5 | −332.50 | 9549.48 | 51.52 | 0.30 |
| 0.6 | −246.67 | 5901.59 | 38.68 | 0.43 |
| 0.7 | −188.89 | 3672.04 | 29.90 | 0.38 |
| 0.8 | −112.41 | 647.18 | 18.27 | 0.52 |
| 0.9 | −17.13 | −3411.03 | 3.94 | 0.41 |
| 1.0 | −197.97 | 4983.43 | 30.70 | 0.47 |

**Table 6.** Model parameters of CNIBS/R-K model for molar ratio solubility of florfenicol in binary solvents.

| T/K | $B_0$ | $B_1$ | $B_2$ | $B_3$ | $B_4$ | $10^6$ RMSD |
|---|---|---|---|---|---|---|
| \multicolumn{7}{c}{Methanol + water} | | | | | | |
| 278.15 | −10.84 | 9.21 | 0.29 | −9.07 | 5.07 | 0.86 |
| 283.15 | −10.77 | 11.31 | −6.39 | −1.24 | 1.87 | 1.68 |
| 288.15 | −10.51 | 9.99 | −1.61 | −7.34 | 4.41 | 1.66 |
| 293.15 | −10.35 | 10.18 | −1.55 | −7.81 | 4.65 | 3.05 |
| 298.15 | −10.17 | 10.15 | −0.12 | −10.22 | 5.69 | 2.64 |
| 303.15 | −9.81 | 9.45 | 1.51 | −12.11 | 6.44 | 2.18 |
| 308.15 | −9.53 | 9.61 | 0.95 | −11.79 | 6.43 | 1.84 |
| 313.15 | −9.19 | 10.24 | −1.73 | −8.76 | 5.30 | 1.94 |
| 318.15 | −8.86 | 12.11 | −9.35 | 0.48 | 1.68 | 1.67 |

**Table 6.** *Cont.*

| T/K | $B_0$ | $B_1$ | $B_2$ | $B_3$ | $B_4$ | $10^6$ RMSD |
|---|---|---|---|---|---|---|
| Ethanol + water | | | | | | |
| 278.15 | −10.88 | 16.50 | −19.06 | 8.80 | −1.17 | 0.97 |
| 283.15 | −10.76 | 17.76 | −22.37 | 11.93 | −2.26 | 1.13 |
| 288.15 | −10.54 | 17.82 | −22.82 | 12.56 | −2.59 | 2.44 |
| 293.15 | −10.27 | 17.72 | −21.99 | 10.62 | −1.51 | 2.18 |
| 298.15 | −10.02 | 18.38 | −23.54 | 11.58 | −1.62 | 1.29 |
| 303.15 | −9.79 | 20.56 | −31.51 | 21.25 | −5.53 | 1.79 |
| 308.15 | −9.52 | 20.61 | −32.18 | 22.54 | −6.28 | 2.07 |
| 313.15 | −9.17 | 20.36 | −31.48 | 21.24 | −5.55 | 1.58 |
| 318.15 | −8.86 | 20.74 | −33.10 | 23.29 | −6.44 | 2.69 |
| 1-Propanol + water | | | | | | |
| 278.15 | −10.88 | 19.40 | −35.74 | 29.69 | −9.64 | 0.84 |
| 283.15 | −10.74 | 21.34 | −42.71 | 39.17 | −14.00 | 1.90 |
| 288.15 | −10.49 | 22.05 | −45.03 | 41.86 | −15.10 | 0.90 |
| 293.15 | −10.27 | 23.11 | −48.76 | 46.64 | −17.22 | 1.96 |
| 298.15 | −10.04 | 24.62 | −55.04 | 55.41 | −21.20 | 0.73 |
| 303.15 | −9.80 | 25.78 | −59.73 | 61.81 | −24.07 | 2.45 |
| 308.15 | −9.54 | 26.50 | −62.14 | 64.34 | −24.94 | 0.99 |
| 313.15 | −9.17 | 26.66 | −64.04 | 67.32 | −26.31 | 0.95 |
| 318.15 | −8.86 | 27.17 | −66.55 | 70.57 | −27.65 | 2.29 |
| Isopropanol + water | | | | | | |
| 278.15 | −10.82 | 17.01 | −29.32 | 22.14 | −6.26 | 1.49 |
| 283.15 | −10.72 | 19.06 | −36.19 | 31.21 | −10.39 | 1.07 |
| 288.15 | −10.49 | 20.04 | −39.52 | 35.31 | −12.14 | 1.26 |
| 293.15 | −10.24 | 20.98 | −43.21 | 40.36 | −14.46 | 1.24 |
| 298.15 | −10.06 | 22.56 | −48.77 | 47.46 | −17.52 | 1.75 |
| 303.15 | −9.78 | 23.06 | −50.73 | 49.82 | −18.47 | 2.60 |
| 308.15 | −9.56 | 24.39 | −55.68 | 56.13 | −21.15 | 2.56 |
| 313.15 | −9.17 | 24.23 | −56.16 | 56.91 | −21.43 | 2.43 |
| 318.15 | −8.86 | 24.74 | −58.55 | 59.89 | −22.61 | 1.84 |

**Table 7.** Model parameters of Jouyban–Acree model for molar ratio solubility of florfenicol in binary solvents.

| Parameters | Methanol + Water | Ethanol + Water | 1-Propanol + Water | Isopropanol + Water |
|---|---|---|---|---|
| $A_0$ | 1150.09 | 1283.14 | 1319.05 | 1311.57 |
| $A_1$ | −55,906.60 | −62,424.52 | −64,161.55 | −63,771.45 |
| $A_2$ | −170.66 | −190.18 | −195.46 | −194.37 |
| $A_3$ | −202.15 | −297.98 | −117.67 | −102.52 |
| $A_4$ | 11,130.77 | 18,497.31 | 11,834.44 | 10,480.10 |

Table 7. Cont.

| Parameters | Methanol + Water | Ethanol + Water | 1-Propanol + Water | Isopropanol + Water |
|---|---|---|---|---|
| $A_5$ | 633.01 | −6878.54 | −15,035.27 | −13,079.80 |
| $A_6$ | −3937.48 | 3236.22 | 14,406.50 | 11,955.51 |
| $A_7$ | 2128.14 | −374.34 | −5287.30 | −4158.02 |
| $A_8$ | 30.53 | 44.60 | 17.81 | 15.56 |
| $10^4$ RMSD | 6.05 | 8.94 | 5.06 | 3.99 |

Table 8. Model parameters of NRTL model for molar ratio solubility of florfenicol in binary solvents.

| Parameters | Methanol + Water | Ethanol + Water | 1-Propanol + Water | Isopropanol + Water |
|---|---|---|---|---|
| $\Delta g_{ij}$ | −11,849.11 | −8212.77 | −4953.91 | −4872.17 |
| $\Delta g_{ik}$ | 16,346.95 | −3890.03 | −614.55 | 4267.26 |
| $\Delta g_{ji}$ | 20,776.62 | 19,191.90 | 14,438.84 | 14,463.19 |
| $\Delta g_{jk}$ | −24,560.44 | −1492.57 | −11,331.24 | −54,803.48 |
| $\Delta g_{ki}$ | 13,303.06 | 21,868.77 | 15,568.84 | 13,816.73 |
| $\Delta g_{kj}$ | 6591.86 | 5907.92 | 15,793.93 | 7239.98 |
| $10^4$ RMSD | 3.33 | 4.50 | 1.55 | 1.01 |

## 5. Conclusions

In this work, the solubility of florfenicol was determined by a gravimetric method in four binary solvents (methanol + water, ethanol + water, 1-propanol + water, and isopropanol + water) with temperatures from 278.15 to 318.15 K under atmospheric pressure. In these four mixed solvents, the solubility of florfenicol increased with the increase in temperature. At the same temperature, the solubility of florfenicol increases with the decrease in the water ratio in the methanol + water mixture solvent. While the solubility of florfenicol increases first and then decreases with the decrease in the ratio of water in ethanol + water, 1-propanol + water, and isopropanol + water mixture solvent, indicating a cosolvency phenomenon occurs in these three binary solvents. In this study, the modified Apelblat model, the CNIBS/R-K model, the Jouyban–Acree model, and the NRTL model were used to correlate the solubility data of florfenicol in four binary solvents. The RMSD values of each model show that the calculated solubility data are in good agreement with the experimental data for all four models, among which the CNIBS/R-K model provides the best fitting result. Most of the previous solubility articles only determined the solubility of florfenicol in pure solvents, and rarely determined the solubility of florfenicol in binary solvents. However, considering the solubilization requirements of florfenicol, especially in the research of preparing nanocrystals by antisolvent crystallization, the solubility data of florfenicol in binary solvents are highly desirable. The four alcohols involved in this study are the most commonly used organic solvents in experimental research and industry, in which Florfenicol has good solubility, so alcohol solvent can be used as an effective proper solvent. Needless to say, water is the most suitable antisolvent for florfenicol. Therefore, the solubility data obtained by this study can be used as fundamental data for research on the florfenicol nanocrystal preparation to achieve florfenicol solubilization.

**Author Contributions:** Data curation, X.Z. and P.C.; Resources, Q.Y.; Supervision, L.Z.; Writing—original draft, X.Z.; Writing—review and editing, P.C., Q.Y. and L.Z. All authors have read and agreed to the published version of the manuscript.

**Funding:** This research was funded by (Tianjin Municipal Natural Science Foundation) grant number [21JCYBJC00600].

**Institutional Review Board Statement:** Not applicable.

**Informed Consent Statement:** Not applicable.

**Data Availability Statement:** Not applicable.

**Acknowledgments:** The authors are grateful for the financial support of the Tianjin Municipal Natural Science Foundation (No. 21JCYBJC00600).

**Conflicts of Interest:** The authors declare no conflict of interest.

## References

1. Bello, A.; Poniak, B.; Smutkiewicz, A.; Świtała, M. The influence of the site of drug administration on florfenicol pharmacokinetics in turkeys. *Poult. Sci.* **2022**, *101*, 101536. [CrossRef] [PubMed]
2. Zhai, Q.; Chang, Z.; Li, J.; Li, J. Effects of combined florfenicol and chlorogenic acid to treat acute hepatopancreatic necrosis disease in Litopenaeus vannamei caused by Vibrio parahaemolyticus. *Aquaculture* **2022**, *547*, 737462. [CrossRef]
3. Zhang, W.; Liu, C.P.; Chen, S.Q.; Liu, M.J.; Zhang, L.; Lin, S.Y.; Shu, G.; Yuan, Z.X.; Lin, J.C.; Peng, G.N.; et al. Poloxamer modified florfenicol instant microparticles for improved oral bioavailability. *Colloids Surf. B Biointerfaces* **2020**, *193*, 111078. [CrossRef] [PubMed]
4. Veach, B.T.; Anglin, R.; Mudalige, T.K.; Barnes, P.J. Quantitation and Confirmation of Chloramphenicol, Florfenicol, and Nitrofuran Metabolites in Honey Using LC-MS/MS. *J. AOAC Int.* **2017**, *101*, 897–904. [CrossRef] [PubMed]
5. Hecker, S.J.; Pansare, S.V.; Glinka, T.W. Florfenicol Prodrug Having Improved Water Solubility. U.S. Patent US20070021387, 17 January 2007.
6. Li, X.; Xie, S.; Pan, Y.; Qu, W.; Tao, Y.; Chen, D.; Huang, L.; Liu, Z.; Wang, Y.; Yuan, Z. Preparation, Characterization and Pharmacokinetics of Doxycycline Hydrochloride and Florfenicol Polyvinylpyrroliddone Microparticle Entrapped with Hydroxypropyl-β-cyclodextrin Inclusion Complexes Suspension. *Colloids Surf. B Biointerfaces* **2016**, *141*, 634–642. [CrossRef] [PubMed]
7. Ma, S.; Shang, X. Preparation and Characterization of Florfenicol-Polyethyleneglycol 4000 Solid Dispersions with Improved Solubility. *Asian J. Chem.* **2012**, *24*, 3059–3063.
8. Periyasamy, S.; Lin, X.; Ganiyu, S.O.; Kamaraj, S.K.; Thiam, A.; Liu, D. Insight into BDD electrochemical oxidation of florfenicol in water: Kinetics, reaction mechanism, and toxicity. *Chemosphere* **2022**, *288*, 132433. [CrossRef] [PubMed]
9. Sun, Z.; Hao, H.; Xie, C.; Xu, Z.; Yin, Q.; Bao, Y.; Hou, B.; Wang, Y. Thermodynamic Properties of Form A and Form B of Florfenicol. *Ind. Eng. Chem. Res.* **2014**, *53*, 13506–13512. [CrossRef]
10. Wang, S.; Chen, N.; Qu, Y. Solubility of Florfenicol in Different Solvents at Temperatures from (278 to 318) K. *J. Chem. Eng. Data* **2011**, *56*, 638–641. [CrossRef]
11. Zhou, J.; Fu, H.; Cao, H.; Lu, C.; Jin, C.; Zhou, T.; Liu, M.; Zhang, Y. Measurement and Correlation of the Solubility of Florfenicol in Binary 1,2-Propanediol + Water Mixtures from 293.15K to 316.25K. *Fluid Phase Equilibria* **2013**, *360*, 118–123. [CrossRef]
12. Lou, Y.; Wang, Y.; Li, Y.; He, M.; Su, N.; Xu, R.; Meng, X.; Hou, B.; Xie, C. Thermodynamic Equilibrium and Cosolvency of Florfenicol in Binary Solvent System. *J. Mol. Liq.* **2018**, *251*, 83–91. [CrossRef]
13. Zhang, P.; Zhang, C.; Zhao, R.; Wan, Y.; Yang, Z.; He, R.; Chen, Q.; Li, T.; Ren, B. Measurement and Correlation of the Solubility of Florfenicol Form A in Several Pure and Binary Solvents. *J. Chem. Eng. Data* **2018**, *63*, 2046–2055. [CrossRef]
14. Li, L.; Zhao, Y.; Hou, B.; Feng, H.; Wang, N.; Liu, D.; Ma, Y.; Wang, T.; Hao, H. Thermodynamic Properties of 1,5-Pentanediamine Adipate Dihydrate in Three Binary Solvent Systems from 278.15 K to 313.15 K. *Crystals* **2022**, *12*, 877. [CrossRef]
15. Apelblat, A.; Manzurola, E. Solubilities of L-Aspartic, DL-Aspartic, DL-Glutamic, P-Hydroxybenzoic, O-Anisic, P-Anisic, and Itaconic Acids in Water from $T = 278$ K to $T = 345$ K. *J. Chem. Thermodyn.* **1997**, *29*, 1527–1533. [CrossRef]
16. Apelblat, A.; Manzurola, E. Solubilities of O-Acetylsalicylic, 4-Aminosalicylic, 3,5-Dinitrosalicylic, and P-Toluic Acid, and Magnesium-DL-Aspartate in Water from $T = (278$ to $348)$ K. *J. Chem. Thermodyn.* **1999**, *31*, 85–91. [CrossRef]
17. Acree, W.E. Mathematical Representation of Thermodynamic Properties: Part 2. Derivation of the Combined Nearly Ideal Binary Solvent (NIBS)/Redlich-Kister Mathematical Representation from A Two-Body and Three-Body Interactional Mixing Model. *Thermochim. Acta* **1992**, *198*, 71–79. [CrossRef]
18. Jouyban-Gharamaleki, A.; Acree, W.E. Comparison of Models for Describing Multiple Peaks in Solubility Profiles. *Int. J. Pharm.* **1998**, *167*, 177–182. [CrossRef]
19. Jouyban, A.; Soltani, S.; Chan, H.K.; Acree, W.E., Jr. Modeling Acid Dissociation Constant of Analytes in Binary Solvents at Various Temperatures Using Jouyban–Acree Model. *Thermochim. Acta* **2005**, *428*, 119–123. [CrossRef]
20. Jouyban, A. Review of the Cosolvency Models for Predicting Solubility of Drugs in Water-Cosolvent Mixtures. *J. Pharm. Pharm. Sci.* **2008**, *11*, 32–58. [CrossRef] [PubMed]
21. Asselineau, L.; Renon, H. Extension de L'equation NRTL Pour La Représentation de L'ensembles des Données D'equilibre Binaire, Liquide—Vapeur et Liquide—Liquide. *Chem. Eng. Sci.* **1970**, *25*, 1211–1223. [CrossRef]

Article

# Image Measurement of Crystal Size Growth during Cooling Crystallization Using High-Speed Imaging and a U-Net Network

Yan Huo, Xin Li * and Binbin Tu

College of Information Engineering, Shenyang University, Shenyang 110044, China
* Correspondence: li_xin@syu.edu.cn

**Abstract:** In this paper, an image measurement method using a high-speed imaging system is proposed for the evolution of crystal population sizes during cooling crystallization processes. Firstly, to resist the negative effect from solution stirring and particle motion during crystallization, a U-net network-based image processing method is established to efficiently detect sufficiently clear crystals from the online captured microscopic images. Accordingly, the crystal size distribution model is analyzed in terms of the counted probability densities of these crystal images. Subsequently, a measurement method of size growth rate based on crystal population distribution is proposed to estimate the growth condition. An experimental case on a crystallization process of β-form LGA is used to show the effectiveness of the proposed strategy.

**Keywords:** crystal image analysis; deep learning; size measurement; crystal growth rate

## 1. Introduction

Crystallization is an important process to obtain crystalline solids from mixed solutions in pharmaceutical and chemical industries [1]. The cooling crystallization process usually includes the formation of a supersaturated solution, nucleation and crystal growth, etc. Crystal size distribution (CSD) is one of key indicators to evaluate the crystallization production quality [2,3]. It is necessary to measure the growth parameters of the crystal population for process optimization and feedback control [4–6]. In recent years, researchers have made significant advances in process analytical technology (PAT) for monitoring crystallization processes, e.g., ATR-FTIR spectroscopy, Raman spectroscopy, focused beam reflectance measurement (FBRM), ultrasound spectroscopy, etc.

With the development of optical imaging sensors, crystallization process detection strategies using image measurement have been promoted for crystal defects, sizes, and shapes [7–12]. Larsen et al. [13] analyzed the high concentration crystal image effectively and processed the needle crystal, making a detailed analysis of various characteristics of the crystal from different application values, striving to achieve a comprehensive description of the crystal. Zhou et al. [14] proposed some parameter optimization approaches for image processing applied to extract useful information from microscopy images regarding the distribution monitoring of particle shape and size. Lins et al. [15] developed a detection method of crystal defects, including crystal contour detection and defect quantification for evaluating and optimizing crystallization processes. For the measurement of crystal agglomeration, Ferreira et al. [16] proposed a novel image analysis technique which combined discriminant factorial analysis to assess the agglomeration of crystals. Lu et al. [17] developed a valid crystal segmentation approach based on background difference and local threshold to overcome the negative effect of particle shadow. In previous work [10], an online image measuring method was presented to analyze two-dimension (2D) crystal sizes during cooling crystallization. Gao et al. [18] developed a valid image analysis technology based on deep learning to detect crystals and measure their sizes. Ma et al. [19] applied a novel online image monitoring system to measure

the growth rate for mean crystal size during LGA crystallization processes. Traditionally, the growth rate was measured based on the mean sizes for β-form L-glutamic acid [20–23], but the changes of the crystal population size distribution were not quantized and estimated roundly in the previous literature.

In this work, in order to eliminate the influence of continuous motion in a stirred reactor, an effective strategy is presented for the growth of crystal population sizes using the imaging measurement method based on a deep learning model. Firstly, for the online crystal images influenced by solution turbulence, uneven illumination, and noise, image preprocessing is used. Secondly, the valid crystals are extracted by an effective image segmentation method using a U-net network model. Thirdly, the CSDs are computed with the measured 2D sizes and probability density function. An indicator for describing the growth of crystal 2D size distribution is estimated with a statics method. Experimental results for the case of the cooling crystallization of β-form L-glutamic acid (LGA) show the effectiveness of the proposed imaging measurement method.

The article is organized as follows. Section 2 introduces the basic algorithms. The Section 3 is dedicated to the experimental set-up. Section 4 demonstrates in detail the method of image measurement based on the U-net network. The experimental case is made for the validity of the method in the Section 5. Finally, conclusions are given in the Section 6.

## 2. Preliminaries

### 2.1. Classical Convolution Neural Network

Convolutional neural network (CNN) is a basic deep neural network with a convolution structure. In 1998, Lecun et al. [24] designed and trained a CNN model (called LeNet-5), which is a classical CNN structure. The basic structure of CNN is composed of an input layer, a convolution layer, a pooling layer, a full connection layer and an output layer, as shown in Figure 1. Generally, a convolution layer is connected with a pooling layer, and the last few layers near the output layer are usually fully connected networks. The training process of CNN is to learn the convolution kernel parameters of convolution layer and the connection weight between layers. In image recognition, the prediction process is mainly based on the input image and network parameters to calculate the category label.

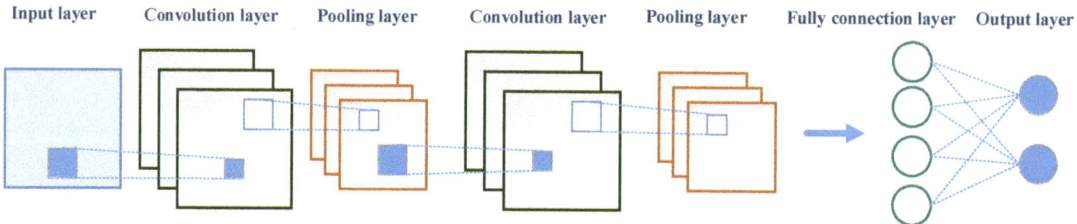

**Figure 1.** Architecture of convolutional neural network.

### 2.2. U-Net Network

U-net network [25] is a full convolutional network improved based on fully convolutional networks (FCN) [25], which is similar to U-type. Compared with other convolutional neural networks, this network requires smaller training set size and has higher segmentation accuracy. A basic U-net network structure consists of a down-sampling path (encoder) and an up-sampling path (decoder). The down-sampling path is used to obtain the context information, and the up-sampling path is used to pinpoint the location. The down-sampling path is in the left of the network, which consists of $3 \times 3$ convolution layers and $2 \times 2$ max pool layers. The activation function $f(x)$ uses ReLU [26], which is defined with $\tau > 0$ as

$$f(x) = \begin{cases} x, & x > 0 \\ \tau(e^x - 1), & x \leq 0 \end{cases} \tag{1}$$

The up-sampling path is in the right of the network. The deconvolution is used to halve the number of channels, then the deconvolution result is spliced with the corresponding feature map, and the spliced feature map is then convolved with a 3 × 3 kernal. The last layer uses a 1 × 1 convolution to map each 2-bit feature vector to the output layer of the network.

## 3. Experimental Set-Up

The experimental setup with an imaging system for measuring crystal size distribution is shown in Figure 2. Experiments were carried out with a crystallizer including a 1 L glass jacketed reactor and a PTFE four-paddle agitator. The temperature control device used a Julabo-CF41 thermostatic circulator (JULABO, Seelbach, Germany). A Pt100 temperature probe was used to measure the solution temperature. A camera device (UI-2280SEC-HQ, IDS, Obersulm, Germany) was adopted to record online crystallization images in cooling crystallization processes. The imaging system includes the following functions: image acquisition, image storage, image compression, image output, etc., which was connected to an industrial personal computer. An LED light and controller (Gardasoft -RT260-20, Gardasoft Vision, Cambridge, UK) were employed to provide illumination.

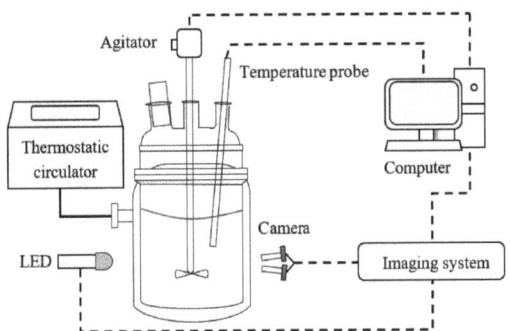

**Figure 2.** Schematic drawing of the experimental setup.

The material used in this experiment was L-glutamic acid (LGA) (Sigma Chemicals, St Louis, MO, USA), which has two forms, the α form and β form. This experiment mainly focused on the study of needle-like β-form LGA. The growth of β-form crystals is statistically analyzed by image analysis in the reactor. The stirring rate was maintained at 200 rpm. Firstly, 0.6 L LGA solution with a concentration of 30 g/L was used in the reactor. The solution was heated to 70 °C and then cooled to 30 °C after 1 h of constant temperature. When the temperature dropped to 55 °C, β-form seeds were added into the reactor, and the online crystal images were recorded with the imaging system.

## 4. Crystallization Measurement Method

The process of the proposed image measurement method is presented in Figure 3. The detailed steps are described in the following subsections.

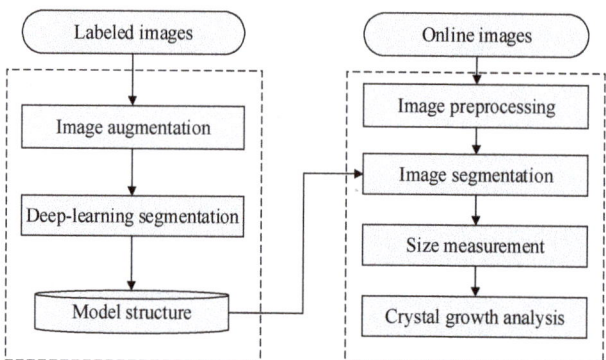

**Figure 3.** Flow chart of the proposed image measurement.

*4.1. Crystal Image Preprocessing*

Due to the disturbance of noise and blur, the quality of crystal images is mostly degraded during online image acquisition. Therefore, guided filtering [27] can be taken as a denoising method. The guided filtering function is considered as an edge-preserving filter, guaranteeing good information preservation around the image edges. Supposing that the pixel of the input image is $m_i$ and the pixel of the output image is $q_i$, which are defined by

$$q_i = \frac{1}{|\omega|} \sum_{i \in \omega_k} \alpha_k m_i + \beta_k \qquad (2)$$

where $k$ is the index of the local square window $\omega_k$ which is taken as 31 × 31 in the input image, $(\alpha_k, \beta_k)$ are the constants in $\omega_k$ which can be obtained by

$$(\alpha_k, \beta_k) = \arg\min_{\alpha_k, \beta_k} \sum_{i \in \omega_k} \left( (\alpha_k m_i + \beta_k - m_i)^2 + \varepsilon \alpha_k^2 \right) \qquad (3)$$

where $\varepsilon$ is the regularization parameter.

*4.2. Crystal Image Segmentation*

Since the calibration method uses clear images to compute pixel equivalent, only clear crystals are able to provide accurate size information. In this work, the U-net network model is improved for online crystal images, being also composed of down-sampling path (encoder) and up-sampling path (decoder) [28], as shown in Figure 4. The improved network consists of 11 layer groups. A parametric rectified linear unit (PReLU) [29] is adopted as the activation function to improve the fitting ability. Each Res Block with two 3 × 3 convolutions in the up-sampling path is connected with the corresponding Res Block of the down-sampling path, as shown in Figure 4. The first convolution layer uses a 1 × 1 convolution kernel to extract the features of the input crystal image. Then the designed layer of Res Block is used to further extract image features and deepen the network processing, as shown in Figure 5, which includes the use of two convolutional operations, batch normalization modules and ELU activation functions. After the Res Block in the down-sample path, a 2 × 2 max pooling layer is used to down-sample the feature map, so that the resolution of the feature map decreases to half of the original one. In the up-sampling path, a bilinear interpolation method is used in an up-interpolation block, and the feature map obtained is splicing by layer hopping connection, so that the network can fully fuse the shallow features and deep semantic information. Following the final 1 × 1 convolution layer, the Sigmoid activation function is employed in the last layer to map the response values to (0, 1) pixel by pixel.

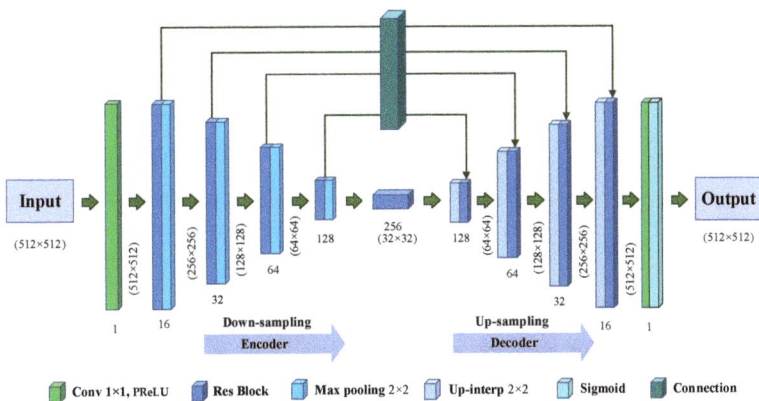

**Figure 4.** Improved U-net network structure.

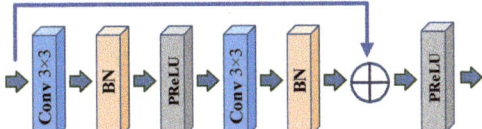

**Figure 5.** Res Block.

To solve the imbalance of positive and negative pixels in the image, the loss function [30] is defined as

$$J_{\text{loss}} = C_{\text{loss}} + D_{\text{loss}} \tag{4}$$

where $C_{\text{loss}}$ is the cross-entropy loss and $D_{\text{loss}}$ is the Dice coefficient loss.

Image augmentation is used to increase the training sample size and strengthen the network generalization ability. Random rotation, scale, and translation are used to enhance the image diversity, and the brightness and contrast of the image are adjusted to reduce the influence of uneven lighting and highlight the edge features.

### 4.3. Crystal Growth Measurement

The two-dimensional sizes (i.e., length and width) of β-form LGA crystals are measured based on the length and width of the minimum enclosing rectangle [31] for crystal imaging, respectively. If $l_a$ is the pixel number of length and $w_a$ is the pixel number of width, the pixel equivalent $\gamma_e$ is obtained with the calibration method [10]. The physical length $x_l$ and the physical width $x_w$ are given by

$$\begin{cases} x_l = \gamma_e l_a \\ x_w = \gamma_e w_a \end{cases} \tag{5}$$

Based on Equation (5), the 2D crystal sizes can be obtained to produce CSD. For a crystal population with a large number of particles showing statistical characteristics, it is meaningful to estimate their size distribution. Generally, the probability density estimation of a log-normal distribution function can be used to smooth the CSD [32] to represent the current size condition of the crystal population.

For length, conforming to the log-normal distribution $LN(\mu_l, \sigma_l^2)$, the likelihood function with length variable $x_l$ is defined as

$$L(\mu_l, \sigma_l^2) = \prod_{i=1}^{n} \frac{1}{\sqrt{2\pi}\sigma_l x_l(i)} \exp\left\{-\frac{(\ln x_l(i) - \mu_l)^2}{2\sigma_l^2}\right\} \tag{6}$$

The likelihood equations are:

$$\begin{cases} \frac{\partial \ln L(\mu_l, \sigma_l^2)}{\partial \mu_l} = \frac{1}{\sigma_l^2} \sum_{i=1}^{n} (\ln x_l(i) - \mu_l) = 0 \\ \frac{\partial \ln L(\mu_l, \sigma_l^2)}{\partial \sigma_l^2} = -\frac{n}{2\sigma_l^2} + \frac{1}{2\sigma_l^4} \sum_{i=1}^{n} (\ln x_l(i) - \mu_l)^2 = 0 \end{cases} \quad (7)$$

By solving Equation (7), the parameters $(\mu_l, \sigma_l^2)$ are estimated as

$$\hat{\mu}_l = \frac{1}{n} \sum_{i=1}^{n} \ln x_l(i) \quad (8)$$

$$\hat{\sigma}_l^2 = \frac{1}{n} \sum_{i=1}^{n} \left( \ln x_l(i) - \frac{1}{n} \sum_{i=1}^{n} \ln x_l(i) \right)^2 \quad (9)$$

Then $LN(\hat{\mu}_l, \hat{\sigma}_l^2)$ can be computed for denoting the length size distribution of the crystal population by using online images in a predefined time window [8].

The growth parameter of crystal population size is an important factor for crystallization detection. Traditionally, the growth rate in the mean size may not denote the size distribution evolution well, due to the noise of size extremes. Then, $x_l^{max}$ in the maximum distribution of $P(x_l)$ is computed as

$$x_l^{max} = \arg \max_{x_l} P(x_l) \quad (10)$$

The growth rate of length $R_l$ is defined as

$$R_l = \frac{\text{Diff}(x_{l,t_1}^{max}, x_{l,t_2}^{max})}{T_{P_1 P_2}} \quad (11)$$

where Diff is the difference function between $x_{l,t_1}^{max}$ of $P_{t_1}$ and $x_{l,t_2}^{max}$ of $P_{t_2}$, and $T_{P_1 P_2}$ is the time interval between point $t_1$ and point $t_2$.

Similar to length, the width size distribution $LN(\mu_w, \sigma_w^2)$ and the growth rate of width $R_w$ are obtained as mentioned above.

## 5. Experiment Results

### 5.1. Deep-Learning Crystal Extraction

The experimental case on the crystallization process of β-form LGA was carried out. Compared with traditional methods [8,10,11,33,34], the method of deep-learning crystal extraction did not require too many tedious procedures (e.g., image preprocessing steps or clear crystal identification, etc.). A typically captured image is demonstrated in Figure 6a. Figure 6b displays the result of image preprocessing. In Figure 6c, the crystals are extracted with the deep leaning-based image processing method applied to the image of Figure 6b. It is seen that the minimum enclosing rectangles of the crystals are obtained in Figure 6d. The 2D sizes were obtained from the fitting rectangles. In the experiment, 40 real-time LGA images were employed as the original training set, which uses image augmentation. Flips, translation and rotation were used for variety, and the brightness and contrast were adjusted to reduce the influence of uneven lighting. Finally, noise interference was added to the training set to further improve the generalization ability of the model.

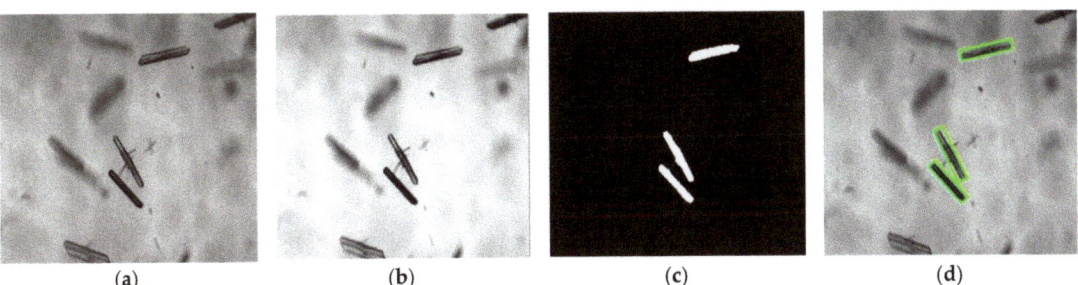

**Figure 6.** Segmentation and measurement results of the LGA image: (**a**) captured image; (**b**) image preprocessing; (**c**) segmented image with an improved U-net; (**d**) rectangle fitting.

In addition, to represent the superiority of the proposed image processing method, the Ostu segmentation method and the Canny method, which are always used in the image processing of crystals [8,10,11,33,34], were performed for the two crystal images in Figure 7a. Figure 7b demonstrates the results with an improved U-net. Figure 7b,c show the results of image preprocessing using the Ostu and Canny segmentation methods, respectively. It can be seen that clear crystals are detected with an improved U-net, whereas the fuzzy crystals may affect the size measurement. Figure 7 shows that the clear crystals are effectively segmented with the proposed online analysis method, while the other two methods make several segmentation mistakes (e.g., overlapping, fuzziness, etc.). Figure 7e shows that the results of the original U-net model are less accurate than the improved one.

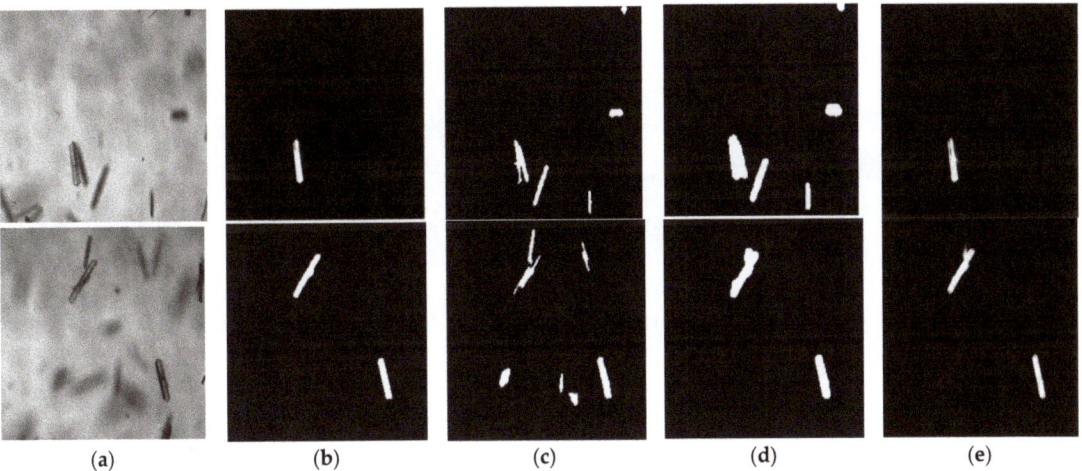

**Figure 7.** Segmentation comparison results of two LGA images: (**a**) captured images; (**b**) improved segmentation results; (**c**) threshold segmentation results; (**d**) edge detection results; (**e**) original U-net results.

### 5.2. Crystal Size Measurement

CSD information is important for production. For crystallization quality control, it is necessary that the information feedback of crystal sizes is provided timely and effectively. In the experiment, an offline measurement method using an electric microscope was utilized to verify the accuracy of the CSD measurement with the proposed method. For the same batch of crystals, the comparison study was made between the proposed online method and the offline method by measuring 2D sizes (i.e., length and width), as shown in Figure 8. It is presented that the measured results between the two methods are very similar. It is noted the online images should be captured immediately after the crystals are added into the reactor to avoid the changes of crystal sizes and shapes.

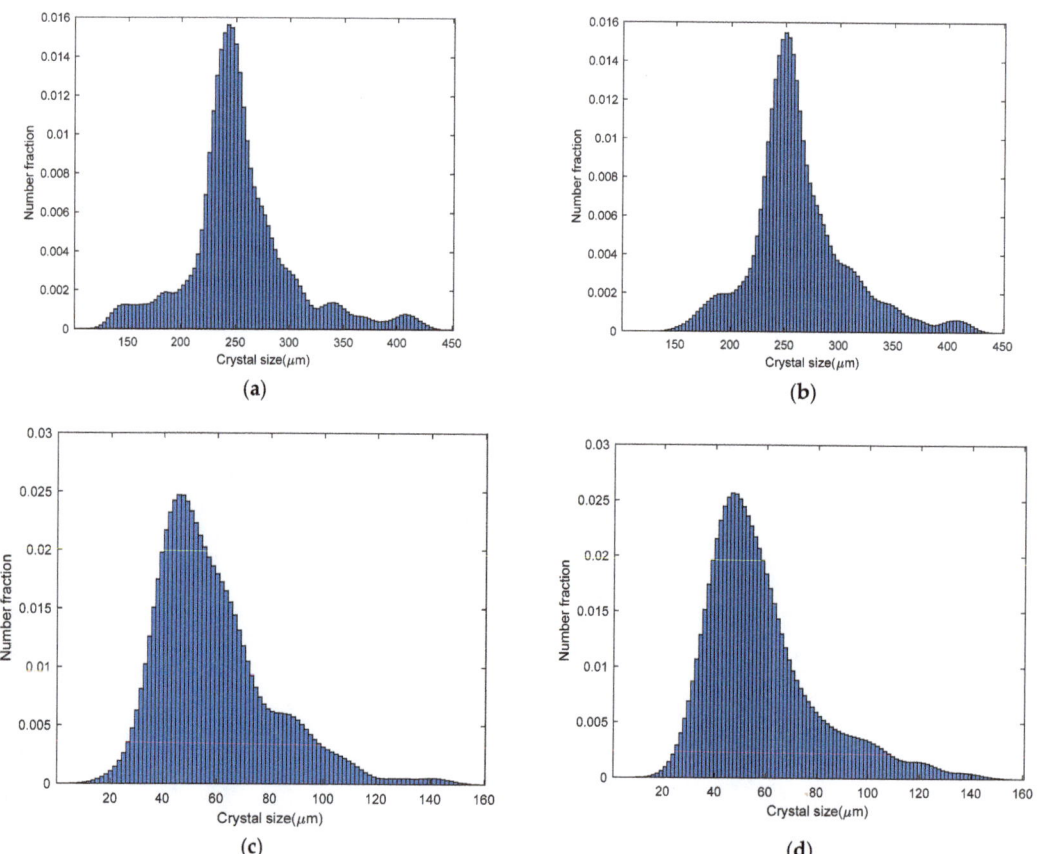

**Figure 8.** Comparisons between online and offline results of LGA crystal products: (**a**) histogram with the online measurement for LGA length; (**b**) histogram with the offline measurement for LGA length; (**c**) histogram with the online measurement for LGA width; (**d**) histogram with offline measurement for LGA width.

In this experiment, the measured CSDs were fitted by the probability density estimation with the lognormal distribution function. In Figure 9, the CSDs are computed with about 200 crystals collected at time points of 0 min, 20 min, 40 min, and 60 min. The predefined time window was set as 30 s. Figure 9 shows that the crystal population size increases with time, but the range of the crystal size distribution becomes much wider. The needle-like crystals may be easier to be broken by the stirring agitator, resulting in small sizes, whereas the LGA crystals are likely to agglomerate, leading to large sizes. Therefore, the size distribution can become wider at the end of LGA crystallization.

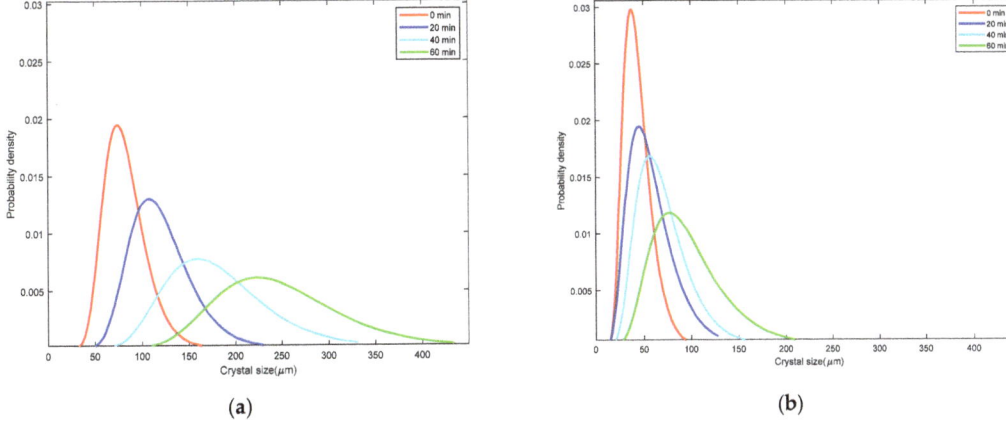

**Figure 9.** Evolution of β-form LGA crystal size distribution at four time points with a log-normal distributed model: (**a**) probability density of crystal length sizes; (**b**) probability density of crystal width sizes.

Traditionally, the measurement of growth rate has been used with mean crystal size. However, the growth rate with mean size may not denote the size distribution evolution well, due to the noise of size extremes. It is significant for characterizing the growth rate of crystal size distribution to crystallization control. The evolution of the β-form LGA crystal population size distribution is exemplified in Table 1. At the three intervals of sample times (i.e., 0, 20, 40, and 60 min), the growth rates $R_l$ and $R_w$ were computed by Equation (11) for crystal size distribution. It is observed that the growth rate gradually increases over time, because solution supersaturation is a main driving force for crystal growth in the cooling crystallization. It is also manifested that the growth rates of β-form LGA population length were much larger than those of β-form LGA population width in Table 1. The growth in length is involved in the growth face (101), and the width is related to the faces (010) and (021) [35]. The behavior of length and width growth rates may be in relation to supersaturation, temperature, etc. It is noted that the growth data for length and width are identified more accurately by increasing the number of sampling time points.

**Table 1.** Measured growth rates for 2D crystal population sizes.

| Time (min) | $R_l$ (µm/min) | $R_w$ (µm/min) |
| --- | --- | --- |
| 20 (0–20) | 1.69 | 0.40 |
| 20 (20–40) | 2.58 | 0.54 |
| 20 (40–60) | 3.26 | 1.11 |

## 6. Conclusions

In this work, an imaging measurement method based on a U-net network was developed to estimate crystal size evolution for β-form LGA using an online non-invasive imaging system. To improve image quality, guided filtering was used for removing the image noise. The deep-learning model with an improved U-net was effectively improved to segment the crystals from the online images. The 2D crystal sizes were measured by using the probability density function. Experimental results showed that the proposed method based on deep learning was effective in obtaining the growth rate of crystal population sizes. The agglomeration condition led to the wrong determination of the crystal sizes. Hence, future work will involve the measurement of crystal agglomeration.

**Author Contributions:** Conceptualization, Y.H.; methodology, B.T.; writing—original draft preparation, Y.H.; writing—review and editing, X.L.; project administration, Y.H. All authors have read and agreed to the published version of the manuscript.

**Funding:** This research was funded by the China Postdoctoral Science Foundation, grant number 2020M680979, and the Doctoral Start-up Foundation of Liaoning Province of China, grant numbers 2021-BS-281 and 2020-BS-263.

**Institutional Review Board Statement:** Not applicable.

**Informed Consent Statement:** Not applicable.

**Data Availability Statement:** Not applicable.

**Conflicts of Interest:** The authors declare no conflict of interest. The funders had no role in the design of the study; in the collection, analyses, or interpretation of data; in the writing of the manuscript, or in the decision to publish the results.

## References

1. Wang, X.; Li, K.; Qin, X.; Li, M.; Liu, Y.; An, Y.; Yang, W.; Chen, M.; Ouyang, J.; Gong, J. Research on mesoscale nucleation and growth processes in solution crystallization: A review. *Crystals* **2022**, *12*, 1234. [CrossRef]
2. Wang, X.Z.; Roberts, K.J.; Ma, C. Crystal growth measurement using 2d and 3d imaging and the perspectives for shape control. *Chem. Eng. Sci.* **2008**, *63*, 1173–1184. [CrossRef]
3. Cardona, J.; Ferreira, C.; Mcginty, J.; Hamilton, A.; Agimelen, O.S.; Cleary, A.; Atkinson, R.; Michie, C.; Marshall, S.; Chen, Y.C. Image analysis framework with focus evaluation for in situ characterisation of particle size and shape attributes. *Chem. Eng. Sci.* **2018**, *191*, 208–231. [CrossRef]
4. Borsos, Á.; Szilágyi, B.; Agachi, P.Ş.; Nagy, Z.K. Real-time image processing based online feedback control system for cooling batch crystallization. *Org. Process Res. Dev.* **2017**, *21*, 511–519. [CrossRef]
5. Gan, C.; Wang, L.; Xiao, S.; Zhu, Y. Feedback control of crystal size distribution for cooling batch crystallization using deep learning-based image analysis. *Crystals* **2022**, *12*, 570. [CrossRef]
6. Wu, Y.; Gao, Z.; Rohani, S. Deep learning-based oriented object detection for in situ image monitoring and analysis: A process analytical technology (pat) application for taurine crystallization. *Chem. Eng. Res. Des.* **2021**, *170*, 444–455. [CrossRef]
7. Liao, C.W.; Yu, J.H.; Tarng, Y.S. On-line full scan inspection of particle size and shape using digital image processing. *Particuology* **2010**, *8*, 286–292. [CrossRef]
8. Ma, C.Y.; Liu, J.J.; Wang, X.Z. Measurement, modelling, and closed-loop control of crystal shape distribution: Literature review and future perspectives. *Particuology* **2016**, *26*, 1–18. [CrossRef]
9. Huo, Y.; Guan, D.; Li, X. In situ measurement method based on edge detection and superpixel for crystallization imaging at high-solid concentrations. *Crystals* **2022**, *12*, 730. [CrossRef]
10. Huo, Y.; Liu, T.; Liu, H.; Ma, C.Y.; Wang, X.Z. In-situ crystal morphology identification using imaging analysis with application to the l-glutamic acid crystallization. *Chem. Eng. Sci.* **2016**, *148*, 126–139. [CrossRef]
11. Zhang, R.; Ma, C.Y.; Liu, J.J.; Wang, X.Z. On-line measurement of the real size and shape of crystals in stirred tank crystalliser using non-invasive stereo vision imaging. *Chem. Eng. Sci.* **2015**, *137*, 9–21. [CrossRef]
12. Li, M.; Zhang, C.; Li, M.; Liu, F.; Zhou, L.; Gao, Z.; Sun, J.; Han, D.; Gong, J. Growth defects of organic crystals: A review. *Chem. Eng. J.* **2022**, *429*, 132350. [CrossRef]
13. Larsen, P.; Rawlings, J.; Ferrier, N. An algorithm for analyzing noisy, in situ images of high-aspect-ratio crystals to monitor particle size distribution. *Chem. Eng. Sci.* **2006**, *61*, 5236–5248. [CrossRef]
14. Zhou, Y.; Lakshminarayanan, S.; Srinivasan, R. Optimization of image processing parameters for large sets of in-process video microscopy images acquired from batch crystallization processes: Integration of uniform design and simplex search. *Chemom. Intell. Lab. Syst.* **2011**, *107*, 290–302. [CrossRef]
15. Lins, J.; Heisel, S.; Wohlgemuth, K. Quantification of internal crystal defects using image analysis. *Powder Technol.* **2021**, *377*, 733–738. [CrossRef]
16. Ferreira, A.; Faria, N.; Rocha, F.; Teixeira, J. Using an online image analysis technique to characterize sucrose crystal morphology during a crystallization run. *Ind. Eng. Chem. Res.* **2011**, *50*, 6990–7002. [CrossRef]
17. Lu, Z.M.; Zhu, F.C.; Gao, X.Y.; Chen, B.C.; Liu, T.; Gao, Z.G. In-situ particle segmentation approach based on average background modeling and graph-cut for the monitoring of l-glutamic acid crystallization. *Chemom. Intell. Lab. Syst.* **2018**, *178*, 11–23. [CrossRef]
18. Gao, Z.; Wu, Y.; Bao, Y.; Gong, J.; Wang, J.; Rohani, S. Image analysis for in-line measurement of multidimensional size, shape, and polymorphic transformation of l-glutamic acid using deep learning-based image segmentation and classification. *Cryst. Growth Des.* **2018**, *18*, 4275–4281. [CrossRef]
19. Ma, C.Y.; Wang, X.Z. Model identification of crystal facet growth kinetics in morphological population balance modeling of l-glutamic acid crystallization and experimental validation. *Chem. Eng. Sci.* **2012**, *70*, 22–30. [CrossRef]

20. Kitamura, M.; Onuma, K. In situ observation of growth process of alpha-l-glutamic acid with atomic force microscopy. *J. Colloid Interface Sci.* **2000**, *224*, 311–316. [CrossRef]
21. Ma, C.Y.; Wang, X.Z.; Roberts, K.J. Multi-dimensional population balance modeling of the growth of rod-like l-glutamic acid crystals using growth rates estimated from in-process imaging. *Adv. Powder Technol.* **2007**, *18*, 707–723. [CrossRef]
22. Hermanto, M.W.; Kee, N.C.; Tan, R.B.H.; Chiu, M.S.; Braatz, R.D. Robust bayesian estimation of kinetics for the polymorphic transformation of l-glutamic acid crystals. *Aiche J.* **2010**, *54*, 3248–3259. [CrossRef]
23. Ochsenbein, D.R.; Schorsch, S.; Vetter, T.; Mazzotti, M.; Morari, M. Growth rate estimation of β l-glutamic acid from online measurements of multidimensional particle size distributions. *Ind. Eng. Chem. Res.* **2014**, *53*, 9136–9148. [CrossRef]
24. Lecun, Y.; Bottou, L. Gradient-based learning applied to document recognition. *Proc. IEEE* **1998**, *86*, 2278–2324. [CrossRef]
25. Long, J.; Shelhamer, E.; Darrell, T. Fully convolutional networks for semantic segmentation. *IEEE Trans. Pattern Anal. Mach. Intell.* **2015**, *39*, 640–651.
26. Clevert, D.-A.; Unterthiner, T.; Hochreiter, S. Fast and accurate deep network learning by exponential linear units (elus). *arXiv* **2015**, arXiv:1511.07289.
27. He, K.; Sun, J.; Tang, X. Guided image filtering. *IEEE Trans. Pattern Anal. Mach. Intell.* **2013**, *35*, 1397–1409. [CrossRef]
28. Ronneberger, O.; Fischer, P.; Brox, T. U-net: Convolutional networks for biomedical image segmentation. In Proceedings of the International Conference on Medical image computing and computer-assisted intervention, Munich, Germany, 5–9 October 2015.
29. He, K.; Zhang, X.; Ren, S.; Sun, J. Delving deep into rectifiers: Surpassing human-level performance on imagenet classification. In Proceedings of the 15th IEEE International Conference on Computer Vision, ICCV 2015, Santiago, Chile, 11–18 December 2015; Institute of Electrical and Electronics Engineers Inc.: Santiago, Chile, 2015; pp. 1026–1034.
30. Huo, Y.; Liu, T.; Jiang, Z.; Fan, J. U-net based deep-learning image monitoring of crystal size distribution during l-glutamic acid crystallization. In Proceedings of the 40th Chinese Control Conference, CCC 2021, Shanghai, China, 26–28 July 2021; IEEE Computer Society: Shanghai, China, 2021; pp. 2555–2560.
31. Wang, W. Image analysis of particles by modified ferret method—Best-fit rectangle. *Powder Technol.* **2006**, *165*, 1–10. [CrossRef]
32. Zhang, B.; Willis, R.; Romagnoli, J.A.; Fois, C.; Tronci, S.; Baratti, R. Image-based multiresolution-ann approach for online particle size characterization. *Ind. Eng. Chem. Res.* **2014**, *53*, 7008–7018. [CrossRef]
33. Calderon De Anda, J.; Wang, X.Z.; Roberts, K.J. Multi-scale segmentation image analysis for the in-process monitoring of particle shape with batch crystallisers. *Chem. Eng. Sci.* **2005**, *60*, 1053–1065. [CrossRef]
34. Wilkinson, M.; Jennings, K.; Hardy, M. Non-invasive video imaging for interrogating pharmaceutical crystallization processes. *Microsc. Microanal.* **2000**, *6*, 996–997. [CrossRef]
35. Wang, X.Z.; Anda, J.; Roberts, K.J. Real-time measurement of the growth rates of individual crystal facets using imaging and image analysis: A feasibility study on needle-shaped crystals of l-glutamic acid. *Chem. Eng. Res. Des.* **2007**, *85*, 921–927. [CrossRef]

Article

# Numerical Simulation of Species Segregation and 2D Distribution in the Floating Zone Silicon Crystals

Kirils Surovovs [1,*], Maksims Surovovs [1], Andrejs Sabanskis [1], Jānis Virbulis [1], Kaspars Dadzis [2], Robert Menzel [2] and Nikolay Abrosimov [2]

[1] Institute of Numerical Modelling, University of Latvia, Jelgavas Street 3, LV-1004 Riga, Latvia
[2] Leibniz-Institut für Kristallzüchtung, Max-Born-Str. 2, 12489 Berlin, Germany
* Correspondence: kirils.surovovs@lu.lv

**Abstract:** The distribution of dopants and impurities in silicon grown with the floating zone method determines the electrical resistivity and other important properties of the crystals. A crucial process that defines the transport of these species is the segregation at the crystallization interface. To investigate the influence of the melt flow on the effective segregation coefficient as well as on the global species transport and the resulting distribution in the grown crystal, we developed a new coupled numerical model. Our simulation results include the shape of phase boundaries, melt flow velocity and temperature, species distribution in the melt and, finally, the radial and axial distributions in the grown crystal. We concluded that the effective segregation coefficient is not constant during the growth process but rather increases for larger melt diameters due to less intensive melt mixing.

**Keywords:** silicon single crystals; floating zone; effective segregation coefficient; numerical modelling

## 1. Introduction

The floating zone (FZ) method is used to produce silicon (Si) single crystals with high purity. In this method, a feed material is supplied as a polycrystalline rod, which is melted by a high-frequency inductor, and the molten Si flows along the open melting front (see the scheme in Figure 1). This forms a volume of melt, which cools and begins to crystallize at some distance from the inductor, thereby, creating the crystallization interface. The single crystal is then pulled downwards. Currently, needle-eye inductors (with the inner diameter smaller than the feed rod diameter) are used for the FZ process because they allow the use of a larger feed rod and achieve a larger crystal diameter.

To obtain the desired electrical resistivity of silicon crystals grown with the FZ method, it is necessary to precisely control the concentration of dopants—typically boron (B) or phosphorus (P) [1]. The level and distribution of various impurities, e.g., carbon (C), is also important in applications, such as the kilogram definition project [2]. A crucial phenomenon that defines the transport of such species is the segregation at the crystallization interface.

Let us consider a one-dimensional distribution of species concentration $C(z)$, where $z$ is perpendicular to the crystallization interface, which is located at $z = 0$; see Figure 2, left. The equilibrium segregation coefficient $k_0$ is theoretically defined as $\frac{C_s(0)}{C_m(0)}$, where $C_s(0)$ is the species concentration in solid silicon and $C_m(0)$ is in molten silicon at the interface.

However, crystal growth experiments with the same species (same $k_0$) can lead to different axial concentration distributions if other process parameters differ [3]. The effective segregation coefficient $k_{\text{eff}} \equiv \frac{C_s(0)}{C_m(z \gg 0)}$ is more practical than $k_0$, because it considers the concentration outside the thin diffusion boundary layer $C_m(z \gg 0)$, which can be measured or evaluated more easily than $C_m(0)$. Therefore, $k_{\text{eff}}$ also depends on the melt flow. An analytical expression for $k_{\text{eff}}$ has been derived in the classical work by Burton, Prim and Slichter (BPS) [4]:

$$k_{\text{eff}} = \frac{k_0}{k_0 + (1-k_0)e^{-\frac{vd}{D}}}, \quad d \sim 1.6 D^{\frac{1}{3}} \eta^{\frac{1}{6}} \omega^{-\frac{1}{2}}, \tag{1}$$

where $v$ is the crystallization velocity, $d$ is the thickness of the diffusion layer in the melt, $D$ is the diffusion coefficient of the impurity in the melt, $\eta$ is the kinematic viscosity, and $\omega$ is the crystal rotation rate. The shape of function $k_{\text{eff}}(d)$ for three different species (carbon, phosphorus and boron) is shown in the right part of Figure 2. According to the BPS model, $k_{\text{eff}} \geq k_0$, and, when the thickness of the diffusion layer increases (for example, due to a decrease in crystal rotation rate), $k_{\text{eff}}$ increases and saturates at $k_{\text{eff}} = 1$ for large $d$. This saturation happens at smaller $d$ for the species with a smaller diffusion coefficient.

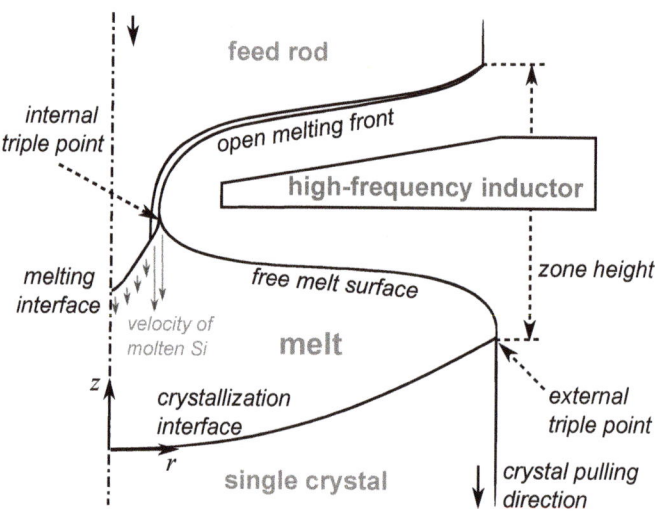

**Figure 1.** Axially symmetrical scheme of the FZ method, with designations of the most important parts and the used coordinate system.

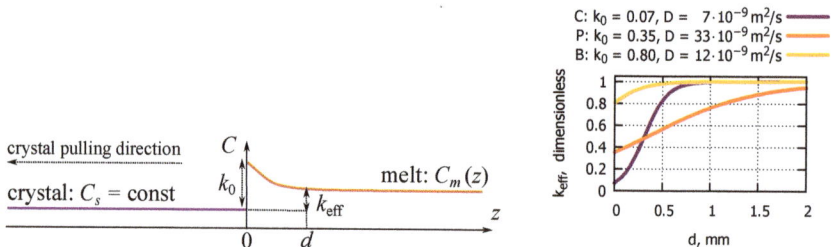

**Figure 2. Left**—a simplified scheme of species distribution in the crystal ($z < 0$) and the molten zone ($z > 0$) near the interface ($z = 0$). **Right**—the dependence of the effective segregation coefficient $k_{\text{eff}}$ on diffusion layer thickness $d$ (e.g., [4]) for different species.

Figure 2 shows that $k_{\text{eff}}(d)$ can vary in a large interval, particularly for carbon. Moreover, the BPS model uses simplified assumptions of melt flow pattern and does not predict $k_{\text{eff}}$ correctly in some cases (e.g., when melt flow is radially converging [5] or when microsegregation is taken into account [6]). Therefore, numerical flow simulations are necessary to obtain a more precise estimation of the effective segregation coefficient for FZ Si crystal growth.

Dopant segregation in the FZ process has been already studied in the literature—an axially symmetrical model of quasi-stationary dopant transport in the melt was described as early as 1997 [7]. Due to the increased availability of computational power, transient 3D simulations became possible around 2010 [8], and these kinds of simulations are still relevant now, e.g., for modelling radial resistivity distribution in 200 mm diameter FZ crystals [9]. However, all the aforementioned publications considered only the radial distribution of dopant concentration, without considering the axial distribution or the effective segregation coefficient.

The literature on $k_{eff}$ of silicon impurities in crucible-free, inductively heated FZ processes was found to be scarce; therefore, we analyse other Si growth processes. An experiment has been reported that considered boron segregation during the Czochralski (CZ) process [10]. In that study, $k_{eff}$ was found to be between 0.75 and 0.98, and it decreased when the melt flow became more intensive. This seemed to agree with the simplified BPS model (1): when the crystal rotation rate increased from 1 to 13 rpm, the diffusion layer became thinner, thus, diminishing $k_{eff}$.

A study where the CZ crystal rotation was 16 rpm [11] reported a slightly smaller $k_{eff} = 0.73$. For crystals grown from a copper crucible using electron beam heating, $k_{eff}$ was found to be 0.8 [12]. In yet another CZ experiment [13], boron $k_{eff}$ approached 1 as the axial magnetic field increased. However, there are studies of oxygen transport in CZ Si that, on the contrary, indicate a lower $k_{eff}$ when magnetic heaters are used [14]. The application of cusp magnetic field also lowered the oxygen content in grown crystals; however, it can be explained not only by the change in $k_{eff}$—oxygen evaporation at the free surface may have played a large role [15].

In this study, we investigated the influence of the melt flow on $k_{eff}$ in the Si FZ growth process and obtained the species distribution $C(r,z)$ in the entire grown crystal including the start cone and end cone growth stages, i.e., not only the radial profile $C(r)$ but also the axial profile $C(z)$. For the numerical simulations of the phase boundaries, electromagnetic field, melt flow, and species transport in the melt, we used previously developed numerical models.

The $k_{eff}$ was then evaluated from the species concentration distribution in the melt. We created a new program that calculates the evolution of the average species concentration in the melt over time, which influences the species distribution along the crystal axis. To cover a wide range of $k_0$, we simulated two species: carbon ($k_0 = 0.07$) and boron ($k_0 = 0.8$).

## 2. Numerical Model

The present section describes the system of numerical models used for the modelling of the FZ crystal growth process. For the previously published models, only a brief summary and references are given in the following subsections.

This system of models is used to describe the following aspects of the FZ process: the shape of phase boundaries, electromagnetic (EM) field, melt flow, species transport in the melt and, finally, species distribution in the grown crystal. Section 2.1 presents a graphical and textual overview of the used models and data. The section explains how the simulated phase boundaries and EM field are used for the 3D melt flow simulations.

The melt flow simulations, in turn, are necessary for obtaining the effective segregation coefficient $k_{eff}$. When $k_{eff}$ is known, we can simulate the evolution of the average species concentration in the melt, which translates to the axial species distribution in the grown crystal. Sections 2.2–2.4 then focus on three main parts of the system of models: the phase boundaries, melt flow and species distribution in the crystal.

## 2.1. Overview of Modelling Scheme

The overall scheme of the models and their interactions is shown in Figure 3. The used data (or the obtained results) are shown in grey boxes, while the models are shown in orange boxes. In the beginning, one needs to specify the process parameters (see grey box No. 1): the zone height, inductor frequency, system geometry, etc. For the modelling of phase boundaries (orange box 2), one of two programs can be used: a quasi-stationary program FZone [16], which is suitable only for the cylindrical phase, or a transient program FZoneT [17], which describes the entire process: start cone, cylinder and end cone.

Both programs assume the axially symmetrical (also denoted by 2D) approximation of the shape of silicon parts. FZone and FZoneT programs are used to calculate melt shape (box 3). FZoneT also allows obtaining the melt volume dynamics (box 8): the history of the change in crystal diameter, melt volume, etc. More information about these programs is given in Section 2.2.

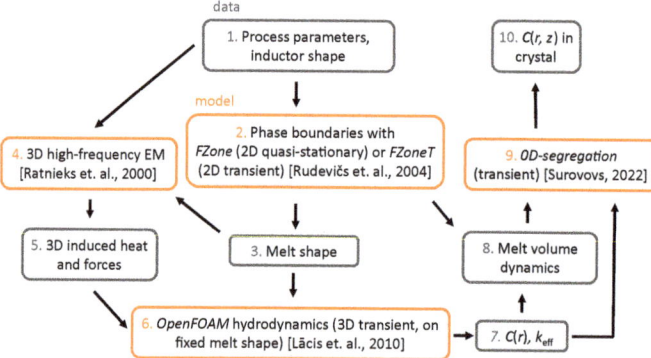

**Figure 3.** The overall scheme of the used numerical models (**orange boxes**) and the data that is being exchanged between them (**grey boxes**)[16–19].

Using the obtained melt shape, a 3D finite volume mesh is created for OpenFOAM [18] hydrodynamic simulation (box 6). This includes transient simulation of the melt flow, heat transfer and species transport; see Section 2.3. To obtain boundary conditions for the melt velocity and temperature at the free surface, a 3D inductor shape is used to calculate the induced heat and Lorentz forces [19] (boxes 4 and 5).

OpenFOAM hydrodynamic simulation produces the radial distribution of the species concentration $C(r)$ in the grown crystal and the effective segregation coefficient $k_{\text{eff}}$. This data, together with the melt volume dynamics, is then used in the transient 0D-segregation program (box 9), which predicts the temporal evolution of the average species concentration in the melt [20]. The complete description of the mathematical model for 0D-segregation is presented in Section 2.4. Finally, the results of 0D-segregation program are interpolated on the point grid that is created in crystal domain (box 10); this process is schematically shown in Figure 4.

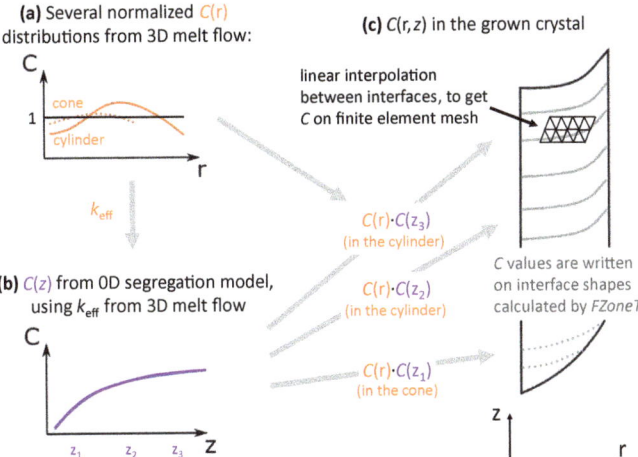

**Figure 4.** The scheme of species concentration interpolation from OpenFOAM melt flow simulations of the cone stage and cylinder stage (**a**) and 0D-segregation simulations (**b**) on the grown crystal mesh (**c**).

*2.2. Phase Boundaries*

The description of the program FZone, which has been used for the calculation of phase boundaries, is provided in [16]. The program considers axially symmetrical approximations of phase boundaries. The melting interface, crystallization interface and open melting front are moved according to the heat flux balance; the influence of the melt flow is neglected. The program FZone operates in quasi-stationary approximation: it describes the cylindrical phase of the process, i.e., it calculates the shape of the phase boundaries for a process stage when they do not change in time.

The transient version of this program that simulates growth process dynamics (e.g., a change in the crystal radius during the cone phase) is called FZoneT [17]. The evolution of the crystal diameter and melt volume is saved in a file, which can be later read by the 0D-segregation program. The species concentration inflow from the melting feed rod (the integral from Equation (3)) is also saved in a file. In the present work, the transient FZoneT program was used to simulate the entire crystal growth (the start cone, cylinder and end cone), and the results from three time instants from the start cone stage were selected for melt flow simulations.

2.2.1. Electromagnetic Field

When quasi-stationary FZone is used, the EM field is calculated in 3D using boundary elements and assuming negligible skin depth [19]. The EM field is iteratively recalculated until the shape of the phase boundaries converges. However, in the transient FZoneT version, only an axially symmetrical (also denoted by 2D) EM field is simulated due to limited computational resources. In the 2D EM simulations, inductor slits are modelled by setting an artificial magnetic field source surface density [16], which allows part of the magnetic field lines to penetrate the inner part of the inductor.

To calculate the 2D EM field distribution, axially symmetrical inductor shape was constructed in a way that ensures the best correspondence to the results obtained with a 3D non-symmetrical shape. Quasi-stationary simulations of the corresponding 3D system were performed to calibrate 2D inductor characteristics so that the differences between the induced power on the silicon surface are the smallest (the results are presented in Section 3.2.2). A similar comparison of different EM field models is described in [21].

The initial simulations were performed using real slit parameters in the 2D model. Then, the width and length of the inductor side slits as well as the width of the main slit were changed to achieve better agreement between 2D and 3D systems.

### 2.3. Species Transport in Melt

The species transport in melt is simulated with the OpenFOAM hydrodynamic solver described in [22]. This assumes fixed phase boundaries and uses the transient, incompressible, laminar Navier–Stokes equation for melt velocity, with the Boussinesq approximation for the description of thermal convection. Boundary conditions for velocity are as follows:

- Marangoni force and the EM force are applied on the free melt surface.
- Fixed velocity (crystal pulling speed and rotation)—on the crystallization interface.
- Fixed velocity (feed rod pushing speed and ring-shaped inflow from the open melting front; see Figure 1)—on the melting interface [18,23].

A standard convection-diffusion equation is used for simulation of the melt temperature with the EM-induced heat and radiation on the free melt surface and fixed temperature on the solid–liquid interfaces. The equation that describes species transport is also of the convection-diffusion type:

$$\frac{\partial C}{\partial t} + (\vec{v}\nabla)C = D\Delta C, \qquad (2)$$

where $C$ is the species concentration, $t$ is the time, $\vec{v}$ is the melt velocity, and $D$ is the species diffusion coefficient. The initial conditions were uniform: $C = 1$ arb.u. (arbitrary unit). The boundary conditions for concentration are as follows:

- On the crystallization interface: $D\frac{\partial C}{\partial n} = v_C(1-k_0)C\cos\theta$, where $n$ is the normal coordinate, $k_0$ is the segregation coefficient, and $\theta$ is the the angle between the horizontal plane and the interface normal vector.
- On the melt free surface: $\frac{\partial C}{\partial n} = 0$ due to the assumption of a pure gas atmosphere and lack of evaporation [24].
- On the melting interface: $C = 1$ arb.u., i.e., the species concentration is normalized to the initial concentration in the feed rod, which is assumed to be homogeneous.

### 2.4. 2D Species Distribution in Crystal

To simulate the axial distribution of species in the crystal, the time-dependence of the average species concentration in the melt $C_m(t)$ was considered. It was assumed (and later supported by simulation results; see Section 3.3.1) that the $\vec{v}$ and $C_m$ in the melt reach a quasi-stationary state sufficiently fast in comparison to the characteristic time of axial changes of concentration in the grown crystal. Therefore, we assumed that the average concentration of species in the crystallizing silicon at any given time $t$ is directly proportional to $C_m(t)$. Thus, the following equation of species mass conservation was proposed:

$$\Delta(C_m V_m) = \int C_F v_F \,\Delta t\, dS - k_{\text{eff}} C_m \Delta V_{\text{out}}, \qquad (3)$$

where $V_m$ is the melt volume, $C_F$ is the species concentration in the feed rod at the melting interface, $v_F$ is the feed rod pushing rate, $\Delta t$ is the simulation time step, $dS$ is the surface element of melting interface and open melting front, $k_{\text{eff}} = C_C/C_m$ is the effective segregation coefficient, $\Delta V_{\text{out}}$ is the volume of the crystal that crystallizes during the time step, and $C_C$ is the average species concentration in the layer that crystallizes during the time step. The value of $k_{\text{eff}}$ is obtained from OpenFOAM simulations (see Section 3.3.1) and can depend on the process parameters, e.g., the crystal diameter $D_C$.

Equation (3) can be rearranged to express $\Delta C_m$, which then is used to calculate $C_m$ iteratively with the time step $\Delta t$:

$$\Delta C_m\, V_m = -C_m\, \Delta V_m + \int C_F v_F \,\Delta t\, dS - k_{\text{eff}} C_m \Delta V_{\text{out}},$$

$$\Delta V_{\text{in}} = \int v_F \Delta t \, dS$$

$$\Delta V_m = \Delta V_{\text{in}} - \Delta V_{\text{out}}$$

$$C_m(t + \Delta t) = C_m + \Delta C_m = C_m + \frac{C_m(\Delta V_{\text{out}} - \Delta V_{\text{in}}) + \int C_F v_F \Delta t \, dS - k_{\text{eff}} C_m \Delta V_{\text{out}}}{V_m}. \quad (4)$$

Equation (4) is implemented using Python language. This program is called 0D-segregation, because it describes transient species segregation without spatial dimensions, i.e., disregarding spatial concentration distribution in the melt. The program source code is published in [20]. Two functionalities of the program are available:

1. Importing the data about process dynamics (time-dependent $V_m$, $C_F$, $\Delta V_{\text{out}}$ and $\Delta V_{\text{in}}$) from transient phase boundary simulations with FZoneT.
2. Creating an approximate description of the cone phase based only on the simplified crystal shape described as $D_C(L)$, where $L$ is the crystal length:

   - Due to the assumption of constant pulling velocity, $L \propto t$.
   - Cone surfaces are approximated as having constant slope, and thus $D_C(t) \propto t$.
   - The free surface height above the external triple point is assumed to be constant even during the cone phases because it is impossible to predict its evolution for an arbitrary crystal shape (without experimental data); therefore, $V_m(t) \propto D_C(t)^2$.
   - The crystallized volume is proportional to the crystal cross-section: $\Delta V_{\text{out}} = v_C S_C \Delta t$, where $v_C$ is the crystal pulling velocity, and $S_C = \frac{\pi}{4} D_C(t)^2$ is the crystal cross-section area. Therefore, $\Delta V_{\text{out}} \propto D_C(t)^2$.
   - Due to silicon mass conservation, $\Delta V_{\text{in}} = \Delta V_m + \Delta V_{\text{out}}$.

After $C_m(t)$ has been calculated, it is combined with $C(r)$, which was obtained during species transport simulations in melt; see the scheme in Figure 4. The final result is $C(r, z)$ distribution in the grown crystal.

## 3. Results

### 3.1. Description of the Experiment

The present work uses process data from a 4″ FZ crystal growth experiment performed at the IKZ (Leibniz Institute for Crystal Growth, Berlin). The main parameters are listed in Table 1. A standard one-turn inductor with three side slits is used; see Figure 5 and our earlier work [25,26]. The shape of the grown crystal as well as the height of the molten zone can be seen in Section 3.2.3, where a comparison to transient simulation results is performed. Process photographs with the visible parts of the phase boundaries were used to validate the simulation of the cone phase in particular. Note that characterization of the grown crystals is not discussed in the present study and will be addressed in further publications.

**Table 1.** Parameters of the experiment.

| Parameter | Value |
|---:|:---|
| Crystal diameter $D_C$ | 102 mm (cylinder phase) |
| Feed rod diameter $D_F$ | 90 mm |
| Crystal pulling rate $v_C$ | 3.5 mm/min |
| Feed rod push rate $v_F$ | 4.5 mm/min (cylinder phase) |
| Crystal rotation rate $\omega_C$ | 6 rpm |
| Feed rod rotation rate $\omega_F$ | −0.8 rpm |
| Zone height $H_Z$ | 27 mm (cylinder phase) |
| Inductor frequency $f$ | 3 MHz |

### 3.2. Phase Boundaries

#### 3.2.1. Quasi-Stationary Simulations

First, quasi-stationary simulations are performed with FZone program to obtain the shape of phase boundaries for the system described in Section 3.1, which are later used in hydrodynamics simulations in Section 3.3. 3D EM simulations are performed to precisely describe the EM field created by the high-frequency inductor. An example of the results of 3D EM simulations (induced heat on silicon free surface) is shown in Figure 5.

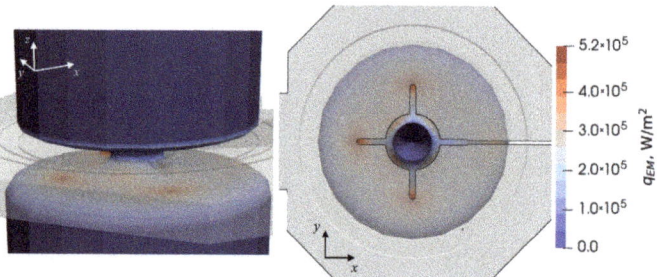

**Figure 5.** 3D high-frequency electromagnetically induced heat sources on silicon surfaces used for 3D melt flow simulations. The one-turn inductor is shown in grey.

3.2.2. Influence of Three-Dimensionality of the EM Field

Three-dimensional melt flow simulations require 3D distributions of induced heat and Lorentz force, which can only be achieved using a 3D model of the high-frequency inductor shape. On the other hand, transient simulations can be performed only with a 2D EM field model, which approximates the 3D features of the asymmetrical EM field (see Section 2.2.1).

Induced current on the open melting front and the free melt surface of silicon are compared between different EM models in Figure 6. For the 3D model, the azimuthally averaged radial distribution is shown. In the case of the 2D model, the effect of inductor slits can be considered only approximately, which results in a significantly different distribution. Various parameter studies were conducted to obtain the optimal inductor parameters (modified slit width and length).

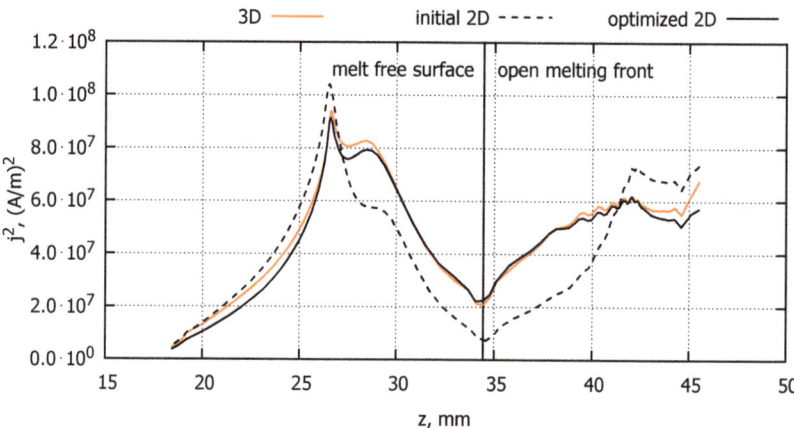

**Figure 6.** Comparison of induced currents on silicon surfaces in the case of 2D and 3D simulations.

The resulting induced electrical current distribution from the optimized 2D model is similar to the 3D model results with the same defined inductor current. The main differences can be observed on the free melt surface and in the outer part of the open melting front. This results in slightly different shapes of the phase boundaries, which are shown in Figure 7. The internal triple point radius is slightly smaller, which results in a different melting interface and free melt surface shape. However, the 2D model is sufficiently precise to describe the melt volume dynamics, because the differences are large only in the central part of the melt.

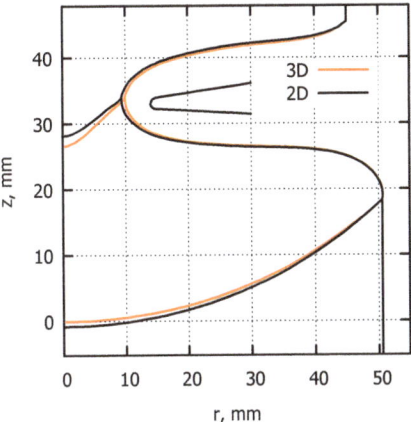

**Figure 7.** Comparison between shapes of the phase boundaries for optimized 2D and 3D EM simulations.

3.2.3. Transient Simulations

While the quasi-stationary approach used in FZone was applied in the present work for the cylindrical growth stage, the transient model should be considered for the initial and final stages of the process when the crystal or feed rod is short and the diameter is changing. Therefore, a transient simulation of the temperature field and phase boundaries has been conducted using FZoneT, starting from a small cone with a diameter of 20 mm and a length of 12 mm at time $t = 0$ min. A rather good agreement with the experiment was obtained for the shape of the phase boundaries during the cone and cylindrical growth stages; see Figure 8.

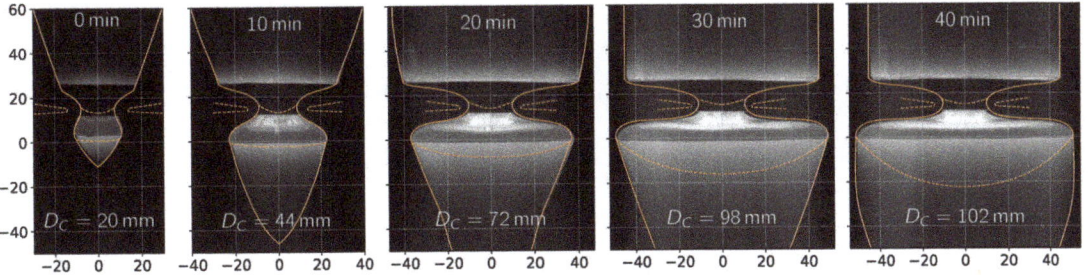

**Figure 8.** Comparison of the shapes of phase boundaries in the transient FZoneT simulation with experimental photos during the cone growth (0–30 min) and cylindrical growth (40 min) stages.

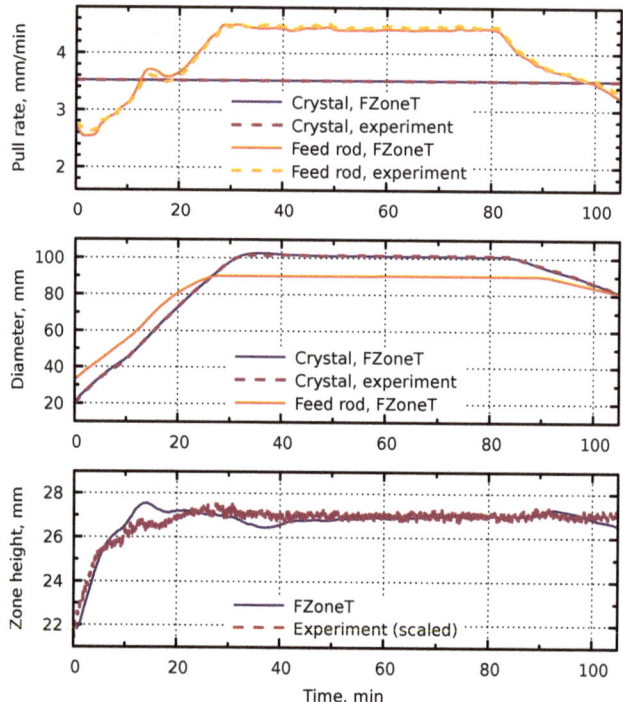

**Figure 9.** Comparison of transient FZoneT simulation results to the experiment during the entire growth process.

The time-dependent inductor current (instead of the power used in the experiment) and feed rod velocity were adjusted to achieve good agreement with the experimentally measured crystal diameter and zone height as demonstrated in Figure 9. A slightly imperfect prediction of the zone height by FZoneT could be caused by differences in input data (inductor current and feed velocity) or by model limitations (e.g., a fluid film model describing the melting of a macroscopically smooth open melting front).

The feed diameter $D_F$ was not explicitly measured in the experiment; therefore, only the curve from FZoneT is plotted in Figure 9. It was verified that $D_F$ during the cylindrical growth stage was 90 mm both in the experiment (from process photos) and FZoneT simulation in accordance with the steady-state mass conservation $D_F = D_C \sqrt{v_C/v_F}$.

An example of the calculated global temperature field and shape of phase boundaries during the entire process is shown in Figure A1 of Appendix A.

### 3.3. Species Transport in Melt

Using the simulated phase boundaries, that were obtained using FZone or FZoneT and presented in the previous section, 3D melt flow simulations were run with OpenFOAM using the model described in Section 2.3. The most important physical and numerical parameters are given in Table 2. An example of the simulation results for the carbon transport during the cylindrical phase is shown in Figure 10. Despite the non-symmetrical high-frequency inductor, the carbon concentration field is mostly axially symmetrical.

Table 2. The physical [27] and numerical parameters that were used for OpenFOAM simulations.

| Parameter | Value |
| --- | --- |
| Silicon density $\rho$ | 2580 kg/m$^3$ |
| Silicon viscosity $\eta$ | $8.6 \cdot 10^{-4}$ Pa·s |
| Silicon heat conductivity $\lambda$ | 67 W/m·K |
| Silicon specific heat capacity $c_p$ | 1000 J/kg·K |
| Silicon thermal expansion coefficient $\beta$ | $10^{-4}$ 1/K |
| Marangoni coefficient $M$ | $-1.3 \cdot 10^{-4}$ N/m·K [26] |
| Carbon diffusion coefficient $D$ | $7 \cdot 10^{-9}$ m$^2$/s [28] |
| Carbon segregation coefficient $k_0$ | 0.07 [3] |
| Boron diffusion coefficient $D'$ | $1.2 \cdot 10^{-8}$ m$^2$/s [29] |
| Boron segregation coefficient $k_0'$ | 0.8 [29] |
| Total number of mesh elements | 614,000 |
| Largest element size | 0.8–1.4 mm (inside the melt) |
| Smallest element thickness | 0.02–0.03 mm (at the crystallization interface) |
| Time step | 2 ms |
| Total simulation time | 350–500 s |
| Averaging interval for $C, T, \vec{v}$ | 100 s |

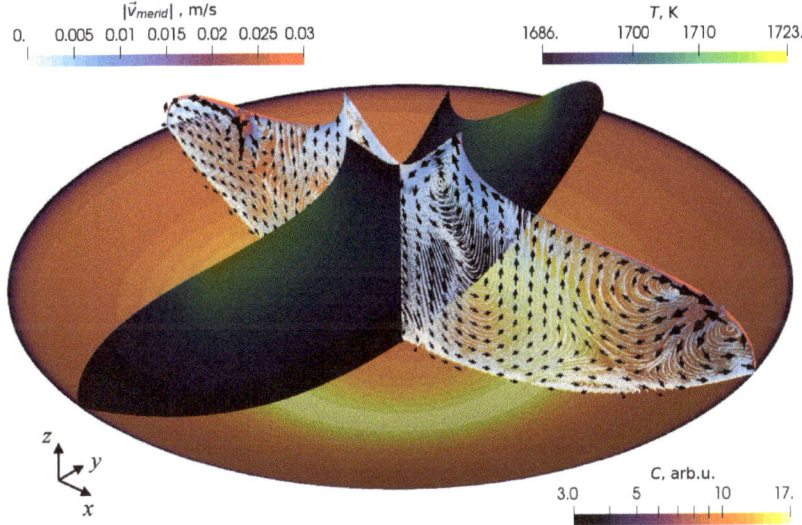

**Figure 10.** The time-averaged meridional melt velocity in the $xz$ plane, temperature in the $yz$ plane and carbon concentration on the crystallization interface, simulated for the cylindrical phase ($D_C = 102$ mm). The direction of the main inductor current suppliers corresponds to the $x$ axis.

The OpenFOAM simulations can be performed only using a fixed melt shape (stationary phase boundaries). The FZone calculation of the cylindrical growth stage with a diameter of 102 mm was used. In addition, the results from FZoneT with several cone diameters were selected to simulate the melt flow in a cone phase: 80, 60 and 45 mm.

The simulated melt velocity in the vertical plane parallel to the main inductor slit is shown in Figure 11, left. Comparison between the EM and Marangoni force densities on the free surface shows that the EM forces dominate; see Figure 12. The EM force is creating an inwards-directed flow at the outer part of the melt. The melt flow is more intensive for the smaller $D_C$ due to larger EM field gradients.

The flow pattern is similar between two stages with larger $D_C$, where the largest velocities (exceeding 1 cm/s) appear only near the free surface and in the neck region.

The decrease in the melt flow intensity during the increase in $D_C$ may be related to various factors. One of them could be the EM force redistribution due to the change in the free surface shape. The influence of crystal rotation on larger melts could also play a role: when the azimuthal melt velocity is higher, it is more difficult to induce a radial flow [7]. Another possible explanation is different surface-to-volume ratios, which decrease the influence of the surface EM force for larger $D_C$.

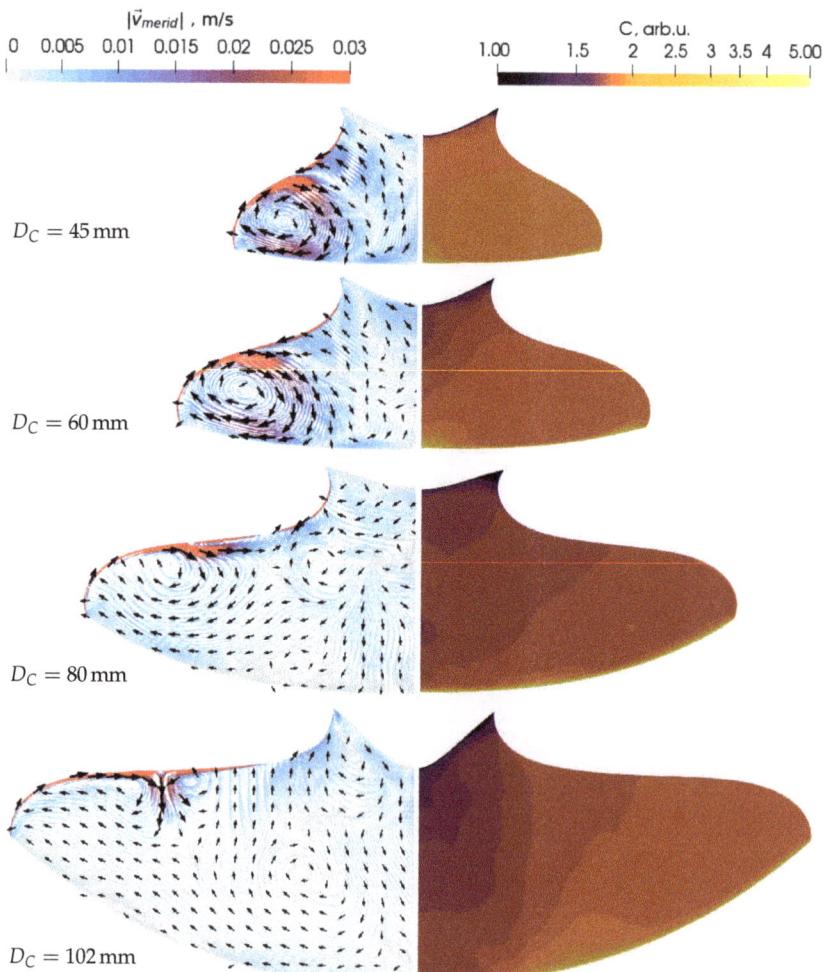

**Figure 11.** The time-averaged melt velocity (**left**) and carbon concentration (**right**) in the vertical cross-section of the melt, simulated for the cylindrical phase ($D_C = 102$ mm) and cone phase ($D_C < 102$ mm).

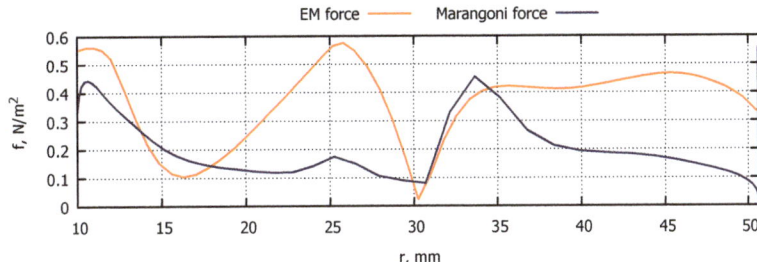

**Figure 12.** Absolute values of the EM and Marangoni force density on the free melt surface during the cylindrical phase ($D_C = 102$ mm).

The carbon concentration distribution is shown in Figure 11, right. Arbitrary units are used, and normalization is made with respect to the concentration in the feed rod, i.e., the $C = 1$ boundary condition is set on the melting interface. Concentration contours, despite the averaging over 100 s, are not always smooth—possibly due to the chaotic temporal behaviour of the melt flow or due to minor numerical effects, e.g., mesh non-orthogonality.

As of segregation with small $k_0$, a thin boundary layer is formed at the crystallization interface; see Figure 13. The width of this layer approximately corresponds to the values calculated using the BPS model, Equation (1): 0.020 mm for carbon and 0.024 mm for boron. The carbon concentration at the crystallization interface is shown in Figure 14. The obtained concentration distributions for large $D_C$ are axially symmetrical due to crystal rotation and due to partial axial symmetry of the boundary conditions.

For smaller $D_C$, only minor deviations from the axial symmetry in the concentration distribution are present, and they do not significantly affect the azimuthally averaged radial distribution. There is no experimental data about such 2D distributions known to the authors.

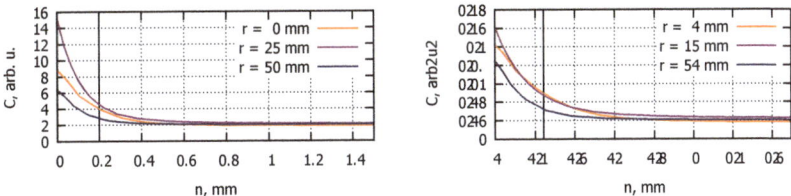

**Figure 13.** The distribution of carbon (**left**) and boron (**right**) concentration in the melt, near the crystallization interface ($D_C = 102$ mm). The normal coordinate is denoted by $n$, and different radial positions $r$ are considered. Boundary layer thicknesses from the BPS model (1) are shown with vertical black lines.

From the species concentration values at the interface, as shown in Figure 14, the radial distribution of species atoms in the grown crystal can be calculated; see Figure 15. For small $D_C$ (60 mm and less), two distinct concentration maxima exist: one in the crystal centre and one near the rim. This effect can be explained by the intensive melt vortex, as shown in the top images in Figure 11. Such a central maximum of species concentration in small crystals ($D_C = 40$ mm) was also observed in experiments [30]. When $D_C$ is large ($D_C = 102$ mm), the species concentration maximum is obtained at $r \approx 20$ mm. It corresponds well to the resistivity maximum that is obtained due to the distribution of phosphorus or other dopant elements in typical 4″ silicon crystals [31].

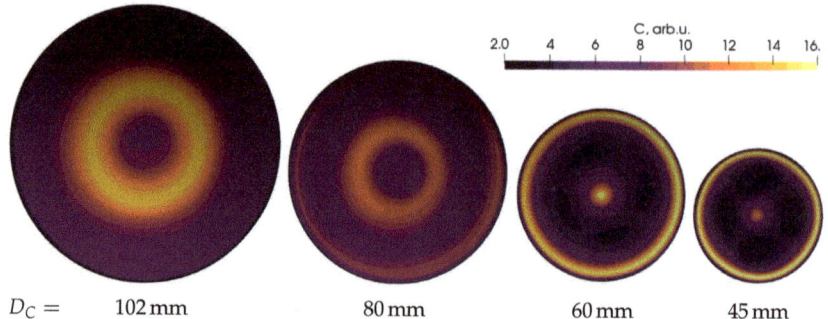

$D_C =$  102 mm  80 mm  60 mm  45 mm

**Figure 14.** The time-averaged carbon concentration on the crystallization interface, simulated for the cone phase ($D_C < 102$ mm) and cylindrical phase ($D_C = 102$ mm).

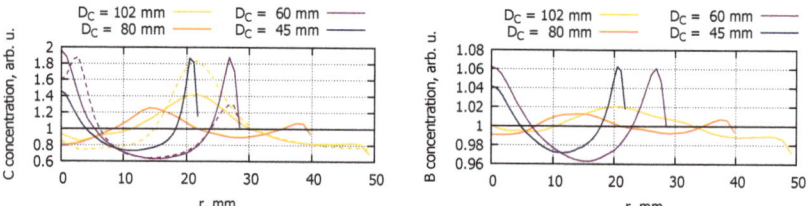

**Figure 15.** Normalized radial distributions of carbon (**left**) and boron (**right**) concentrations in the grown crystal (yellow line for the cylindrical phase, violet and orange lines for the cone phase), using crystal rotation rates 6 rpm (solid lines) and 12 rpm (dashed lines).

### 3.3.1. Effective Segregation Coefficient

From OpenFOAM melt flow simulations, the effective segregation coefficient can be calculated:

$$k_{\text{eff}} = \frac{C_C}{C_m} = \frac{k_0\, C_{m,\,\text{crys. inter.}}}{C_m}, \quad (5)$$

where $C_C$ is the concentration of species in the crystal, $C_m$ is the average species concentration in the melt, and $C_{m,\,\text{crys. inter.}}$ is the average species concentration on the crystallization interface, calculated by averaging the concentration values from the faces of OpenFOAM calculation cells.

Values of $k_{\text{eff}}$ converged relatively rapidly—after less than 200 s, it did not change more than ±2%. This means that, until the saturation of $k_{\text{eff}}(t)$ curve, less than 12 mm of crystal is grown, which is a small part in comparison to total crystal length (400 mm and more). The obtained values of $k_{\text{eff}}$ are summarized in Figure 16. The effective segregation coefficient monotonously increases as the crystal diameter increases. This can be explained by considering a simplified scheme in Figure 2: when $D_C$ is larger, the melt motion is less intensive, the boundary layer remains undisturbed, and the drop of species concentration in the melt is sharper. Therefore, $C_m(z \gg 0)$ becomes closer to $C_s$, and $k_{\text{eff}} \equiv \frac{C_s}{C_m(z \gg 0)}$ increases.

The obtained range of carbon $k_{\text{eff}}$ variation only partly intersects with the range calculated using the BPS model (1), where $k_{\text{eff}}$ is in the interval from 0.3 to 0.5. This could possibly be explained by the simplifications during the derivation of the BPS model, which does not capture all features of the 3D melt flow in the FZ process. It should also be noted that the used carbon diffusion coefficient $D$ is rather uncertain (its values are given in [28] with ±30% precision). The range of boron $k_{\text{eff}}$, in turn, agrees with Equation (1) rather well.

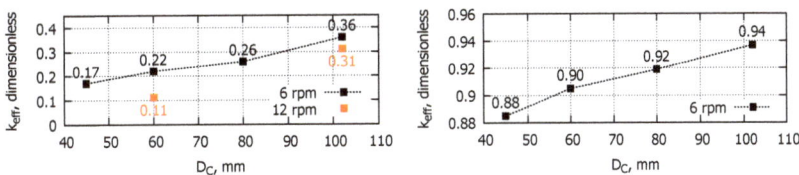

**Figure 16.** The effective segregation coefficient $k_{eff}$ of carbon (**left**) and boron (**right**) obtained for different crystal diameters $D_C$ and crystal rotation rates of 6 rpm (black) and 12 rpm (orange).

The obtained $k_{eff}$ value may depend on the used mesh due to the combination of extremely low $D$ and low $k_0$ for carbon. During the research, mesh influence studies were performed, and it was shown that, in order to obtain consistent results, the cell size at the crystallization interface should not be larger than 0.03 mm in the normal direction. This criterion was ensured in all meshes that were used for the melt flow simulations described above.

### 3.3.2. Increased Crystal Rotation Rate

The crystal rotation rate $\omega_C$ was increased from 6 to 12 rpm in two stages—for $D_C = 102$ mm and $D_C = 60$ mm—in order to investigate if it is possible to decrease carbon $k_{eff}$ and thus improve the crystal purification. The comparison of meridional velocity fields revealed only small differences; see Figure 17. Vertical motion (downwards in the centre of the melt and upwards at the middle of the radius) became slightly stronger when $\omega_C$ was increased to 12 rpm.

Therefore, the central maximum in the radial distribution of the carbon concentration became more pronounced; see Figure 15 (dashed lines). As shown in Figure 16, an increase in $\omega_C$ from 6 to 12 rpm led to a decrease in $k_{eff}$ from 0.36 to 0.31 for $D_C = 102$ mm and decreased $k_{eff}$ by half for $D_C = 60$ mm. This means that the high $\omega_C$ could help to achieve a slightly higher average crystal purity; however, an increase in the $C(r)$ profile maximum is predicted, as shown in Figure 15, which will make the 2D $C(r, z)$ distribution in the grown crystal less homogeneous.

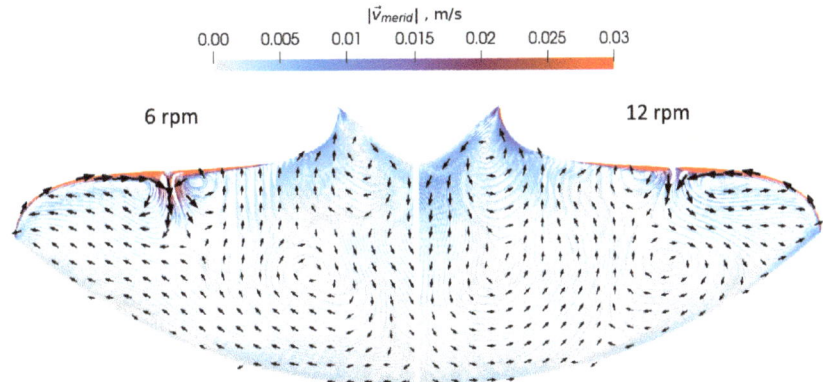

**Figure 17.** The time-averaged melt velocity in the vertical cross-section of the melt, simulated for the cylindrical phase with $D_C = 102$ mm and crystal rotation rates of 6 rpm (**left**) and 12 rpm (**right**).

### 3.4. Species Distribution in Crystal

The developed program `0D-segregation`, which calculates the axial distribution of impurities (Section 2.4), was verified using an analytical solution for a theoretical "cylindrical crystal"—a crystal with constant diameter, disregarding the cone phases. The results of the verification are presented in Appendix B.

A proper comparison with analytical expressions is only possible for the fully cylindrical system. In reality, the crystal diameter is not constant and includes cone stages at the beginning and the end of the growth process—the crystal diameter evolution is shown in Figure 9. For the cases when precise process data (either experimental logs or the output of the transient FZoneT program) is not available, the 0D-segregation program has a setting that creates a simplified crystal shape $D_C(L)$ as described in the end of Section 2.4. Figure 18 (left) shows this simplified shape for a 102 mm crystal: the crystal diameter and melt volume are functions of the crystal length and are compared with the shape simulated by FZoneT.

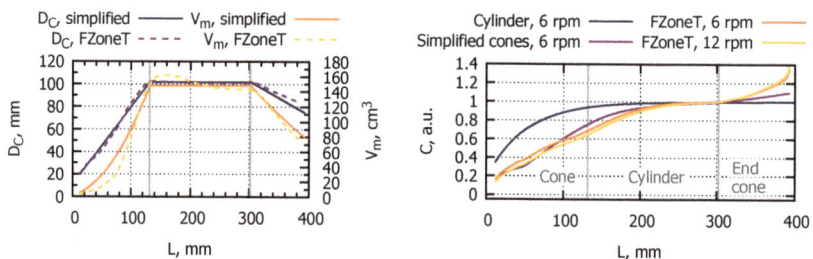

**Figure 18.** The crystal diameter and melt volume as a function of the crystal length $L$ for the considered $D_C = 102$ mm system with cone stages (**left**) and comparison of the axial carbon concentration in the crystal using different crystal shapes with $k_{\text{eff}}(D_C)$ from the melt flow simulation results (**right**). Vertical grey lines represent the boundaries between the start cone, cylinder and end cone stages.

Changes in crystal diameter have a significant impact in the resulting species distribution in the crystal as shown in Figure 18 (right). In all cases, uniform carbon concentration in the feed rod is assumed. During the start cone stage, the melt volume increases, which causes a slower impurity build-up than in the analytically solved cylindrical crystal case.

The opposite effect can be seen during the end cone stage when the impurity concentration exceeds the initial values in the feed rod. Another reason for the difference between the cylindrical crystal and the real crystal is the $k_{\text{eff}}$ dependence on the crystal radius, which was shown in Figure 16. The simplification of the crystal shape significantly impairs the precision of the obtained $C(L)$ distribution, which is likely due to the imprecise description of the melt volume.

The axial concentration distribution, which was obtained using the precise crystal shape simulated by FZoneT, was combined with radial distributions at corresponding diameters (shown in Figure 15), and linear interpolation was used between them. As a result, the 2D $C(r, z)$ distribution in the entire crystal is shown in Figure 19.

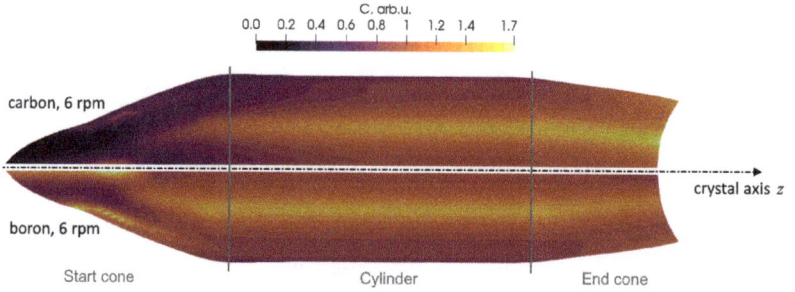

**Figure 19.** *Cont.*

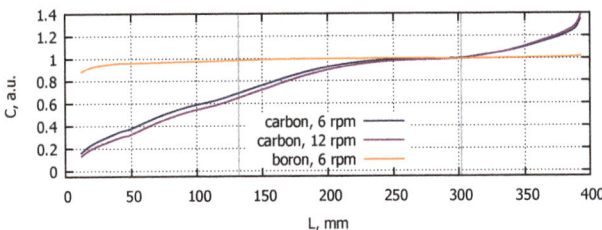

**Figure 19.** An example of the carbon and boron distribution in the grown crystal (**top**) and the axial distribution of radially averaged concentration of these species (**bottom**).

## 4. Discussion

A new model system was proposed to describe the transient FZ growth of silicon, including the shapes of the phase boundaries and the resulting species concentration in the crystal influenced by melt flow.

Simulations of the shapes of phase boundaries, both transient and quasi-stationary, corresponded well to the experimental data (process photographs, zone height and crystal diameter). The differences usually did not exceed 1–2 mm, which is sufficient to use the obtained phase boundaries and the calculated 3D electromagnetic field for further melt flow simulations.

The melt flow simulations demonstrated significant differences between stages with small ($D_C \leq 65$ mm) and large ($D_C \geq 80$ mm) crystal diameters during the cone phase. In the small melts, the vortex induced by EM force occupied more than a half of the melt. In the large melts, on the contrary, intensive melt motion occurred only in the thin layer near the free melt surface.

Differences in the melt flow influence species transport significantly: two maximums in the radial concentration profile occurred for small crystal diameters but only one for large crystal diameters. The radial variation of concentration reached ±50% for carbon and ±5% for boron. The profile shapes were similar for both elements with larger variations obtained for small crystal diameters. The differences between the simulations and analytically predicted axial distribution are considerably large due to the effects of cone phases. It was also observed that the precise description of melt volume evolution is important for obtaining axial species distribution in the grown crystal.

The simulated values of the effective segregation coefficient $k_{eff}$ qualitatively agree with the classical BPS model (the BPS model predicted 0.38–0.48 for carbon) and experimental data from other silicon growth processes (experimental results were 0.73–0.98 for boron). However, numerical simulations showed that $k_{eff}$ increases for larger melt diameters due to less intensive melt motion. The increase in the crystal rotation rate from 6 to 12 rpm reduces $k_{eff}$, particularly for small crystal diameters. These results are crucial for the accurate prediction of the species distribution in FZ Si crystals.

**Author Contributions:** Conceptualization, J.V. and K.D.; methodology, J.V., K.D. and K.S.; software, K.S. and A.S.; validation, K.S. and M.S.; investigation, N.A., R.M., K.S., M.S. and A.S.; writing—original draft preparation, K.S., M.S. and A.S.; writing—review and editing, J.V. and K.D.; supervision, J.V. All authors have read and agreed to the published version of the manuscript.

**Funding:** This research was funded by University of Latvia, "Strengthening of the capacity of doctoral studies at the University of Latvia within the framework of the new doctoral model", grant number No. 8.2.2.0/20/I/006. K. Dadzis has received funding from the European Research Council (ERC) under the European Union's Horizon 2020 research and innovation programme (grant agreement No 851768).

**Data Availability Statement:** Not applicable.

**Acknowledgments:** We gratefully acknowledge Daniela Eppers, Horst Bettin and all members of the project team at the Physikalisch-Technische Bundesanstalt (PTB, Braunschweig) for the many helpful discussions about the impurities in FZ Si growth. We also thank Lucas Vieira (IKZ) for reviewing the manuscript.

**Conflicts of Interest:** The authors declare no conflict of interest. The funders had no role in the design of the study; in the collection, analyses, or interpretation of data; in the writing of the manuscript; or in the decision to publish the results.

## Appendix A

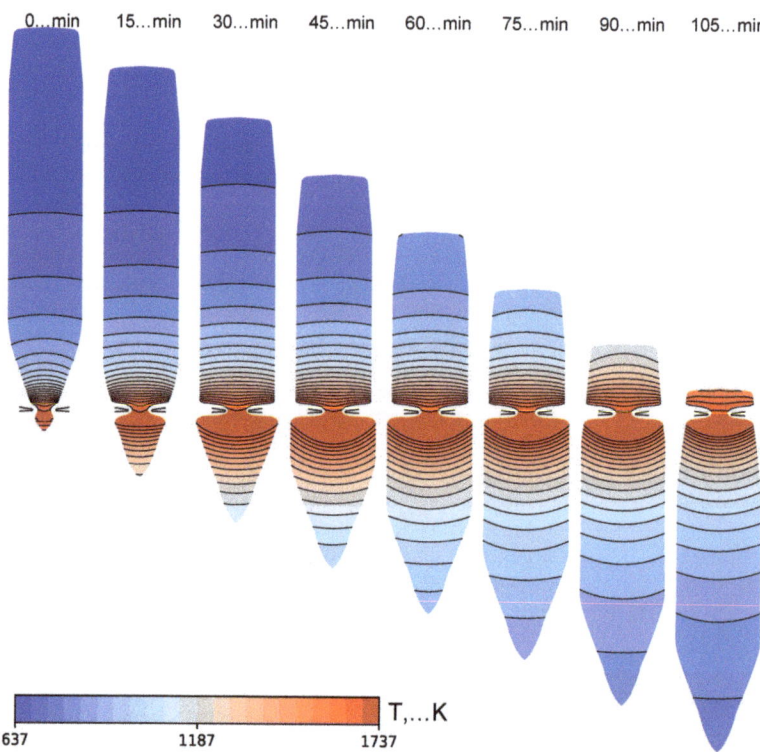

**Figure A1.** The time-dependence of the shape of the phase boundaries and global temperature field in Si plotted each 15 min. The spacing between isotherms is 50 K.

## Appendix B

A simple cylindrical system was considered to verify the developed program 0D-segregation for axial distribution of impurities (Section 2.4). The system consists of a crystal with a constant 102 mm diameter, ignoring the cone stages at the beginning and end of the growth process. The axial distribution of impurities in such a system can be described by an analytical formula [32]:

$$\frac{C(a)}{C_0} = 1 - (1 - k_{\text{eff}})e^{-k_{\text{eff}}a}, \tag{A1}$$

where $C_0$ is the initial impurity concentration and $a = L/H$ is the crystal length $L$ expressed in units of molten zone height $H$. The model results precisely agree with the analytical formula.

We compare the axial species concentration in the crystal with different segregation coefficient values; see Figure A2. The model results are in agreement with the analytical formula for all considered $k_{\text{eff}}$ values, both $k_{\text{eff}} > 1$ and $k_{\text{eff}} < 1$.

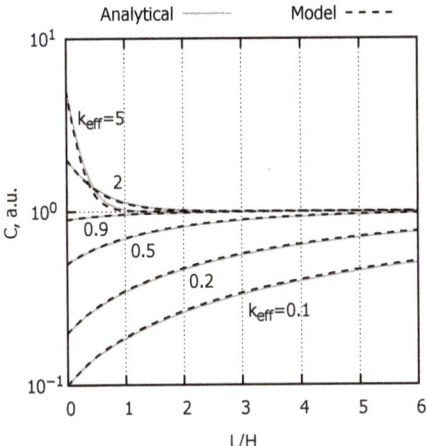

**Figure A2.** Comparison of model results with analytical formula for species concentration in the cylindrical crystal (without start and end cones) with different segregation coefficient values $k_{eff}$ ($L$ represents the crystal length, and $H$ represents the height of the molten zone).

## References

1. Mühlbauer, A.; Muižnieks, A.; Raming, G.; Riemann, H.; Lüdge, A. Numerical modelling of the microscopic inhomogeneities during FZ silicon growth. *J. Cryst. Growth* **1999**, *198*, 107–113. [CrossRef]
2. Abrosimov, N.V.; Aref'ev, D.G.; Becker, P.; Bettin, H.; Bulanov, A.D.; Churbanov, M.F.; Filimonov, S.V.; Gavva, V.A.; Godisov, O.N.; Gusev, A.V.; et al. A new generation of 99.999% enriched 28Si single crystals for the determination of Avogadro's constant. *Metrologia* **2017**, *54*, 599–609. [CrossRef]
3. Sze, S.; Lee, M. *Semiconductor Devices: Physics and Technology*; John Wiley & Sons, Inc.: Hoboken, NJ, USA, 2012; p. 361.
4. Burton, J.A.; Prim, R.C.; Slichter, W.P. The distribution of solute in crystals grown from the melt. Part I. Theoretical. *J. Chem. Phys.* **1953**, *21*, 1987–1991. [CrossRef]
5. Priede, J.; Gerbeth, G. Breakdown of Burton-Prim-Slichter approach and lateral solute segregation in radially converging flows. *J. Cryst. Growth* **2005**, *285*, 261–269. [CrossRef]
6. Wilson, L.O. Analysis of microsegregation in crystals. *J. Cryst. Growth* **1980**, *48*, 363–366. [CrossRef]
7. Mühlbauer, A.; Muižnieks, A.; Virbulis, J. Analysis of the dopant segregation effects at the floating zone growth of large silicon crystals. *J. Cryst. Growth* **1997**, *180*, 372–380. [CrossRef]
8. Lācis, K.; Muižnieks, A.; Jēkabsons, N.; Rudevičs, A.; Nacke, B. Unsteady 3D and analytical analysis of segregation process in floating zone silicon single crystal growth. *Magnetohydrodynamics* **2009**, *45*, 549–556. [CrossRef]
9. Han, X.F.; Liu, X.; Nakano, S.; Kakimoto, K. Numerical analysis of dopant concentration in 200 mm (8 inch) floating zone silicon. *J. Cryst. Growth* **2020**, *545*, 125752. [CrossRef]
10. Sim, B.C.; Kim, K.H.; Lee, H.W. Boron segregation control in silicon crystal ingots grown in Czochralski process. *J. Cryst. Growth* **2006**, *290*, 665–669. [CrossRef]
11. Hong, Y.H.; Sim, B.C.; Shim, K.B. Distribution coefficient of boron in Si crystal ingots grown in cusp-magnetic Czochralski process. *J. Cryst. Growth* **2008**, *310*, 83–90. [CrossRef]
12. Mei, P.R.; Moreira, S.P.; Cortes, A.D.S.; Cardoso, E.; Marques, F.C. Determination of the effective distribution coefficient (K) for silicon impurities. *J. Renew. Sustain. Energy* **2012**, *4*, 043118. [CrossRef]
13. Series, R.W.; Hurle, D.T.; Barraclough, K.G. Effective distribution coefficient of silicon dopants during magnetic Czochralski Growth. *IMA J. Appl. Math.* **1985**, *35*, 195–203. [CrossRef]
14. Jeon, H.J.; Park, H.; Koyyada, G.; Alhammadi, S.; Jung, J.H. Optimal cooling system design for increasing the crystal growth rate of single-crystal silicon ingots in the Czochralski process using the crystal growth simulation. *Processes* **2020**, *8*, 1077. [CrossRef]
15. Ding, J.; Li, Y.; Liu, L. Effect of cusp magnetic field on the turbulent melt flow and crystal/melt interface during large-size Czochralski silicon crystal growth. *Int. J. Therm. Sci.* **2021**, *170*, 107137. . [CrossRef]
16. Ratnieks, G.; Muižnieks, A.; Mühlbauer, A. Modelling of phase boundaries for large industrial FZ silicon crystal growth with the needle-eye technique. *J. Cryst. Growth* **2003**, *255*, 227–240. [CrossRef]
17. Rudevičs, A.; Muižnieks, A.; Ratnieks, G.; Mühlbauer, A.; Wetzel, T. Numerical study of transient behaviour of molten zone during industrial FZ process for large silicon crystal growth. *J. Cryst. Growth* **2004**, *266*, 54–59. [CrossRef]
18. Lācis, K.; Muižnieks, A.; Rudevičs, A.; Sabanskis, A. Influence of DC and AC magnetic fields on melt motion in FZ large Si crystal growth. *Magnetohydrodynamics* **2010**, *46*, 199–218.

19. Ratnieks, G.; Muiznieks, A.; Buligins, L.; Raming, G.; Mühlbauer, A.; Lüdge, A.; Riemann, H. Influence of the three dimensionality of the HF electromagnetic field on resistivity variations in Si single crystals during FZ growth. *J. Cryst. Growth* **2000**, *216*, 204–219. [CrossRef]
20. Surovovs, K. A Program for Calculating Dopant Concentration Distribution in a Crystal Grown in Float-Zone Process. Available online: https://git.lu.lv/ks10172/zero-d (accessed on 30 September 2022).
21. Muižnieks, A.; Rudevics, A.; Riemann, H.; Lacis, U. Comparison between 2D and 3D modelling of HF electromagnetic field in FZ silicon crystal growth process. In Proceedings of the International Scientific Colloquium Modelling for Material Processing, Riga, Latvia, 16–17 September 2010; pp. 61–65.
22. Muižnieks, A.; Lacis, K.; Nacke, B. 3D unsteady modelling of the melt flow in the FZ silicon crystal growth process. *Magnetohydrodynamics* **2007**, *43*, 377–386. [CrossRef]
23. Sabanskis, A.; Surovovs, K.; Virbulis, J. 3D modeling of doping from the atmosphere in floating zone silicon crystal growth. *J. Cryst. Growth* **2017**, *457*, 65–71. [CrossRef]
24. Ribeyron, P.; Durand, F. Oxygen and carbon transfer during solidification of semiconductor grade silicon in different processes. *J. Cryst. Growth* **2000**, *210*, 541–553. [CrossRef]
25. Rost, H.J.; Menzel, R.; Luedge, A.; Riemann, H. Float-Zone silicon crystal growth at reduced RF frequencies. *J. Cryst. Growth* **2012**, *360*, 43–46. [CrossRef]
26. Surovovs, K.; Muiznieks, A.; Sabanskis, A.; Virbulis, J. Hydrodynamical aspects of the floating zone silicon crystal growth process. *J. Cryst. Growth* **2014**, *401*, 120–123. [CrossRef]
27. Mills, C.; Courtney, L. Thermophysical Properties of Silicon. *ISIJ Int.* **2000**, *40*, 130–138. [CrossRef]
28. Eremenko, V.N.; Gnesin, G.G.; Churakov, M.M. Dissolution of polycrystalline silicon carbide in liquid silicon. *Test Methods Prop. Mater.* **1972**, *11*, 471–474. [CrossRef]
29. Garandet, J.P. New determinations of diffusion coefficients for various dopants in liquid silicon. *Int. J. Thermophys.* **2007**, *28*, 1285–1303. [CrossRef]
30. Kolbesen, B.O.; Mühlbauer, A. Carbon in silicon: Properties and impact on devices. *Solid-State Electron.* **1982**, *25*, 759–775. [CrossRef]
31. Menzel, R. Growth Conditions for Large Diameter FZ Si Single Crystals. Ph.D. Thesis, Technischen Universität Berlin, Germany, 2013.
32. Milliken, K.S. Simplification of a molten zone refining formula. *J. Met.* **1955**, *7*, 838. [CrossRef]

Article

# Long-Term Stability of Novel Crucible Systems for the Growth of Oxygen-Free Czochralski Silicon Crystals

Felix Sturm [1,*], Matthias Trempa [1], Gordian Schuster [2], Rainer Hegermann [2], Philipp Goetz [2], Rolf Wagner [3], Gilvan Barroso [3], Patrick Meisner [4], Christian Reimann [1] and Jochen Friedrich [1]

[1] Fraunhofer IISB, Schottkystrasse 10, 91058 Erlangen, Germany
[2] CVT GmbH & Co. KG, Romantische Strasse 18, 87642 Halblech, Germany
[3] Rauschert Heinersdorf-Pressig GmbH, Bahnhofstrasse 1, 96332 Pressig, Germany
[4] SGL Carbon GmbH, Drachenburgstrasse, 53170 Bonn, Germany
* Correspondence: felix.sturm@iisb.fraunhofer.de

**Abstract:** The replacement of the silica glass crucible by oxygen-free crucible materials in silicon Czochralski (Cz) growth technology could be a key factor to obtaining Cz silicon, with extremely low oxygen contamination $< 1 \times 10^{17}$ at/cm$^3$ required for power electronic applications. So far, isostatic pressed graphite or nitrogen-bonded silicon nitride (NSN) crucible material, in combination with a chemical vapor deposited silicon nitride (CVD-Si$_3$N$_4$) surface coating, could be identified as promising materials by first short-term experiments. However, for the evaluation of their potential for industrial scale Cz growth application, the knowledge about the long-term behavior of these crucible setups is mandatory. For that purpose, the different materials were brought in contact with silicon melt up to 60 h to investigate the infiltration and dissolution behavior. The chosen graphite, as well as the pore-sealed NSN material, revealed a subordinated infiltration-depth of ≤1 mm and dissolution of ≤275 µm by the silicon melt, so they basically fulfilled the general safety requirements for Cz application. Further, the highly pure and dense CVD Si$_3$N$_4$ crucible coating showed no measurable infiltration as well as minor dissolution of ≤50 µm and may further acts as a nucleation site for nitrogen-based precipitates. Consequently, these novel crucible systems have a high potential to withstand the stresses during industrial Cz growth considering that more research on the process side relating to the particle transport in the silicon melt is needed.

**Keywords:** Czochralski growth; silicon; crucible; oxygen concentration

## 1. Introduction

For some types of microelectronic devices, like insulated-gate bipolar transistors (IGBT), silicon wafers with significantly low oxygen contamination are indispensable [1,2]. Therefore, silicon crystals used for this application are typically grown by the Floating zone (Fz) technique, revealing an oxygen content far below $1 \times 10^{16}$ at/cm$^3$ [3,4]. However, from an economical point of view, Czochralski (Cz) material would be preferred by electronic device manufacturers due to its lower process cost and the availability of larger crystal diameter up to 300 mm. The major drawback of the Cz technique is basically the limitation to silicon material with a significant increased oxygen content, typically higher than $1 \times 10^{17}$ at/cm$^3$. These examples of intense oxygen incorporation, even by application of an additional magnetic field (MCz), are mainly caused by melting up the silicon feedstock in the state of the art silica glass (SiO$_2$) crucibles [4,5]. Hence, one promising approach would be the replacement of SiO$_2$ by oxygen-free crucible materials.

Only a few approaches to reduce the oxygen concentration during the Cz growth process by optimization of the crucible concept and materials were reported in literature during the 1980s. By application of a dense CVD-Si$_3$N$_4$ coating on a SiO$_2$ glass crucible, Doi et al. could demonstrate a reduction of oxygen concentration to $5 \times 10^{16}$ atoms/cm$^3$ [6].

Oxygen-free crucible systems were achieved by Watanabe et al. [7] and Matsuo et al. [8]. By graphite and $Si_3N_4$-based crucibles in combination with a CVD-$Si_3N_4$ surface coating, the oxygen concentration in monocrystalline silicon ingots could be reduced below $2 \times 10^{16}$ atoms/cm$^3$. Nevertheless, due to the $Si_3N_4$ coating dissolution by the silicon melt, it becomes saturated by nitrogen, which cannot be degassed from the melt. Consequently, nitrogen-related precipitates like $Si_3N_4$ needles were formed by exceeding the nitrogen solubility limit of the melt [9]. By transportation of these particles to the solid–liquid interface and incorporation into the growing silicon ingot, an increase in dislocation density or even multi-crystalline growth occurred.

More recently, several new concepts for graphite [10–12] and $Si_3N_4$-based [13–18] crucibles for the growth of multi-crystalline silicon ingots were presented, which underlines the potential to use these materials in crucible setups during Si crystal growth.

In our previous work [19], we reconsidered this approach by evaluating several isostatic pressed graphite materials and nitride-bonded silicon nitride (NSN) in combination with a chemical vapor-deposited (CVD) $Si_3N_4$ inner surface coating, according to their general applicability as crucible materials for Cz silicon growth. The crucible materials were characterized according to their performance and interaction with liquid silicon in lab scale melting experiments with moderate melt holding times up to 4 h.

However, considering the fact that, in an industrial Cz process the crucible is typically in contact with the silicon melt at much longer timescales ranging from ~30 h to >100 h, the knowledge about the long-term behavior and stability of the crucible setups is mandatory. For that purpose, so called dipping experiments as well as directional solidification (DS) runs were carried out with extended melt holding time up to 60 h. By detailed investigation of the dissolution behavior as well as the infiltration depth in combination with the resulting material morphology after the processes, the capability of the novel crucible setups to safely contain the silicon melt over long times scales was evaluated. Further, the minimum required thickness of the CVD-$Si_3N_4$ coating should be discussed. Additionally, the DS crystallization experiments allowed the evaluation of the crucible behavior in contact with larger Si melt volumes over longer time scales and the Si ingot quality with respect to the formation of $Si_3N_4$ based precipitates. In conclusion, the experimental results obtained in this work will help to further evaluate the use of these novel crucible setups during industrial Cz silicon growth processes and to identify which developments will still be necessary during perspective research.

## 2. Materials and Methods

For the dipping experiments, bare shaped NSN and graphite (blank and coated with CVD $Si_3N_4$ coated) samples ($65 \times 40 \times 10$ mm$^3$) were brought in direct contact with liquid silicon at defined process conditions for various time scales.

The NSN samples were achieved by the slip casting of a water-based slurry containing silicon powder (provided by Wacker Chemie AG, Munich, Germany) and $Si_3N_4$ powder (provided by Alzchem Trostberg GmbH, Trostberg, Germany), followed by a heat treatment in nitrogen atmosphere. This method allows the formation of a $Si_3N_4$ material with sufficient mechanical strength at relatively low temperatures about 1450 °C, but with an increased porosity up to 60%. For that reason, an additional pore sealing coating based on a polysilazane slurry [20,21] was developed and applied on the inner surface of NSN crucibles, finalized by a second nitridation step. For graphite samples, isostatic pressed material with low porosity and high mechanical strength (SGL Carbon GmbH, Wiesbaden, Germany) was used; hence, it has already shown excellent stability against the silicon melt in our previous tests [19].

The setup of the dipping experiments is described in Figure 1. In a first step, high purity silicon feedstock was melted in uncoated graphite crucibles (Ø 97/87 × 122 mm). The ratio of weighted silicon portion to the dipped surface of the NSN and graphite samples was chosen in that way, that it corresponds to the ratio of silicon volume to crucible contact area in an industrial scale 32-inch Cz crucible setup. For CVD $Si_3N_4$-coated graphite samples, the ratio was doubled due to the only one-side coating.

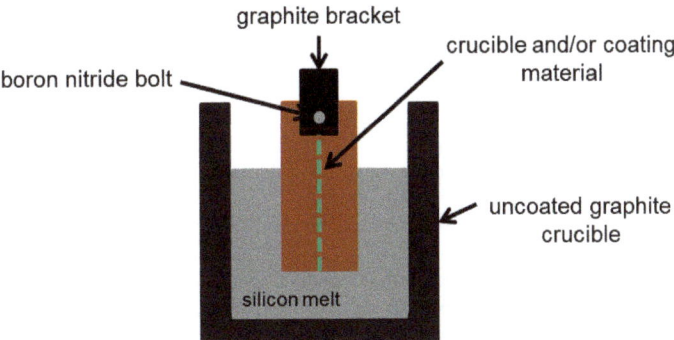

**Figure 1.** Schematic drawing of the experimental setup for investigating the dissolution of crucible and coating materials in liquid silicon. The dashed green line marks the preparation position of the vertical cross-section.

Furthermore, the chosen experimental setup excludes any kind of nitrogen contamination of the silicon melt, besides from the dipped materials itself. After a short melt homogenization period, the different samples were dipped into the silicon melt for 4 h, 8 h 16 h, 25 h and 40 h. After the extraction from the melt, the liquid silicon in the crucible was crystallized in a directional solidification process. The dipped samples were treated with a solution of HF, $HNO_3$ and $CH_3COOH$ (volume ratio 6:81:13) to remove solidified silicon residues. After the etching and cleaning procedure, the dipped samples were embedded in epoxy resin, and vertical cross-sections were prepared. For the investigation of the dissolution of uncoated graphite vertical cross-sections were additionally cut from the crucibles after the dipping experiment. The resulting decrease of material/coating thickness and the infiltration depth of liquid silicon was determined by optical microscopy.

Besides the systematic investigation of the interaction with liquid silicon at small sample scale, the novel crucible setups were also tested in laboratory scaled DS silicon growth experiments. Therefore, crucibles (Ø 121.5/106.5 × 110 mm) with a spherical bottom design in various configurations were prepared to evaluate their potential for Cz application (see Figure 2).

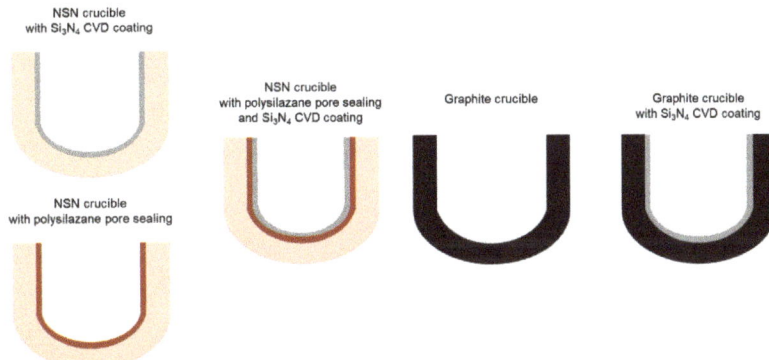

**Figure 2.** Overview of investigated crucible setups for oxygen-free Cz silicon ingot growth.

NSN-based crucibles were produced analogously to the NSN dipping samples. Additionally, the inner surface of the crucibles was covered either with a CVD $Si_3N_4$ coating, a pore sealing coating $Si_3N_4$/Si/polysilazane (70/18/12 wt.%) or a combination of both. Further, isostatic pressed graphite crucibles, which have already shown good resistance against melt infiltration in our previous investigations [19], were used in uncoated and CVD-coated variants.

The averaged CVD $Si_3N_4$ coating thickness on graphite crucibles was determined to be 150 µm at the crucible wall to >400 µm at the bottom, while for NSN crucibles, it was 170 µm at the walls to >700 µm at the bottom. The increasing coating thickness towards the crucible bottom is caused by the high reactive precursor gases and the geometric limitation of gas flux conditions during CVD application. The CVD coating reveals a good wear resistance, which allows the handling of the crucible without delamination or chipping of the coating.

The crucibles were heated up in vacuum conditions, and the silicon was melted within a period of 6 h. Afterwards, the liquid silicon was contained up to 60 h at 20 mbar in the different crucible setups, followed by a 7 h crystallization step (growth rate = 1–1.5 cm/h). After silicon solidification, 2 mm thick vertical slices were cut out of the center of the crucible as well as from the Si crystal. The crucible parts were etched analogously to the dipping samples to remove attached silicon and to expose the remaining CVD coating. The characterization of the structure and morphology of the CVD coating was carried out by optical and scanning electron microscopy (SEM).

Furthermore, the silicon samples were investigated by infrared (IR) transmission microscopy to observe the particles and precipitates formed in the silicon crystal.

## 3. Results and Discussion

### 3.1. Dipping Experiments

For investigation of the infiltration behavior of liquid silicon in the coating and crucible materials at long time scale, dipping experiments with a variation of contact time to the silicon melt (4 h up to 40 h) were performed. Among others, this is important, hence the Si infiltration in C-based materials results in a silicon carbide (SiC) formation inducing a volume increase, which can lead to an intense crack formation in graphite materials with minor mechanical strength [19,22]. The resulting averaged infiltration depth, which was measured at vertical cross-sections through the various kinds of dipped samples, is summarized in Figure 3.

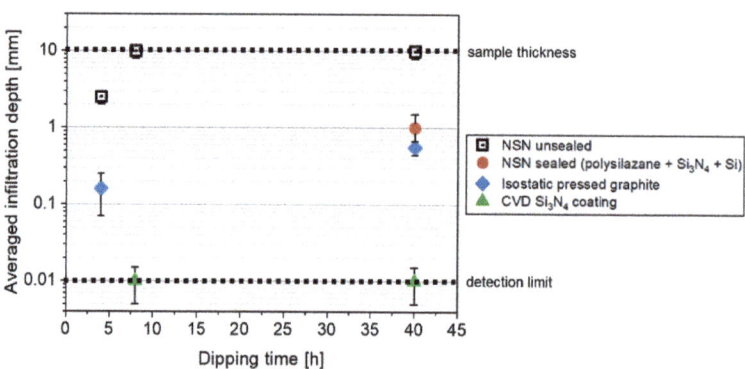

**Figure 3.** Averaged infiltration depth of liquid silicon after dipping in silicon melt between 4 h and 40 h.

First, the relatively dense graphite material (~11% porosity) shows only a minor infiltration depth of about 600 µm, even after 40 h in contact with liquid silicon. In combination with a sufficient mechanical strength of the chosen graphite material, this prohibits any crack formation according to the occurring SiC formation within the infiltrated zone. The NSN material shows a significant higher affinity to silicon infiltration due to its high porosity of ~60 % combined with a good wettability by liquid silicon. In consequence, a 10 mm thick NSN sample was completely infiltrated by silicon melt, already after a dipping of 8 h. Therefore, the NSN material cannot be used as Cz crucible material in this untreated condition. However, if a polysilazane-based sealing coating is applied, a significant reduction of the infiltration depth after 40 h from >10 mm to about 1 mm could be achieved, which is close to the result of the graphite sample. Finally, in case of the CVD $Si_3N_4$ coating on graphite, no infiltration in the range of the detection limit (<10 μm) could be observed at all, even at long time scale.

Beside the infiltration, also the dissolution behavior of the crucible (NSN, graphite) and coating (CVD $Si_3N_4$) materials was examined by the dipping experiments. Residues of solidified silicon were removed from the dipped sample surface by an etching process and the averaged dissolved layer thickness was determined by vertical cross-sections. The results of the averaged dissolved layer thickness in dependence on the dipping time are shown in Figure 4.

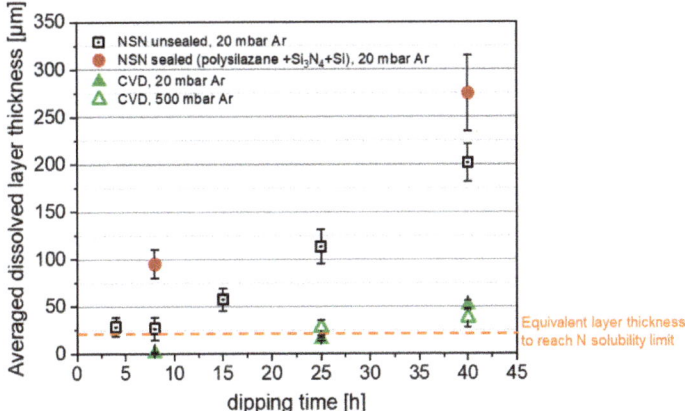

**Figure 4.** Measurement of dissolved layer thickness of NSN with and without polysilazane sealing as well as CVD $Si_3N_4$.

For unsealed NSN samples, an averaged dissolved layer thickness of about 25 μm can be observed after a 4 h dipping, which nearly corresponds to the N solubility limit in the given silicon melt volume. With an increase of dipping time above 15 h also the averaged dissolved layer thickness increases. After 40 h nearly 200 μm of the NSN material was dissolved in the silicon melt. NSN samples with an additional pore sealing (polysilazane + $Si_3N_4$ + Si) reveal a higher dissolved layer thickness than the uncoated NSN material. This indicates that the sealing coating has a lower resistance to dissolution in liquid silicon than the NSN material itself. Nevertheless, the application of a pore sealing coating is basically indispensable to avoid complete crucible infiltration in case of CVD coating failure in terms of cracks or delamination.

In contrast to the NSN material, the CVD $Si_3N_4$ coating exhibits a significant weaker material dissolution (see Figure 5), which predestines the CVD coating as ideal crucible functional surface coating. In case of 8 h of melt contact, the averaged dissolved layer thickness is in the range of the detection limit of 10 μm and increases with increasing dipping time to 25 μm for 25 h and 50 μm for 40 h. This decelerated dissolution behavior compared to NSN can be connected to the high density and exceptionally low open porosity of the CVD $Si_3N_4$, of which properties have already been proposed to reduce dissolution in silicon melt [23]. Also at higher Ar pressures, in this case at 500 mbar (see Figure 4), the dissolution process of CVD $Si_3N_4$ is rather slow. This could be beneficial if the Ar pressure during the Cz process will be set to higher values (combined with the addition of nitrogen in the atmosphere), in order to increase the chemical stability of the $Si_3N_4$ [17].

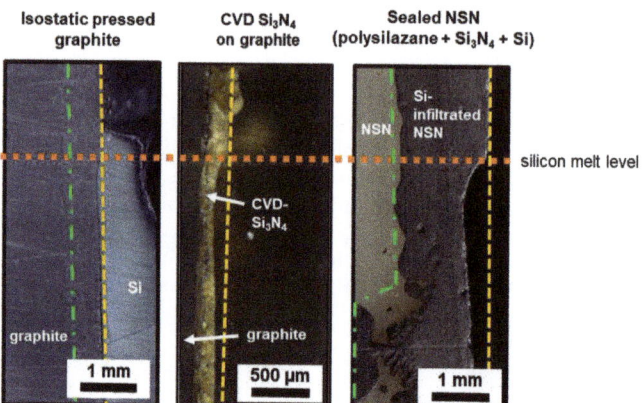

**Figure 5.** Vertical cross-section of isostatic pressed graphite samples, uncoated as well as with CVD $Si_3N_4$ coating, and sealed NSN dipped in liquid silicon for 40 h. The orange dotted line marks the silicon melt level. The yellow dashed line represents the original sample surface and indicates the dissolution of material during melt contact. The green dot dashed line marks the silicon infiltration zone.

The dissolution of the uncoated isostatic graphite material during the dipping procedure was significantly lower than for CVD $Si_3N_4$ coated graphite or NSN samples. In these cases, no dissolution, except directly below the Si melt level (see Figure 5), could be observed, meaning the averaged dissolved layer thickness must be below the detection limit of about 10 μm. This could be mainly attributed to the formation of a SiC layer [22,24], which leads to the passivation of the surface against further dissolution [25].

Besides the global material dissolution, which happens along the dipped sample surface, a more pronounced dissolution was observed at the region where silicon melt, graphite/$Si_3N_4$ and the argon atmosphere coincide. This could be clearly seen at vertical cross-sections, which are exemplarily shown in Figure 5 for uncoated and CVD $Si_3N_4$-coated graphite as well as uncoated, sealed NSN samples after 40 h of melt contact.

The isostatic pressed graphite material shows the smallest maximum dissolved layer thickness in the triple area of about 100 μm, followed by the CVD $Si_3N_4$ coating (150 μm) and sealed NSN (600 μm). As expected, this trend is in good correlation to the measurements of the averaged dissolved layer thickness over the whole dipped sample surface (see Figure 4).

This observation is mainly important for the prediction of the necessary CVD coating thickness. To ensure a closed CVD coating layer over the whole crucible surface for the entire silicon crystal growth process, the coating thickness must exceed the maximum dissolved layer thickness at the melt surface.

The phenomena of enhanced dissolution of different ceramic crucible and coating materials directly below the melt level are basically described in literature for iron containing melt systems [26] and also for quartz glass crucibles in contact with liquid silicon during a standard Cz growth process [27]. This effect is mainly caused by the Marangoni convection. By wetting the coating or crucible material surface by the silicon melt, the dissolution process is initiated, and a local change of melt composition and surface tension occurs. These gradients of concentration and surface tension are the driving forces for the Marangoni convection, and an enhanced mass flow advances the further dissolution of the crucible or coating material [26,28]. Furthermore, the depth of the resulting groove and, consequently, the dissolution rate of the crucible and coating materials seem also to depend on the material properties. The dissolution rate $V_{dissolution}$ [mm/h] by the Marangoni enhanced dissolution can be described with the following Equation (1) [26].

$$V_{dissolution} = 360 \cdot \frac{\rho_{melt}}{\rho_{crucible}} \cdot \beta \cdot (C_s - C_0) \tag{1}$$

where $\rho_{melt}$ is the density of the silicon melt, $\rho_{crucible}$ the density of graphite, NSN or CVD $Si_3N_4$, $\beta$ the mass transfer constant, $C_s$ the saturation concentration of the dissolved phase (C or N) and $C_0$ the initial concentration of the dissolved phase in the melt. This also explains basically the smallest maximum dissolved layer thickness in the triple area for CVD $Si_3N_4$ compared to NSN. The CVD coating exhibits a significant higher density than the NSN material and, consequently, $V_{dissolution}$ is reduced for the CVD coating. For graphite materials it must be considered that according to the use of a graphite crucible for the dipping experiments $C_0$ of C may significantly higher than it is the case for N. This results in a decrease of the dissolution rate for the graphite samples.

Summarized, graphite and NSN with a pore sealing coating as well as CVD $Si_3N_4$ showed material loss by dissolution and silicon melt infiltration only in a µm range during the dipping experiments. Due to the fact a Cz crucible typically exhibits a wall thickness in a range of 10 to 35 mm, the observed interaction of the crucible and coating materials with the liquid silicon is no crucial drawback for Cz application.

*3.2. Long Term Crystallization Experiments in G0 Scale*

3.2.1. Performance and Durability of the Oxygen-Free Crucible Systems

In Figure 6, the top views on the crucible systems containing the resulting silicon crystal after the experiment are shown. For all crucible systems no melt leakage could be observed, and regular crystallization of silicon melt could be achieved. Further, all ingots stick to the crucible/coating system as expected due to the wettability of the graphite, NSN and CVD $Si_3N_4$ materials towards the silicon melt. As expected, Fourier transformed infrared spectroscopy (FTIR) confirmed that no oxygen was incorporated into silicon ingots during the DS process.

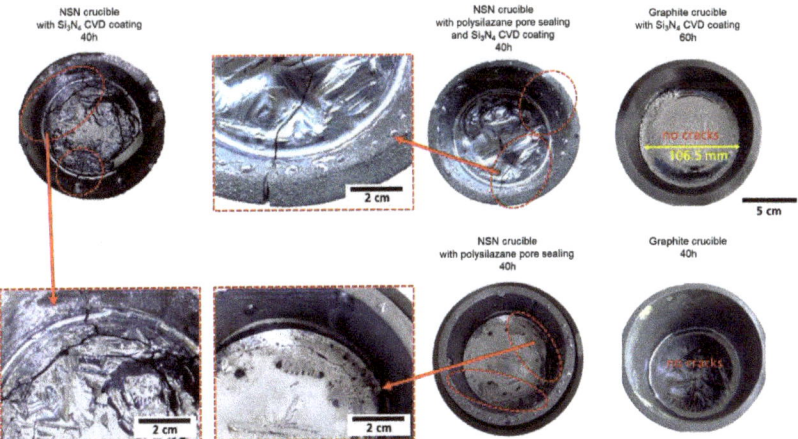

**Figure 6.** NSN and graphite-based crucible systems with silicon crystals after DS crystallization experiments with melt holding times of 40 h and 60 h. The red marks indicate crack formation in the crucible and the crystal.

NSN Crucibles

In case of NSN material, a significant crack formation in the crucibles and in the silicon crystals occurred, independent of the presence of a CVD $Si_3N_4$ coating. Due to the absence of melt leakage, and the fact that no strong Si infiltration was observed in cross-sections of the NSN crucibles (compare Figure 7), it can be assumed that the cracking was caused by thermal stresses occurring during the cooling after silicon crystallization. The stresses are induced by the difference in thermal expansion of solid silicon (CTE~$2.5 \times 10^{-6}$ K$^{-1}$) and the CVD coating/NSN crucible (CTE~$3.3 \times 10^{-6}$ K$^{-1}$). So basically, the mechanical strength of the NSN is not sufficient to absorb the thermal stresses resulting from the CTE mismatch. Although this may be in general no ideal basic condition for crucible application, it has to be considered that, in Cz growth, typically only a small portion of the residual melt solidifies in the crucible after the pulling of the Cz crystal is finished and therefore the CTE mismatch should not be such a severe issue.

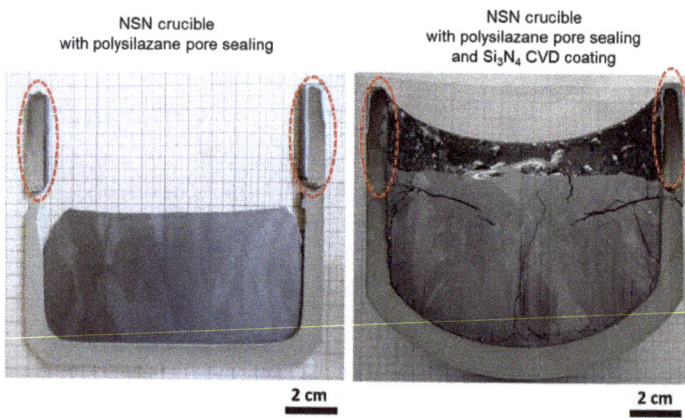

**Figure 7.** Vertical cross section of CVD coated NSN crucible bottom part after 40 h melt contact and silicon crystallization. In the marked areas silicon melt infiltration locally occurred.

By preparation of vertical cross-sections through the center of the crucible/crystal compound, the silicon infiltration in the different NSN crucible setups were investigated. For unsealed NSN crucible with CVD coating, no infiltration in coating or crucible could be detected. This underlines the ability of the CVD coating to inhibit direct melt contact with the crucible also on long time scales. Furthermore, no significant infiltration of liquid silicon in the two sealed NSN crucibles (with and even without CVD coating) could be observed below the silicon melt surface level (see Figure 7). The resulting infiltration depth on this site was even lower than it was previously observed for the bare shaped samples in the dipping experiments. This indicates that the pore sealing can be an effective tool to protect the NSN material from intense melt infiltration.

Above the melt level, a locally increased silicon infiltration depth was found at the top part of the crucibles in both tested NSN crucible configurations. This could be correlated to some micro-defects (cracks or voids) of the CVD coating and/or the sealing layer, which should be avoided by further optimization of the coating procedure.

Graphite Crucibles

As shown in Figure 6, the visual inspection after the crystallization process of the graphite-based crucibles, with and without CVD coating, shows no obvious crack formation or deformation. Despite, the graphite material has a higher CTE (~$4.1 \times 10^{-6}$ K$^{-1}$) than the NSN/CVD materials, and in consequence, the CTE mismatch between crucible and the solidified Si is larger in this case; the mechanical strength of the used graphite material is sufficient to withstand the thermal stresses.

Silicon melt infiltration was not observed in the CVD-coated variant, showing again that the CVD coating acts as an effective barrier. In the uncoated case, only a small silicon melt infiltration of some hundreds of μm appears, as it was already observed in the dipping experiments. However, the additional mechanical stresses, which occurs due to the SiC formation within the Si infiltrated zone, are obviously not high enough to induce crack formation in the graphite material. So, the chosen graphite material is suitable for the application as Cz crucible material, even in case of coating dissolution, delamination or failure. But it has to be noted that not all isostatic presses graphite types reveal high enough mechanical strength, as it was already presented in [19].

CVD Coating

To evaluate the dissolution of CVD $Si_3N_4$ coating during the crystallization experiments, a vertical cross-section of one coated graphite crucible (60 h melt contact) was prepared from various positions along the crucible wall and bottom. After removing the attached silicon by an etching process, the thickness of the still present CVD coating was measured (Figure 8).

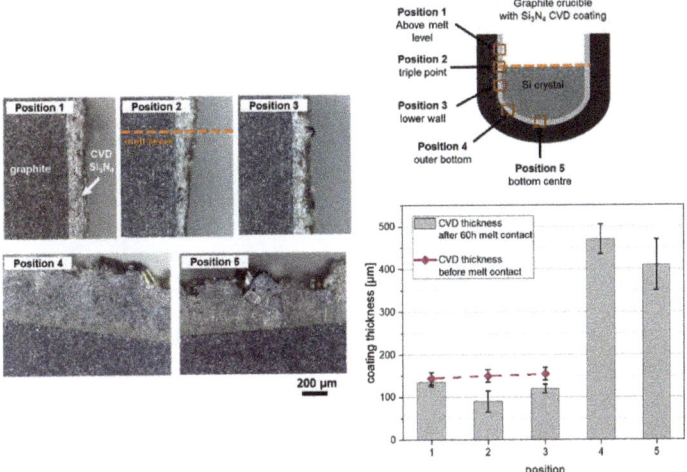

**Figure 8.** CVD coating thickness at different positions of a graphite crucible after 60 h melt contact and silicon crystallization. Because the measurement of coating thickness was only achievable by destructive methods, the CVD thickness before melt contact was determined at a crucible, which was coated with identical process condition as the crucible, used for silicon melting experiment.

Above the silicon melt surface level, a coating thickness of about 135 ± 10 µm is measured. Analogously to the dipping experiments, the lowest coating thickness (90 µm ± 30 µm) occurred directly below the contact point of the coating with the Si melt surface. Moving along the crucible wall downwards, the CVD thickness increases again up to 120 µm ± 10 µm, before it further increases in the spherical bottom region to >400 µm. Since the exact local CVD coating thickness before the melting experiment was unknown, values from destructive measurements of previously coated graphite crucibles were used for a rough estimation of the dissolved layer thickness. The results show that, in the region without Si melt contact, almost no dissolution exists, while in the region of the melt surface level the strongest dissolution of about 50 µm has occurred. In the lower wall region, the coating thickness is only reduced by 30–40 µm, which is on a similar level as in the dipping experiments (40–50 µm for 40 h, compare Figure 4). For the bottom region no reliable investigation of the CVD dissolution process could be achieved. The geometric limitation of flux, in combination with the high reactivity of the precursor gases, results in an increased variation of resulting CVD coating thickness at the bottom for each coating process.

Based on these results, a CVD Si$_3$N$_4$ coating layer thickness of at least 100 μm should be applied for application in Cz growth to be sure that the coating is not completely dissolved.

3.2.2. Impact of Crucible/Coating Systems on Precipitate Formation in Grown Crystals

Within the frame of previous done short-term experiments (4 h melt holding time) [19], the application of a CVD Si$_3$N$_4$ coating on the crucible has led to a significant reduction of Si$_3$N$_4$ precipitates in the crystal volume in comparison to an experiment without the use of the CVD coating. This gives the hint that the coating could act as nucleation site for the Si$_3$N$_4$ precipitates formed in the silicon melt supersaturated with nitrogen.

The same phenomenon could be also found in the new long-term experiments within uncoated and coated graphite crucibles. The observation in crystals grown within the NSN-crucibles was not possible due to the extended crack formation.

Corresponding investigations by IR transmission microscopy (Figure 9) reveals that, in the case of the non-coated graphite crucible (here, the nitrogen comes from Si$_3$N$_4$ powder, which was intentionally added to the Si feedstock), the resulting crystal shows a significant formation of Si$_3$N$_4$ precipitates in the central crystal volume and the top region.

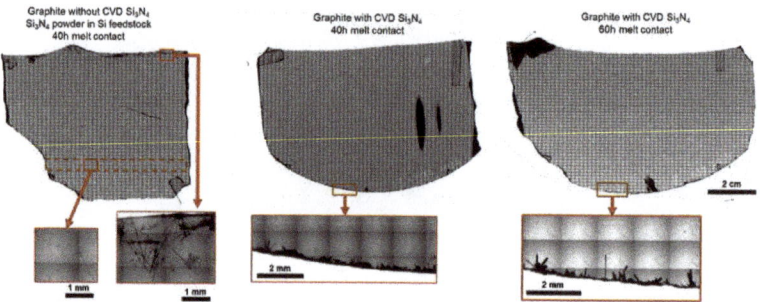

**Figure 9.** IR transmission microscopy images of vertical center slices from silicon crystals grown in graphite crucible with different nitrogen sources (CVD coating and Si$_3$N$_4$ powder addition to feedstock).

In contrast, for both crystals grown in graphite crucibles with CVD coating (40 h and 60 h of melt contact), Si$_3$N$_4$ needle-like structures were mostly detected directly at the interface between CVD Si$_3$N$_4$ coating and silicon crystal. In the crystal volume no precipitates could be found, despite FTIR measurements also here show N contents closely above the solubility limit and therefore Si$_3$N$_4$ precipitate formation could be expected. In this case, the heterogeneous nucleation on the CVD coating surface seems to be favored according to a reduced nucleation energy compared to the homogeneous nucleation of Si$_3$N$_4$ precipitates in the silicon melt. Consequently, the CVD Si$_3$N$_4$ coating could act as a nucleation site for nitrogen-related precipitates. A similar effect of preferred precipitate formation at a Si$_3$N$_4$ based crucible coating used for DS growth of mc-Si ingots was already observed by Trempa et al. [29].

Additionally, a significant change in the resulting CVD Si$_3$N$_4$ coating morphology after long time scale experiment (60 h) could be observed (Figure 10).

Besides the change in surface morphology, the Si$_3$N$_4$ crystal size is also increased. This may indicate that, on the one hand, a recrystallization process of amorphous coating elements and/or a transition of Si$_3$N$_4$ phase occurred during the silicon melt contact. On the other hand, also a deposition of parasitic precipitates on the CVD coating surface could take place due to the dissolution of Si$_3$N$_4$ coating in liquid silicon, followed by an oversaturation of nitrogen in the melt [17].

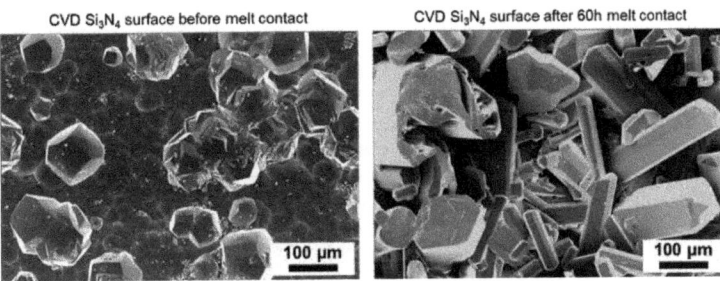

**Figure 10.** Morphology of CVD Si$_3$N$_4$ surface before and after 60 h melt contact.

Nevertheless, even if all the experimental data support the thesis of the beneficial impact of the CVD Si$_3$N$_4$ crucible coating on precipitate formation in lab scale directional solidification processes, the proof of concept must also be further demonstrated for real Cz growth process conditions.

## 4. Conclusions

The basic evaluation of the long-term stability of oxygen-free crucible systems for the application in industrial Cz growth has shown that graphite and NSN-based crucible materials, in combination with a dense CVD Si$_3$N$_4$ surface coating, have promising application potential. According to dipping experiments up to 40 h the CVD coating shows no infiltration and only weak dissolution by the silicon melt, which makes it attractive as a protection coating for the crucible base materials. But also, if the CVD coating would locally fail during process, the tested crucible base materials itself, such as isostatic pressed graphite with low porosity and appropriate mechanical strength as well as pore sealed NSN, have shown an excellent stability at this time scale, which is an important safety issue for industrial application.

These results are confirmed by laboratory scaled DS crystallization experiments with extended melt holding period up to 60 h. Also in this case, both crucible systems show a high robustness against the silicon melt.

With respect to the nitrogen contamination of the silicon melt, which is also a critical issue in Cz growth related to Si$_3$N$_4$ precipitate formation and structure loss events, it could be found that the Si melt is always saturated with nitrogen by the partly dissolved CVD coating. However, all ingots grown in CVD-coated crucibles reveal no precipitates in the ingot volume. This indicates that the CVD Si$_3$N$_4$ coating acts as nucleation site, which could enable the pulling of single crystalline Si ingots out of these crucible systems.

**Author Contributions:** Conceptualization, F.S. and M.T.; methodology, F.S. and M.T.; investigation, F.S., M.T., G.S., R.H., R.W., G.B. and P.M.; resources, G.S., P.G., R.W., G.B. and P.M.; writing—original draft preparation, F.S. and M.T.; writing—review and editing, G.S., R.H., P.G., R.W., G.B., P.M., C.R. and J.F.; visualization, F.S.; supervision, J.F.; project administration, M.T. and C.R.; funding acquisition, M.T. and C.R. All authors have read and agreed to the published version of the manuscript.

**Funding:** These research and development activities are carried out within the "X-treme" project funded by the Bavarian Research Foundation in Germany under contract number AZ-1317-17.

**Data Availability Statement:** Not applicable.

**Acknowledgments:** The project partners Alzchem Trostberg GmbH and Wacker Chemie AG are acknowledged for providing the Si$_3$N$_4$- and Si-powders.

**Conflicts of Interest:** The authors declare no conflict of interest.

## References

1. Schulze, H.J.; Öfner, H.; Niedernostheide, F.J.; Laven, J.G.; Felsl, H.P.; Voss, S.; Schwagmann, A.; Jelinek, M.; Ganagona, N.; Susiti, A.; et al. Use of 300 mm magnetic Czochralski wafers for the fabrication of IGBTs. In Proceedings of the 28th International Symposium on Power Semiconductor Devices and ICs (ISPSD), Prague, Czech Republic, 12–16 June 2016; pp. 355–358.
2. Kajiwara, K.; Harada, K.; Torigoe, K.; Hourai, M. Oxygen Precipitation Properties of Nitrogen-Doped Czochralski Silicon Single Crystals with Low Oxygen Concentration. *Phys. Status Solidi (A)* **2019**, *216*, 1900272. [CrossRef]
3. Hourai, M.; Nagashima, T.; Nishikawa, H.; Sugimura, W.; Ono, T.; Umeno, S. Review and Comments for the Development of Point Defect-Controlled CZ-Si Crystals and Their Application to Future Power Devices. *Phys. Status Solidi (A) Appl. Mater. Sci.* **2019**, *216*, 1800664. [CrossRef]
4. Kiyoi, A.; Kawabata, N.; Nakamura, K.; Fujiwara, Y. Influence of oxygen on trap-limited diffusion of hydrogen in proton-irradiated n -type silicon for power devices. *J. Appl. Phys.* **2021**, *129*, 025701. [CrossRef]
5. Schulze, H.-J.; Öfner, H.; Niedernostheide, F.-J.; Lükermann, F.; Schulz, A. Fabrication of Medium Power Insulated Gate Bipolar Transistors Using 300 mm Magnetic Czochralski Silicon Wafers. *Phys. Status Solidi (A)* **2019**, *216*, 1900235. [CrossRef]
6. Doi, H.; Kikuchi, N.; Oosawa, Y. Chemical vapour deposition coating of crystalline $Si_3N_4$ on a quartz crucible for nitrogen-doped Czochralski silicon crystal growth. *Mater. Sci. Eng. A* **1988**, *105/106*, 465–480. [CrossRef]
7. Watanabe, M.; Usami, T.; Muraoka, H.; Matsuo, S.; Imanishi, Y.; Nagashima, H. Oxygen-Free Silicon Single Crystal Grown from Silicon Nitride Crucible. In *Semiconductor Silicon 1981: Proceedings of the 4th International Symposium on Silicon Materials Science and Technology*; Huff, H.R., Ed.; The Electrochemical Society: Pennington, NJ, USA, 1981; pp. 126–137.
8. Matsuo, S.; Imanishi, Y.; Nagashima, H.; Watanabe, M.; Usami, T.; Muraoka, H. Device made of silicon nitride for pulling single crystal made of silicon and method of manufacturing the same. patent EP 0065122, 24 April 1982.
9. Nakajima, K.; Morishita, K.; Murai, R.; Usami, N. Formation process of $Si_3N_4$ particles on surface of Si ingots grown using silica crucibles with $Si_3N_4$ coating by noncontact crucible method. *J. Crys. Growth* **2014**, *389*, 112–119. [CrossRef]
10. Huguet, C.; Dechamp, C.; Camel, D.; Drevet, B.; Eustathopoulos, N. Study of interactions between silicon and coated graphite for application to photovoltaic silicon processing. *J. Mater. Sci.* **2019**, *54*, 11546–11555. [CrossRef]
11. Camel, D.; Cierniak, E.; Drevet, B.; Cabal, R.; Ponthenier, D.; Eustathopoulos, N. Directional solidification of photovoltaic silicon in re-useable graphite crucibles. *Sol. Energy Mater. Sol. Cells* **2020**, *215*, 110637. [CrossRef]
12. Hendawi, R.; Arnberg, L.; Di Sabatino, M. Novel coatings for graphite materials in PV silicon applications: A study of the surface wettability and interface interactions. *Sol. Energy Mater. Sol. Cells* **2022**, *234*, 111422. [CrossRef]
13. Schneider, V.; Reimann, C.; Friedrich, J.; Müller, G. Nitride bonded silicon nitride as a reusable crucible material for directional solidification of silicon. *Cryst. Res. Technol.* **2016**, *51*, 74–86. [CrossRef]
14. Bellmann, M.P.; Noja, G.; Ciftja, A. Eco-Solar Factory: Utilisation of Kerf-Loss from Silicon Wafer Sawing for the Manufacturing of Silicon Nitride Crucibles. In Proceedings of the 35th European Photovoltaic Solar Energy Conference, Brussels, Belgium, 24–28 September 2018; pp. 501–502.
15. Hendawi, R.; Ciftja, A.; Stokkan, G.; Arnberg, L.; Di Sabatino, M. The effect of preliminary heat treatment on the durability of reaction bonded silicon nitride crucibles for solar cells applications. *J. Crys. Growth* **2020**, *542*, 125670. [CrossRef]
16. Lan, A.; Liu, C.E.; Yang, H.L.; Yu, H.T.; Liu, I.T.; Hsu, H.P.; Lan, C.W. Silicon ingot casting using reusable silicon nitride crucibles made from diamond wire sawing kerf-loss silicon. *J. Crys. Growth* **2019**, *525*, 125184. [CrossRef]
17. Knerer, D. Verfahren und Vorrichtung zum Ziehen eines Einkristalls und Halbleiterscheibe aus Silizium. DE 102018210286 A1, 25 June 2018.
18. Hendawi, R.; Søndenå, R.; Ciftja, A.; Stokkan, G.; Arnberg, L.; Di Sabatino, M. Microstructure and electrical properties of multicrystalline silicon ingots made in silicon nitride crucibles. *AIP Conf. Proc.* **2022**, *2487*, 130005. [CrossRef]
19. Sturm, F.; Trempa, M.; Schuster, G.; Götz, P.; Wagner, R.; Barroso, G.; Meisner, P.; Reimann, C.; Friedrich, J. Material evaluation for engineering a novel crucible setup for the growth of oxygen free Czochralski silicon crystals. *J. Cryst. Growth* **2022**, *584*, 126582. [CrossRef]
20. Mirkhalaf, M.; Yazdani Sarvestani, H.; Yang, Q.; Jakubinek, M.B.; Ashrafi, B. A comparative study of nano-fillers to improve toughness and modulus of polymer-derived ceramics. *Sci. Rep.* **2021**, *11*, 6951. [CrossRef] [PubMed]
21. Barroso, G.; Li, Q.; Bordia, R.K.; Motz, G. Polymeric and ceramic silicon-based coatings—A review. *J. Mater. Chem. A* **2019**, *7*, 1936–1963. [CrossRef]
22. Israel, R.; Voytovych, R.; Protsenko, P.; Drevet, B.; Camel, D.; Eustathopoulos, N. Capillary interactions between molten silicon and porous graphite. *J. Mater. Sci.* **2010**, *45*, 2210–2217. [CrossRef]
23. Chaney, R.E.; Varker, C.J. The Erosion of Materials in Molten Silicon. *J. Electrochem. Soc.* **1976**, *123*, 846–852. [CrossRef]
24. Eustathopoulos, N.; Israel, R.; Drevet, B.; Camel, D. Reactive infiltration by Si: Infiltration versus wetting: Viewpoint set no. 46 "Triple Lines". *Scr. Mater.* **2010**, *62*, 966–971. [CrossRef]
25. Hoseinpur, A.; Safarian, J. Mechanisms of graphite crucible degradation in contact with Si–Al melts at high temperatures and vacuum conditions. *Vacuum* **2020**, *171*, 108993. [CrossRef]
26. Lian, P.; Huang, A.; Gu, H.; Zou, Y.; Fu, L.; Wang, Y. Towards prediction of local corrosion on alumina refractories driven by Marangoni convection. *Ceram. Int.* **2018**, *44*, 1675–1680. [CrossRef]
27. Chaney, R.E.; Varker, C.J. The dissolution of fused silica in molten silicon. *J. Cryst. Growth* **1976**, *33*, 188–190. [CrossRef]

28. Mukai, K. Wetting and Marangoni Effect in Iron and Steelmaking Processes. *ISIJ Int.* **1992**, *32*, 19–25. [CrossRef]
29. Trempa, M.; Reimann, C.; Friedrich, J.; Müller, G. The influence of growth rate on the formation and avoidance of C and N related precipitates during directional solidification of multi crystalline silicon. *J. Crys. Growth* **2010**, *312*, 1517–1524. [CrossRef]

**Disclaimer/Publisher's Note:** The statements, opinions and data contained in all publications are solely those of the individual author(s) and contributor(s) and not of MDPI and/or the editor(s). MDPI and/or the editor(s) disclaim responsibility for any injury to people or property resulting from any ideas, methods, instructions or products referred to in the content.

Article

# Influence of Foreign Salts and Antiscalants on Calcium Carbonate Crystallization

Raghda Hamdi [1,*] and Mohamed Mouldi Tlili [2]

[1] Department of Chemistry, College of Science and Humanities in Al-Kharj, Prince Sattam bin Abdulaziz University, Al-Kharj 11942, Saudi Arabia
[2] Laboratory of Desalination and Natural Water Valorisation (LADVEN), Water Researches and Technologies Center (CERTE), Techno-Park Borj Cedria, BP 273, Soliman 8020, Tunisia
* Correspondence: hamdi.raghda@gmail.com

**Abstract:** For more than a century, crystallization has remained a chief research topic. One of the most undesirable crystallization phenomena is the formation of calcium carbonate scale in drinking and industrial water systems. In this work, the influence of chemical additives on $CaCO_3$ formation—in either nucleation, crystal growth, or inhibition processes—is investigated by using the $CO_2$-degasification method. Chemical additives are foreign salts ($MgCl_2$, $Na_2SO_4$ and $MgSO_4$) to the calco-carbonic system and antiscalants (sodium polyacrylate 'RPI' and sodium-tripolyphosphate 'STPP'). The results show that additives affects both crystallization kinetics and the $CaCO_3$ microstructure. Sulfate and magnesium ions, added separately at constant ionic strength, influence the nucleation step more than the growth of the formed crystallites. Added simultaneously, their effect was accentuated on both nucleation and the growth of $CaCO_3$. Furthermore, antiscalants RPI and STPP affect the crystallization process by greatly delaying the precipitation time and largely increasing the supersaturation coefficient. It was also shown that the calco-carbonic system with additives prefers the heterogeneous nucleation to the homogeneous one. X-ray diffraction patterns show that additives promote the formation of a new crystal polymorph of calcium carbonate as aragonite, in addition to the initial polymorphs formed as calcite and vaterite.

**Keywords:** calcium carbonate; foreign salts; antiscalants; crystallization

**Citation:** Hamdi, R.; Tlili, M.M. Influence of Foreign Salts and Antiscalants on Calcium Carbonate Crystallization. *Crystals* **2023**, *13*, 516. https://doi.org/10.3390/cryst13030516

Academic Editor: José L. Arias

Received: 28 February 2023
Revised: 13 March 2023
Accepted: 15 March 2023
Published: 17 March 2023

**Copyright:** © 2023 by the authors. Licensee MDPI, Basel, Switzerland. This article is an open access article distributed under the terms and conditions of the Creative Commons Attribution (CC BY) license (https:// creativecommons.org/licenses/by/ 4.0/).

## 1. Introduction

Crystallization in the calco-carbonic system $CaCO_3$-$CO_2$-$H_2O$ and the parameters that significantly control the polymorph's formation have been studied for more than a century by several researchers in different scientific disciplines ranging from industrial crystallization [1,2] to biomineralization [3–5].

Calcium carbonate exists in three crystalline forms, which are calcite, aragonite and vaterite, in the order of stability at earth surface conditions. These crystalline phases have different crystal structures and morphologies. Calcite, aragonite and vaterite crystals have rhombohedral, orthorhombic, and hexagonal structures, respectively. In addition, crystalline monohydrocalcite ($CaCO_3.H_2O$) and ikaite ($CaCO_3.6H_2O$) are known. The observation of an intermediated or transient amorphous phase (amorphous calcium carbonate ACC) has also been reported [6–9]. In accordance with the Ostwald's step rule [10], the transformation of the unstable phase (ACC) to a metastable phase (vaterite or aragonite) is followed by the formation of the more stable phase (calcite) through dissolution–crystallization reactions of various degrees of complexity.

The nucleation and growth of these forms are generally related to kinetic and thermodynamic factors such as the degree of supersaturation, solvent, organic, and inorganic additives. $CaCO_3$ crystallization is intensely affected by the presence of such additives, which delay or inhibit nucleation and growth, and change the polymorph morphology [11–13]. The additives' impact is detected in the nucleation and growth steps [14–18]. In the current work, the effect of different types of chemical additives such as magnesium, sulfate, and antiscalants on the $CaCO_3$ crystallization process was studied.

The presence of dissolved ions, for instance magnesium and sulfate, is supposed to produce modification in the kinetic and the driving force of calcium carbonate precipitation [19,20]. They have long been considered as additives in calcium carbonates, usually known to affect calcite growth and dissolution. Magnesium ions have a significant role in the formation and transformation phases of $CaCO_3$ crystals. They inhibit calcite formation [18,21] by decelerating its growth rate [22] and favoring the appearance of aragonite [23] and monohydrocalcite [24], which are less stable phases. They extend the duration of the amorphous phase life [25], and determine the polymorph formation. Magnesium interacts directly with calcium carbonate crystals and affects their morphologies [26]. The presence of magnesium and sulfate, separately and together, modifies the shape of the calcite crystal [27]. Sulfate affects the nucleation even at lower concentrations, by suppressing the formation of $CaCO_3$ crystals and modifying their morphology [28]. It reduces the crystallization rate and favors the formation of the aragonite form [29].

The addition of antiscalants, also known as chemical inhibitors, is an effective method to inhibit the formation of scale [30]. Indeed, antiscalants delay calcium carbonate crystallization through growth sites blocking [31]. They have three mechanisms of action, which are dispersion, chelation, and crystal modification, consisting of the distortion of the crystal to become irregular and less adhesive, thus inhibiting crystal growth at calcite surfaces [32]. The polymer adsorption on the crystal growing surface suppresses the growth rate and displaces the precipitation mode from the most thermodynamic stable phase (calcite) to the metastable phases (aragonite and vaterite) [33]. The addition of the inhibitor resulted in (i) an important inhibition of crystallization, and (ii) a porous or an unconsolidated consistency of precipitates [34]. Therefore, the incorporation of the chemical additives in the structure of the crystal phase form modified its thermodynamic properties.

The present paper aims to improve our understanding of the effect of chemical additives on the $CaCO_3$ crystallization. The changes induced in the kinetic, thermodynamic, and morphology of $CaCO_3$ phases formed after the use of diverse types of chemical additives will be evaluated to explore the complex behaviors of the crystallization described in the calco-carbonic system. Consequently, the experimental process for $CaCO_3$ crystallization based on the fast controlled precipitation (FCP) method [35,36] will be used. The chemical additives employed are three foreign salt ($MgCl_2$, $Na_2SO_4$, and $MgSO_4$) and two antiscalants (sodium polyacrylate and sodium-tripolyphosphate).

## 2. Experimental Section

### 2.1. Solution Preparation

The pure calco-carbonic water (PCCW) was obtained by the dissolution of reagent grade calcium carbonate solid in distilled water under $CO_2$ bubbling, as follows:

$$CaCO_3 + CO_2 + H_2O \leftrightarrows Ca^{2+} + 2HCO_3^- \tag{1}$$

The initial pH value was fixed at 5.9, which corresponds to a supersaturation coefficient equal to 0.25 (calculated using Equation (2)) to maintain the solution undersaturated.

The degree of supersaturation ($\Omega$) with respect to calcite, the most $CaCO_3$ stable form, is defined as follows:

$$\Omega_{CaCO_3} = \frac{[Ca^{2+}] \cdot [CO_3^{2-}] \cdot \gamma_{Ca^{2+}} \cdot \gamma_{CO_3^{2-}}}{K_s(\text{calcite})} \tag{2}$$

where $[i]$, $\gamma_i$ and $K_s$ (calcite) are the ions' concentrations, the activity coefficients and the solubility product of calcite, respectively.

All the additives used were introduced to the PCCW before the precipitation test. Studied foreign salts ($MgCl_2$, $Na_2SO_4$, and $MgSO_4$) quantities were calculated according to the solution ionic strengths (IS) of about 0.012 ($IS_0$), 0.024 ($IS_1$) and 0.036 mol $L^{-1}$

(IS$_2$). The antiscalants used were organic sodium salt of polyacrylate (C$_3$H$_3$NaO$_2$)$_n$ and inorganic sodium tripolyphosphate (Na$_5$P$_3$O$_{10}$), thereafter called, respectively, RPI and STPP. The corresponding quantities used are small and do not affect the ionic strength and the resistivity.

## 2.2. Fast Controlled Precipitation Set Up

The experimental set up of the fast controlled precipitation (FCP) method was presented in Figure 1. Thermostatic water bath was used to maintain the temperature of the FCP test at 30 °C. A 0.5 L volume of pure calco-carbonic water (PCCW) was filled in a polytetrafluoroethylene (PTFE) cell and was stirred at 800 rpm. The solution pH and resistivity values were constantly measured each 5 min, using a pH meter (Hanna HI 110, Hanna Instruments, Woonsocket, RI, USA) and a conductivity meter (Meter Lab CDM210, Radiometer Analytical's, Villeurbanne, France). The measuring electrode positions in the round bottom cell were well controlled. The calcium ion concentration was measured using EDTA complexometric titration.

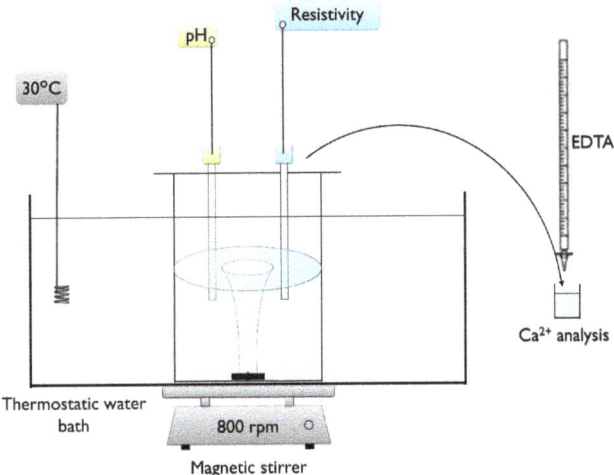

**Figure 1.** Fast controlled precipitation experimental set up.

In the FCP vessel, CaCO$_3$ can precipitate on the cell wall and in the bulk solution. After the completion of the experiment, homogeneously formed precipitate m$_b$ can be recuperated by filtration using cellulose nitrate membrane with 0.45 μm porosity. The amount of Ca$^{2+}$ residual in the solution was measured, and then the total precipitated calcium carbonate m$_t$ was calculated. The heterogeneous mass m$_w$ deposited on the cell wall can be deduced (m$_w$ = m$_t$ − m$_b$), and its rate is calculated as follows:

$$\%_{hete} = \frac{m_w}{m_t} \times 100 \qquad (3)$$

Figure 2 represents a model of an FCP test reporting the evolution of pH and resistivity as a function of time of PCCW throughout the crystallization procedure. The figure emphasizes the limit between three different zones of precipitation. During zone 1, the resistivity is roughly constant and pH values increase up to reach the prenucleation pH "pH$_{prenuc}$" at t$_{prenuc}$. In zone 2, the pH continues rising up with a small change in the resistivity variation speed before precipitation time "t$_{prec}$". The medium is considered to be supersaturated, and calcium carbonate precipitation is thermodynamically possible. In zone 3, the slope of the resistivity temporal evolution diverges significantly, which highlights the precipitation beginning. After attaining the precipitation threshold at "pH$_{prec}$", the pH decreases considerably. Therefore, the precipitation is rapid and followed by a crystal growth stage.

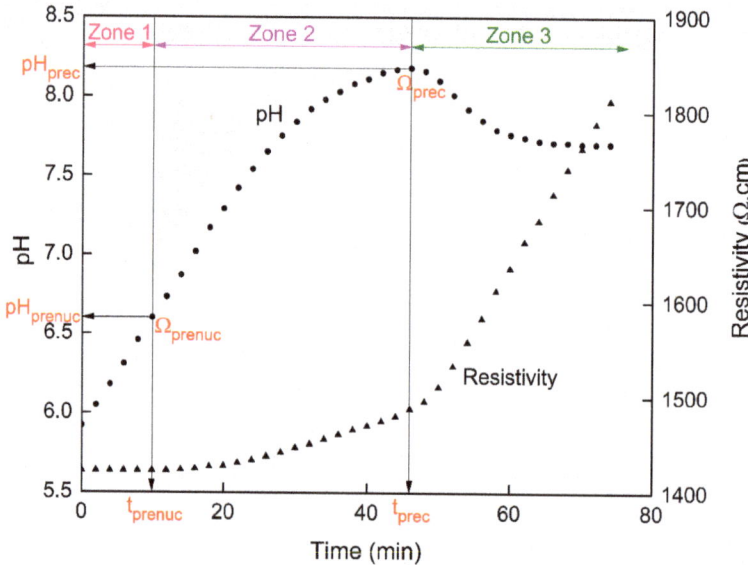

**Figure 2.** Temporal evolution of pH and resistivity curves of a PCCW 400 mg L$^{-1}$ at 30 °C, 800 rpm.

The reproducibility of the experimental measurements was verified for three FCP tests using 400 mg L$^{-1}$ of PCCW, as seen in Figure 3.

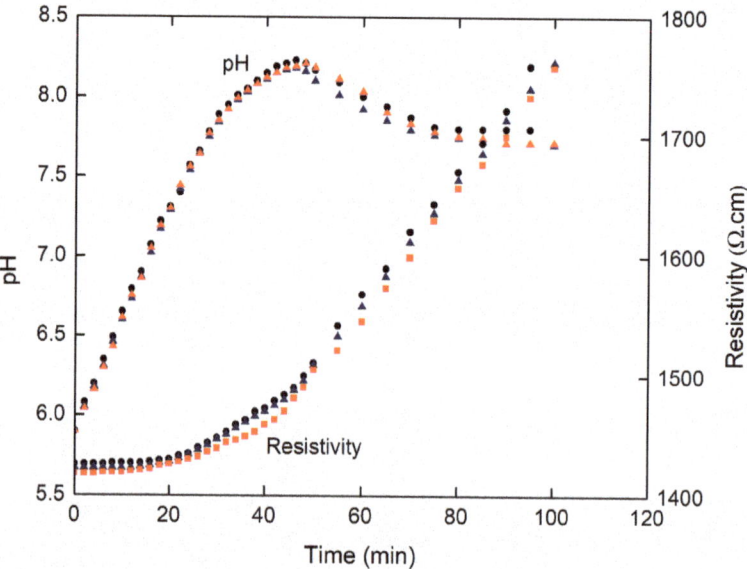

**Figure 3.** Reproducibility of the temporal evolution of pH and resistivity curves for three FCP tests at 30 °C, 800 rpm and PCCW 400 mg L$^{-1}$.

## 2.3. X-ray Diffraction Method for Solid Precipitate Characterization

The X-ray Diffraction (XRD) method is a very efficient and universal tool for determining the structure of crystals. At the end of each FCP test, the precipitate was recuperated by solution filtration through cellulose nitrate membrane with 0.45 µm porosity. The collected samples were dried at ambient conditions before XRD analysis. The analyses were performed using a diffractometer Philips X'PERT PRO (PANalytical, Philips, Amsterdam, The Netherlands) in step-scanning mode using Cu Kα (1.54 Å) radiation. The XRD patterns were chronicled in the angular range 2θ = 10–60°, with a slight step size of 2θ = 0.017° and a fixed count time of 4 s. The software 'X-Pert HighScore Plus' was used to determine the XRD reflection positions. The XRD patterns of the formed precipitates were compared to the joint committee on powder diffraction standards data.

## 3. Results and Discussion

### 3.1. Effect of Initial Calcium Carbonate Concentration

The calcium carbonate concentrations vary in natural waters in a range of 100 to 600 mg L$^{-1}$. Examples include geothermal water (100 mg L$^{-1}$), tap water (200 to 400 mg L$^{-1}$), and saline water (600 mg L$^{-1}$) [37]. Therefore, three solutions with different dissolved CaCO$_3$ content levels (200, 400, 600 mg L$^{-1}$) were prepared to study the effect of initial calcium carbonate amount on the nucleation threshold. Figure 4 reports the temporal evolution of pH and $\Delta_{Resistivity}$ curves during the FCP tests by varying the initial concentration of calcium carbonate. The main results are presented in Table 1.

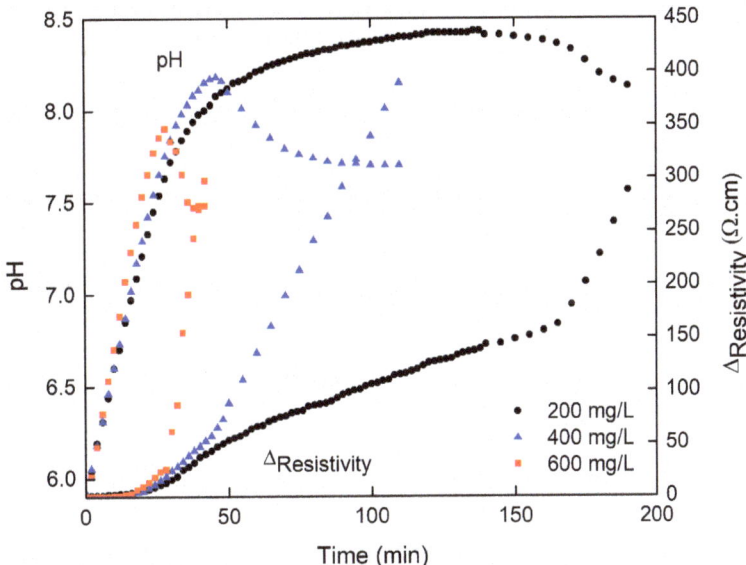

**Figure 4.** pH and $\Delta_{Resistivity}$ vs. time curves by varying the initial amount of calcium carbonate.

**Table 1.** FCP tests results by varying the initial amount of calcium carbonate.

| [CaCO$_3$]$_i$ (mg L$^{-1}$) | $t_{prenuc}$ (min) | $\Omega_{prenuc}$ | $t_{prec}$ (min) | $\Omega_{prec}$ | Δt (min) | %$_{hete}$ |
|---|---|---|---|---|---|---|
| 200 | 18 | 1.1 | 138 | 24 | 120 | 56 |
| 400 | 12 | 1.7 | 46 | 48 | 34 | 45 |
| 600 | 8 | 2.0 | 28 | 51 | 20 | 34 |

As seen in Figure 4, the prenucleation ($t_{prenuc}$) and the precipitation ($t_{prec}$) time decrease as the $CaCO_3$ amount increases. Indeed, $t_{prenuc}$ and $t_{prec}$ are roughly divided by two and by five, respectively, as the initial concentration increases from 200 to 600 mg $L^{-1}$ (Table 1). Ben Amor et al. [38] proved that the precipitation time was delayed from 90 to 7 min after increasing the water hardness from 20 to 50 °F by using a $CO_2$ degasification test. Additionally, Fathi et al. [39] found that the increase in $Ca(HCO_3)_2$ content initially dissolved from 30 to 50 °F accelerates the nucleation time from 20 to 7 min.

Thus, the nucleation step ($\Delta t = t_{prec} - t_{prenuc}$) is shorter as the initial calcium carbonate concentration increases. According to Table 1, this nucleation phase lasts 120 min for 200 mg $L^{-1}$ and only 20 min for 600 mg $L^{-1}$. This can be explained by the fact that the medium saturation state is affected by the initial concentration. According to Raffaella et al. [8], the lifetime of a cluster, before the growth phase, becomes longer when the concentration decreases as a result of the reduction in collision frequency. Indeed, the variation in the solution concentration modifies the frequency of collisions between the different species present (ions, ions pairs, and clusters). Thus, the maximum size that a germ can reach varies according to the different equilibriums between the association and dissociation processes.

From a thermodynamic point of view, the prenucleation threshold is attained at low $\Omega$ values less than two despite the difference in time to reach it (8 min for 600 mg $L^{-1}$ and 18 min for 200 mg $L^{-1}$). However, at the precipitation threshold, the supersaturation coefficients increase from 24 to 51 and the time decreases from 138 to 28 min as the initial concentration of calcium carbonate increases from 200 mg $L^{-1}$ to 600 mg $L^{-1}$, respectively.

Table 1 also reveals that the initial concentration of calcium carbonate $[CaCO_3]_i$ influences the heterogeneous precipitation rate ($\%_{hete}$) of $CaCO_3$. Indeed, $\%_{hete}$ decreases from 56 to 34% as $[CaCO_3]_i$ increases from 200 to 600 mg $L^{-1}$. This is in agreement with Fathi et al. [39], who found that the heterogeneous precipitation rate, which was around 90% for 30 and 40 °F $CaCO_3$ solutions, reached 63% for 50 °F solution. The same observation can be made for Ben Amor et al. [38], who proved that when the water hardness increases from 20 to 50 °F, the $\%_{hete}$ decreases from 98 to 73%. Consequently, the adhesion of tartar on the cell walls is favored for low levels of calcium carbonate initially dissolved, and therefore for low nucleation thresholds. This can be due to the slow aggregation of ion pairs and clusters in low concentrated solutions, so the pre-nucleus remain at a small size for a longer time. Thus, it is more likely that they reach the surface and cling before they become large enough to not adhere to the cell wall.

### 3.2. Effect of Foreign Salts

#### 3.2.1. Influence on the Nucleation Threshold

The most abundant salts found in natural waters are $MgCl_2$, $Na_2SO_4$ and $MgSO_4$. Their effect on the nucleation process of $CaCO_3$ will be investigated. The results of FCP tests obtained in PCCW-$MgCl_2$, PCCW-$Na_2SO_4$ and PCCW-$MgSO_4$ solutions are compared to the results of PCCW solution. The pH and resistivity time curves are represented in Figures 5 and 6.

As shown in Table 2, the prenucleation time $t_{prenuc}$ varies from 12 to 60 min as a function of the solution composition. During this time, the resistivity remains invariable, corresponding to the prenucleation threshold frontier. As expected, the increase of $Mg^{2+}$ and $SO_4^{2-}$ ions concentrations highly delays the nucleation time for both ionic strengths (Table 2). Any precipitation time obtained in the presence of foreign ions was more than the 46 min precipitation time of $CaCO_3$ in a PCCW solution. The $MgSO_4$ ion greatly delayed the precipitation of $CaCO_3$ compared to $Mg^{2+}$ and $SO_4^{2-}$ ions. Indeed, $t_{prec}$ increased from 46 to 214 min when $MgSO_4$ concentration increased from 0 ($IS_0$) to 828 ($IS_2$) mg $L^{-1}$. As to $Mg^{2+}$ and $SO_4^{2-}$, they increased the precipitation time from 46 min (PCCW) to only 102 and 64 min, respectively. Moreover, the ion pairs' formation ($t_{prenuc}$) and aggregation ($t_{prec}$) times were remarkably affected by the addition of each foreign ion. Indeed, the duration of the stable nuclei formation ($\Delta t = t_{prec} - t_{prenuc}$) was extended after the addition of these ions to a PCCW solution. At the same ionic strength ($IS_2$), magnesium had more of an influence on the formation of $CaCO_3$ stable nuclei than sulfate since $\Delta t$ was equivalent to 74 and 40 min for $Mg^{2+}$ and $SO_4^{2-}$ ions, respectively.

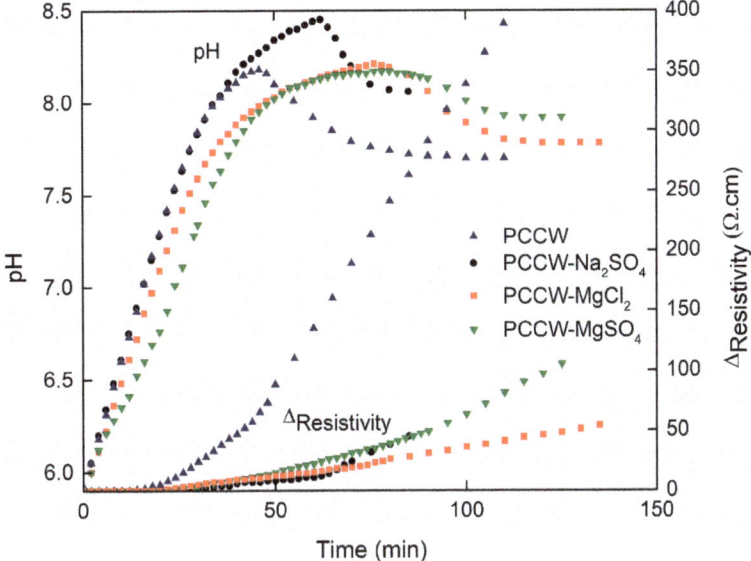

**Figure 5.** pH and $\Delta_{Resistivity}$ vs. time curves as a function of the added salt for $IS_0$ and $IS_1$.

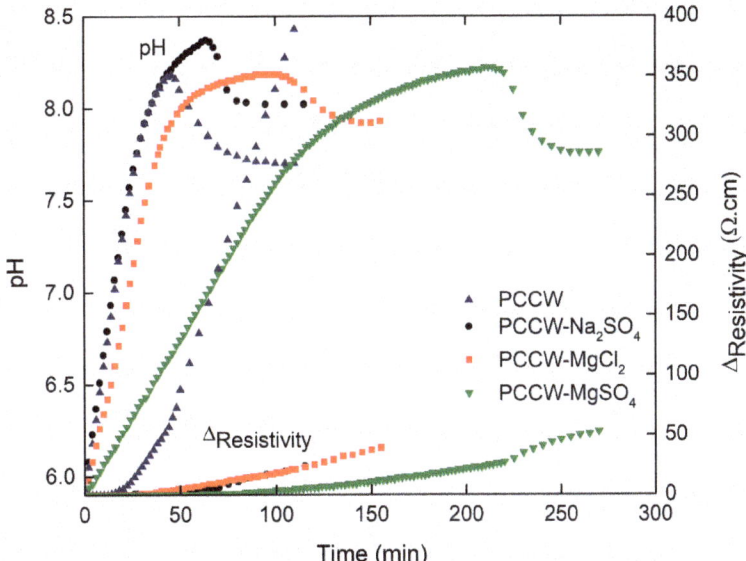

**Figure 6.** pH and $\Delta_{Resistivity}$ vs. time curves as a function of the added salt for $IS_0$ and $IS_2$.

Table 2. FCP tests results by varying the added salt amount.

| IS (mol L$^{-1}$) | Solution | $t_{prenuc}$ (min) | $\Omega_{prenuc}$ | $t_{prec}$ (min) | $\Omega_{prec}$ | $\tau_{prec}$ (%) | %$_{hete}$ |
|---|---|---|---|---|---|---|---|
| IS$_0$ = 0.012 | PCCW | 12 | 1.7 | 46 | 48 | 45 | 45 |
| IS$_1$ = 0.024 | PCCW-Na$_2$SO$_4$ | 20 | 5.0 | 62 | 74 | 27 | 77 |
| | PCCW-MgCl$_2$ | 20 | 3.2 | 76 | 43 | 40 | 80 |
| | PCCW-MgSO$_4$ | 22 | 1.7 | 80 | 35 | 36 | 72 |
| IS$_2$ = 0.036 | PCCW-Na$_2$SO$_4$ | 24 | 8.7 | 64 | 55 | 26 | 72 |
| | PCCW-MgCl$_2$ | 28 | 4.4 | 102 | 35 | 38 | 79 |
| | PCCW-MgSO$_4$ | 60 | 1.6 | 214 | 33 | 35 | 71 |

Furthermore, Table 2 shows that the supersaturation coefficient values of the prenucleation threshold $\Omega_{prenuc}$ were very low and varied from 1 to 9. By contrast, the precipitation threshold was reached at high $\Omega_{prec}$, especially after adding Na$_2$SO$_4$ to the PCCW solution. In fact, at the same ionic strength IS$_1$, SO$_4^{2-}$ ion created a large supersaturation coefficient, which reached up to 1.55 times higher than the PCCW test (reference), while Mg$^{2+}$ ion created a small supersaturation coefficient, which was 0.9 times lesser than that of PCCW. By combining these two ions to obtain MgSO$_4$, $\Omega_{prec}$ went down to 0.7 times less than that of PCCW. Consequently, sulfate ion had more of an impact on the pH of prenucleation and precipitation. As for magnesium ion, it influenced the nucleation time. Thus, the presence of magnesium and sulfate ions affected the nucleation and growth in different ways. Indeed, according to Waly et al. [40], two mechanisms can occur in the presence of magnesium and sulfate ions individually or together. The first mechanism involved is complexation, affecting the activity coefficients of calcium and carbonate ions by decreasing the total present ions for precipitation. The second implicated mechanism is inhibition, where the formed nuclei are inhibited from growing more by blocking the active growth sites. Therefore, both ions can attach to the recently formed calcium carbonate nuclei, causing a decrease in the nuclei growth rate due to the reduction in the available sites for growth by the blockage of the growth steps [41]. The growth can be suppressed entirely if the presence of foreign ions is dominant compared to the presence of calcium and carbonate ions [42].

3.2.2. Influence on the Surface Scaling

The effect of foreign ions on CaCO$_3$ formation has been investigated systematically in this paper, including aspects of kinetics, thermodynamics and behaviors. However, their influence on scale surface deposition has not yet been studied. For this, the total precipitation rate and the heterogeneous precipitation percentage are calculated using the weight method [38] in the absence and presence of foreign ions. As illustrated by Table 2, the total precipitation rate $\tau_{prec}$ decreases as the ionic strength increases. Indeed, $\tau_{prec}$ decreases for IS$_1$ by 5% for magnesium, 18% for sulfate and 9% for the two combined ions (MgSO$_4$) compared to the PCCW solution test. Moreover, $\tau_{prec}$ is practically invariant by increasing the ionic strength for each salt. Furthermore, in the presence of magnesium or sulfate ions or combined magnesium sulfate ions, more than 70% of the CaCO$_3$ precipitates are formed on the different surfaces in contact with water (cell-wall and probes) for both ionic strengths, whereas this heterogeneous percentage does not exceed 45% for pure calco-carbonic water. Subsequently, the introduction of one of these two ions, or both at once, leads to an increase in the precipitation on the walls for both ionic strengths studied. At the macroscopic scale, it would seem that both magnesium and sulfate ions, although presenting opposite charges, have the same effect on the orientation of CaCO$_3$ precipitation to the heterogeneous precipitation, regardless of their modes of action in solution. However, in the study of Chen et al. [43], it was proved that Mg$^{2+}$ ions apparently inhibit both CaCO$_3$ bulk precipitation, regarded as a homogenous process, and CaCO$_3$ surface deposition, regarded as a heterogeneous process. It is known that Mg will bond with water and form a complex at room temperature. This complex prohibits the nucleation and growth of calcite structure carbonate [44].

3.2.3. Influence on the Solid Precipitate Microstructure

The effect of foreign salts addition on the solid precipitate was studied. At the end of each FCP experiment, the precipitate obtained was characterized using the X-ray diffraction analyses. Figures 7–9 present, respectively, the XRD patterns of the precipitate obtained in the absence and presence of $MgCl_2$, $Na_2SO_4$ and $MgSO_4$ for $IS_1$ and $IS_2$. $CaCO_3$ precipitated in the PCCW is mainly vaterite, with a little fraction of calcite. The formation of these calcite crystals is inhibited when $Mg^{2+}$ ions (from $IS_1$) and $SO_4^{2-}$ ions are separately added (Figures 7 and 8). Nevertheless, the presence of $Mg^{2+}$ promotes the aragonite (Figure 7), while $SO_4^{2-}$ has no effect and vaterite remains the main $CaCO_3$ polymorphous (Figure 8). The action mode of magnesium ions can be by (i) the substitution of calcium ions, or (ii) the insertion in or (iii) adsorption on the $CaCO_3$ lattice [45]. If substitution or insertion leads to magnesian calcite, the adsorption at the earlier stage of nucleation and growth affects the $CaCO_3$ crystalline structure, which is the most probable action mode for our case. Indeed, the interactions between Mg and $CaCO_3$ nuclei are strong because they are improved by electrostatic interactions [46]. Moreover, $Mg^{2+}$ competes with $Ca^{2+}$ due to having the same charge, which favors the magnesium adsorption on the surface of calcite and vaterite [47]. The amount of $Mg^{2+}$ being adsorbed onto $CaCO_3$ crystal active growth sites can be influenced by the precipitation rate and the local solution chemistry [22]. In the presence of $MgSO_4$ at $IS_1$, the vaterite is completely transformed into aragonite (Figure 9) as for the precipitate obtained in $MgCl_2$-solution (Figure 7). This confirms the role of magnesium ions in favoring the aragonite instead of vaterite. However, at such IS and in the presence of $SO_4^{2-}$, $Mg^{2+}$ ions could not inhibit the crystallization of the calcite form (Figure 9). By increasing the ionic strength ($IS_2$) in the case of $MgSO_4$, calcite is gradually inhibited.

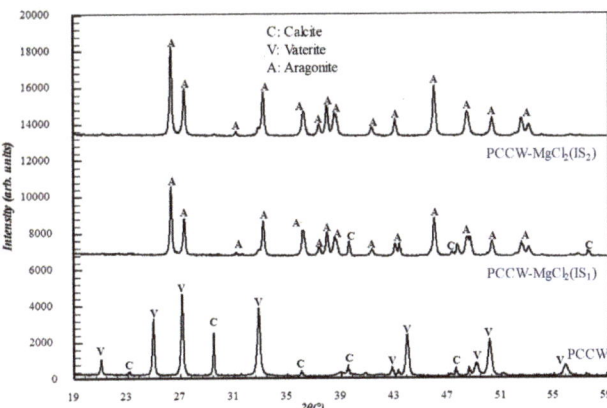

**Figure 7.** XRD patterns of the precipitate formed in the presence of $MgCl_2$ by varying IS.

*3.3. Effect of Antiscalant*

3.3.1. Influence on the Nucleation Threshold

Calcium carbonate scale deposits in natural water installations are a stimulating issue, mainly by obstructing the water flow. The formation of such undesirable deposits can be inhibited by the addition of antiscalants such as organic sodium salt of polyacrylate (RPI) and inorganic sodium tripolyphosphate (STPP). The results of FCP tests obtained in PCCW-RPI and PCCW-STPP solutions were compared to the results of the PCCW solution.

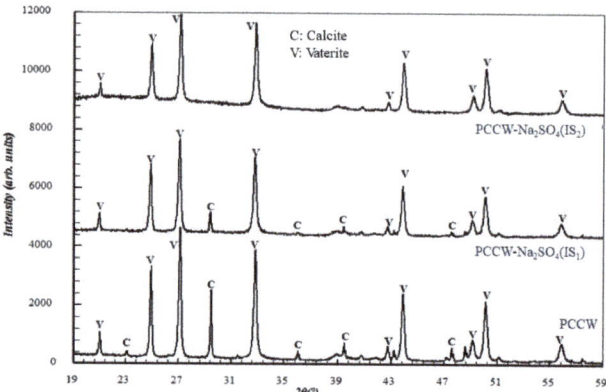

**Figure 8.** XRD patterns of the precipitate formed in the presence of $Na_2SO_4$ by varying IS.

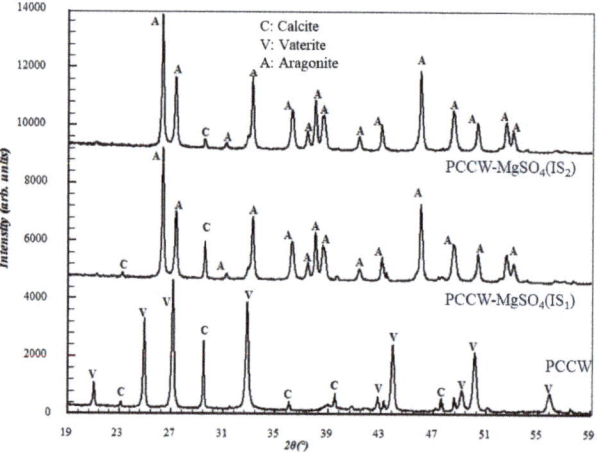

**Figure 9.** XRD patterns of the precipitate formed in the presence of $MgSO_4$ by varying IS.

Figures 10 and 11 report the temporal evolution of pH and $\Delta_{Resistivity}$ as a function of RPI and STPP antiscalants amount. They show that the presence of both antiscalants RPI and STPP significantly influences the kinetics of the nucleation process. Indeed, $t_{prenuc}$ increased from 12 min for PCCW to 20 min in the presence of 4 ppm RPI and 0.8 ppm STPP, as seen in Table 3. During this time, the resistivity remained invariable corresponding to the prenucleation threshold limit. In addition, the increase of the antiscalant amount added to PCCW led to a higher precipitation time. In addition, STPP greatly delayed the precipitation of $CaCO_3$ compared to RPI. In fact, $t_{prec}$ increased from 46 to 210 min and to 365 min for 4 ppm RPI and 0.8 ppm STPP, respectively. Furthermore, the duration of the stable nuclei formation ($\Delta t = t_{prec} - t_{prenuc}$) was longer as the antiscalant amount was larger. Therefore, the distinction between the prenucleation and precipitation thresholds became easier. Moreover, the prenucleation threshold was always attained at $\Omega_{prenuc}$ values less than 7. Conversely, the precipitation threshold was reached at very high $\Omega_{prec}$, especially after adding STPP to PCCW solution, which went up to 353 at 0.8 ppm. Consequently, both antiscalants affected the kinetics and thermodynamics of calcium carbonate prenucleation and precipitation with different manners.

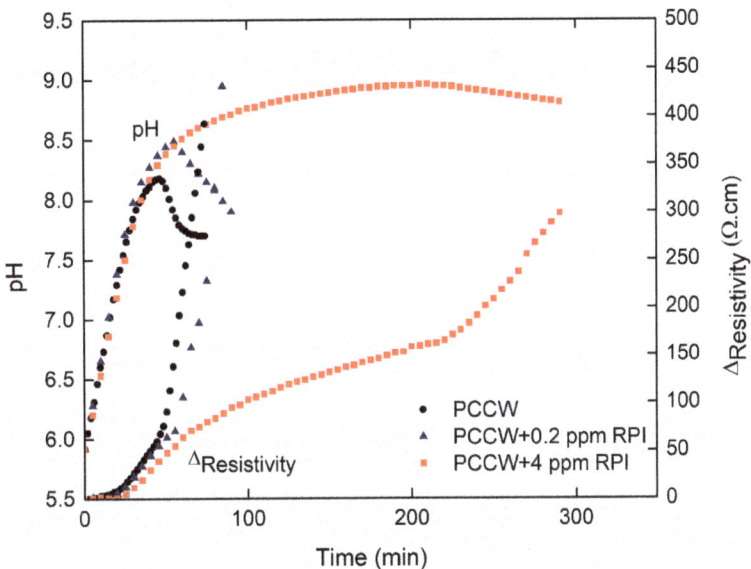

**Figure 10.** pH and $\Delta_{Resistivity}$ vs. time curves as a function of RPI amount.

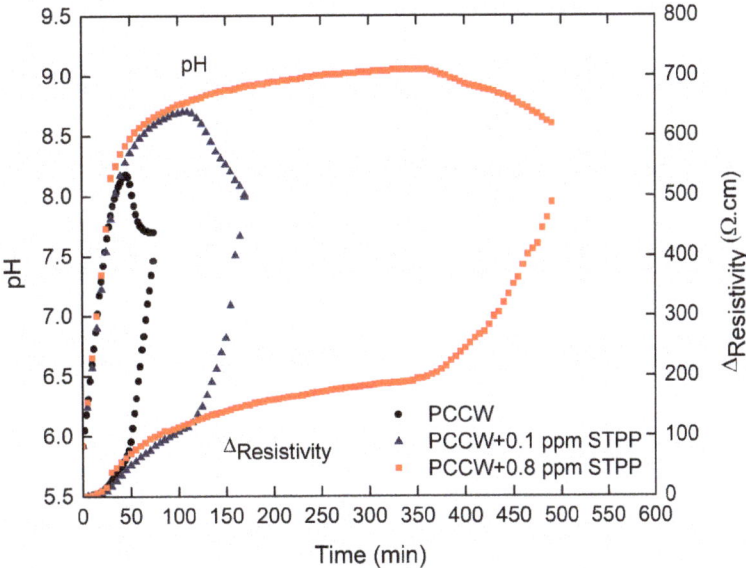

**Figure 11.** pH and $\Delta_{Resistivity}$ vs. time curves as a function of STPP amount.

3.3.2. Influence on the Surface Scaling

The influence of the two antiscalants on scale surface deposition was studied by calculating the total precipitation rate and the heterogeneous precipitation percentage using the weight method. As seen in Table 3, after the addition of each antiscalant, the total precipitation rate $\tau_{prec}$ decreased remarkably from 45% for PCCW to 29% and 33% in the presence of RPI (4 ppm) and STPP (0.8 ppm), respectively. Moreover, for the same antiscalant, $\tau_{prec}$ decreased by increasing its amount added to PCCW. Furthermore, both antiscalants greatly affected the scale adherence to the cell wall and probes. Indeed, %$_{hete}$ increased from 45%

for PCCW to 79% and to 70% after adding 4 ppm of RPI and 0.8 ppm of STPP, respectively. Despite RPI being organic and STPP being inorganic antiscalants, they both favorably promote precipitation on the cell wall detrimentally to the bulk solution scaling.

Table 3. FCP test results by varying the antiscalant amount.

| Solution | $t_{prenuc}$ (min) | $\Omega_{prenuc}$ | $t_{prec}$ (min) | $\Omega_{prec}$ | $\tau_{prec}$ | %$_{hete}$ |
|---|---|---|---|---|---|---|
| PCCW | 12 | 1.7 | 46 | 48 | 45 | 45 |
| PCCW-RPI (0.2 ppm) | 15 | 3 | 55 | 97 | 34 | 74 |
| PCCW-RPI (4 ppm) | 20 | 5 | 210 | 286 | 29 | 79 |
| PCCW-STPP (0.1 ppm) | 15 | 2.5 | 110 | 157 | 37 | 66 |
| PCCW-STPP (0.8 ppm) | 20 | 7 | 365 | 353 | 33 | 70 |

3.3.3. Influence on the Solid Precipitate Microstructure

The X-ray diffraction patterns of the precipitate formed in the presence of RPI and STPP are presented in Figures 12 and 13. The patterns of deposit scale obtained in PCCW reveal the formation of calcite with majority vaterite. After the addition of 4 ppm RPI, the crystallization was orientated towards the formation of calcite and to the complete inhibition of vaterite. In the presence of 0.8 ppm STPP, the phases formed remained vaterite and calcite, with the appearance of a new aragonite phase. Consequently, the two antiscalants acted on the $CaCO_3$ precipitation in different ways. XRD characterizations proved that the presence of antiscalant can affect the morphology and the crystal shape of calcium carbonate [48].

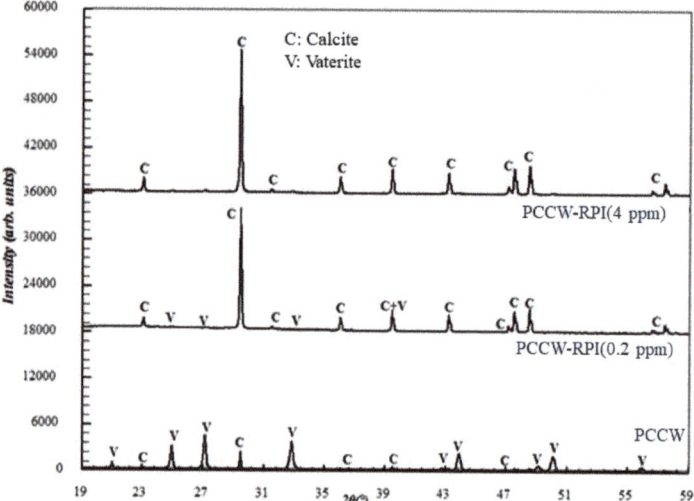

Figure 12. XRD patterns of the precipitate formed in the presence of RPI.

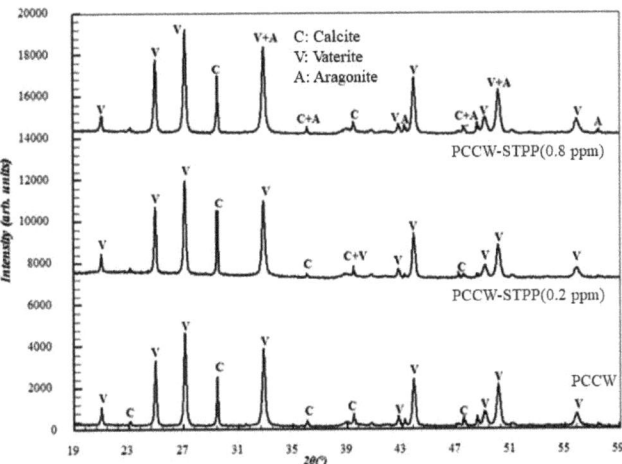

Figure 13. XRD patterns of the precipitate formed in the presence of STPP.

## 4. Conclusions

Chemical additives play a significant role in several stages of the formation and transformation of $CaCO_3$ crystals. In the present paper, the fast controlled precipitation method was used to study the influence of chemical additives on the $CaCO_3$ crystallization process. The additives employed were three foreign salts ($MgCl_2$, $Na_2SO_4$ and $MgSO_4$) and two antiscalants (sodium polyacrylate and sodium-tripolyphosphate). The results showed that the nucleation time was retarded after the addition of each foreign salt, regardless of its ionic strength value. The $MgSO_4$ ion greatly delayed the precipitation of $CaCO_3$ compared to $Mg^{2+}$ and $SO_4^{2-}$ ions. Moreover, the sulfate ion had more impact on the pH of prenucleation and precipitation. As for magnesium ion, it influenced the nucleation time. Thus, the presence of magnesium and sulfate ions affected nucleation and growth through different ways. At the macroscopic scale, it would seem that both magnesium and sulfate ions, although presenting opposite charges, had the same effect on the orientation of $CaCO_3$ precipitation to the heterogeneous precipitation, regardless of their modes of action in solution. Furthermore, both antiscalants RPI and STPP had a great effect on the crystallization kinetic by greatly delaying the precipitation time. Indeed, the prenucleation stage duration ($\Delta t = t_{prec} - t_{prenuc}$) was longer when the antiscalant concentration was larger. Therefore, the distinction between the prenucleation and precipitation thresholds became easier. In addition, the presence of each antiscalant affected the precipitation threshold remarkably, leading to a large variation in the supersaturation coefficient. In addition, the chemical inhibition favorably promoted precipitation on the cell wall, detrimental to the bulk solution scaling. Finally, the X-ray diffraction patterns of deposit scale formed in the absence and presence of each chemical additive revealed the formation of different crystal forms of calcium carbonate, which were calcite, vaterite and aragonite.

**Author Contributions:** Conceptualization, R.H. and M.M.T.; methodology, R.H. and M.M.T.; software, R.H.; validation, R.H. and M.M.T.; formal analysis, R.H. and M.M.T.; investigation, R.H.; resources, R.H. and M.M.T.; data curation, R.H.; writing—original draft preparation, R.H.; writing—review and editing, R.H. and M.M.T.; visualization, R.H.; supervision, R.H. and M.M.T.; project administration, R.H. and M.M.T.; funding acquisition, R.H. and M.M.T. All authors have read and agreed to the published version of the manuscript.

**Funding:** This research was funded by the Deputyship for Research and Innovation, Ministry of Education in Saudi Arabia, Project number (IF2/PSAU/2022/01/21605).

**Institutional Review Board Statement:** Not applicable.

**Informed Consent Statement:** Not applicable.

**Data Availability Statement:** Data are available on reasonable request.

**Acknowledgments:** The authors extend their appreciation to the Deputyship for Research and Innovation, Ministry of Education in Saudi Arabia for funding this research work through the Project number (IF2/PSAU/2022/01/21605).

**Conflicts of Interest:** The authors declare no conflict of interest.

## References

1. Lu, J.; Ruan, S.; Liu, Y.; Wang, T.; Zeng, Q.; Yan, D. Morphological characteristics of calcium carbonate crystallization in $CO_2$ pre-cured aerated concrete. *RSC Adv.* **2022**, *12*, 14610–14620. [CrossRef] [PubMed]
2. Feng, Z.; Yang, T.; Dong, S.; Wu, T.; Jin, W.; Wu, Z.; Wang, B.; Liang, T.; Cao, L.; Yu, L. Industrially synthesized biosafe vaterite hollow $CaCO_3$ for controllable delivery of anticancer drugs. *Mater. Today Chem.* **2022**, *24*, 100917. [CrossRef]
3. Lu, H.; Huang, Y.-C.; Hunger, J.; Gebauer, D.; Cölfen, H.; Bonn, M. Role of Water in $CaCO_3$ Biomineralization. *J. Am. Chem. Soc.* **2021**, *143*, 1758–1762. [CrossRef] [PubMed]
4. Kontrec, J.; Tomašić, N.; Mlinarić, N.M.; Kralj, D.; Džakula, B.N. Effect of pH and Type of Stirring on the Spontaneous Precipitation of $CaCO_3$ at Identical Initial Supersaturation, Ionic Strength and $a(Ca^{2+})/a(CO_3^{2-})$ Ratio. *Crystals* **2021**, *11*, 1075. [CrossRef]
5. Dobberschütz, S.; Nielsen, M.R.; Sand, K.K.; Civioc, R.; Bovet, N.; Stipp, S.L.S.; Andersson, M.P. The mechanisms of crystal growth inhibition by organic and inorganic inhibitors. *Nat. Commun.* **2018**, *9*, 1578. [CrossRef]
6. Lázár, A.; Molnár, Z.; Demény, A.; Kótai, L.; Trif, L.; Béres, K.A.; Bódis, E.; Bortel, G.; Aradi, L.E.; Karlik, M.; et al. Insights into the amorphous calcium carbonate (ACC) → ikaite → calcite transformations. *CrystEngComm* **2023**, *25*, 738–750. [CrossRef]
7. Dong, L.; Xu, Y.-J.; Sui, C.; Zhao, Y.; Mao, L.-B.; Gebauer, D.; Rosenberg, R.; Avaro, J.; Wu, Y.-D.; Gao, H.-L.; et al. Highly hydrated paramagnetic amorphous calcium carbonate nanoclusters as an MRI contrast agent. *Nat. Commun.* **2022**, *13*, 5088. [CrossRef]
8. Raffaella, D.; Paolo, R.; Julian, D.G.; David, Q.; Denis, G. Stable prenucleation mineral clusters are liquid-like ionic polymers. *Nat. Commun.* **2011**, *2*, 590.
9. Gebauer, D.; Völkel, A.; Cölfen, H. Stable prenucleation calcium carbonate clusters. *Science* **2008**, *322*, 1819–1822. [CrossRef]
10. Ostwald, W. Studies on formation and transformation of solid materials (Studien uber die bildung und umwandlung fester korper). *Z. Phys. Chem.* **1897**, *22*, 289–330. [CrossRef]
11. Hamdi, R.; Tlili, M.M. Influence of foreign salts on the $CaCO_3$ pre-nucleation stage: Application of the conductometric method. *CrystEngComm* **2022**, *24*, 3256–3267. [CrossRef]
12. Wolf, S.L.P.; Jähme, K.; Gebauer, D. Synergy of $Mg^{2+}$ and poly(aspartic acid) in additive-controlled calcium carbonate precipitation. *CrystEngComm* **2015**, *17*, 6857–6862. [CrossRef]
13. Song, R.-Q.; Cölfen, H. Additive controlled crystallization. *CrystEngComm* **2011**, *13*, 1249–1276. [CrossRef]
14. Raj, K.S.; Devi, M.N.; Palanisamy, K.; Subramanian, V.K. Individual and synergetic effect of EDTA and NTA on polymorphism and morphology of $CaCO_3$ crystallization process in presence of barium. *J. Solid State Chem.* **2021**, *302*, 122026.
15. Sancho-Tomás, M.; Fermani, S.; Durán-Olivencia, M.A.; Otálora, F.; Gómez-Morales, J.; Falini, G.; García-Ruiz, J.M. Influence of Charged Polypeptides on Nucleation and Growth of $CaCO_3$ Evaluated by Counterdiffusion Experiments. *Cryst. Growth Des.* **2013**, *13*, 3884–3891. [CrossRef]
16. Ihli, J.; Kim, Y.-Y.; Noel, E.H.; Meldrum, F.C. The Effect of Additives on Amorphous Calcium Carbonate (ACC): Janus Behavior in Solution and the Solid State. *Adv. Funct. Mater.* **2013**, *23*, 1575–1585. [CrossRef]
17. Xiang, L.; Xiang, Y.; Wang, Z.G.; Jin, Y. Influence of chemical additives on the formation of super-fine calcium carbonate. *Powder Technol.* **2002**, *126*, 129–133. [CrossRef]
18. Davis, K.J.; Dove, P.M.; De Yoreo, J.J. The Role of $Mg^{2+}$ as an Impurity in Calcite Growth. *Science* **2000**, *290*, 1134–1137. [CrossRef]
19. Loste, E.; Wilson, R.M.; Seshadri, R.; Meldrum, F.C. The role of magnesium in stabilising amorphous calcium carbonate and controlling calcite morphologies. *J. Cryst. Growth* **2003**, *254*, 206–218. [CrossRef]
20. Chong, T.H.; Sheikholeslami, R. Thermodynamics and kinetics for mixed calcium carbonate and calcium sulfate precipitation. *Chem. Eng. Sci.* **2001**, *56*, 5391–5400. [CrossRef]
21. Chen, T.; Neville, A.; Yuan, M. Assessing the effect of $Mg^{2+}$ on $CaCO_3$ scale formation–bulk precipitation and surface deposition. *J. Cryst. Growth* **2005**, *275*, e1341–e1347. [CrossRef]
22. Zhang, Y.; Dawe, R.A. Influence of $Mg^{2+}$ on the kinetics of calcite precipitation and calcite crystal morphology. *Chem. Geol.* **2000**, *163*, 129–138. [CrossRef]
23. Niedermayr, A.; Köhler, S.J.; Dietzel, M. Impacts of aqueous carbonate accumulation rate, magnesium and polyaspartic acid on calcium carbonate formation (6–40 °C). *Chem. Geol.* **2013**, *340*, 105–120. [CrossRef]
24. Blue, C.R.; Giuffre, A.; Mergelsberg, S.; Han, N.; De Yoreo, J.J.; Dove, P.M. Chemical and physical controls on the transformation of amorphous calcium carbonate into crystalline $CaCO_3$ polymorphs. *Geochim. Cosmochim. Acta* **2017**, *196*, 179–196. [CrossRef]
25. Rodriguez-Blanco, J.D.; Shaw, S.; Bots, P.; Roncal-Herrero, T.; Benning, L.G. The role of Mg in the crystallization of monohydrocalcite. *Geochim. Cosmochim. Acta* **2014**, *127*, 204–220. [CrossRef]

26. Zhang, A.; Xie, H.; Liu, N.; Chen, B.-L.; Ping, H.; Fu, Z.-Y.; Su, B.-L. Crystallization of calcium carbonate under the influences of casein and magnesium ions. *RSC Adv.* **2016**, *6*, 110362–110366. [CrossRef]
27. Nielsen, M.R.; Sand, K.K.; Rodriguez-Blanco, J.D.; Bovet, N.; Generosi, J.; Dalby, K.N.; Stipp, S.L.S. Inhibition of Calcite Growth: Combined Effects of $Mg^{2+}$ and $SO_4^{2-}$. *Cryst. Growth Des.* **2016**, *16*, 6199–6207. [CrossRef]
28. Sheikholeslami, R.; Ong, H.W.K. Kinetics and thermodynamics of calcium carbonate and calcium sulfate at salinities up to 1.5 M. *Desalination* **2003**, *157*, 217–234. [CrossRef]
29. Tang, Y.; Zhang, F.; Cao, Z.; Jing, W.; Chen, Y. Crystallization of $CaCO_3$ in the presence of sulfate and additives: Experimental and molecular dynamics simulation studies. *J. Colloid Interface Sci.* **2012**, *377*, 430–437. [CrossRef]
30. Hamdi, R.; Tlili, M.M. Investigation of scale inhibitors effect on calcium carbonate nucleation process. *Desalination Water Treat.* **2019**, *160*, 14–22. [CrossRef]
31. Jain, T.; Sanchez, E.; Owens-Bennett, E.; Trussell, R.; Walker, S.; Liu, H. Impacts of antiscalants on the formation of calcium solids: Implication on scaling potential of desalination concentrate. *Environ. Sci. Water Res. Technol.* **2019**, *5*, 1285–1294. [CrossRef]
32. Bai, S.; Naren, G.; Nakano, M.; Okaue, Y.; Yokoyama, T. Effect of polysilicic acid on the precipitation of calcium carbonate. *Colloids Surf. A Physicochem. Eng. Asp.* **2014**, *445*, 54–58. [CrossRef]
33. Chhim, N.; Haddad, E.; Neveux, T.; Bouteleux, C.; Teychené, S.; Biscans, B. Performance of green antiscalants and their mixtures in controlled calcium carbonate precipitation conditions reproducing industrial cooling circuits. *Water Res.* **2020**, *186*, 116334. [CrossRef] [PubMed]
34. Eichinger, S.; Boch, R.; Leis, A.; Baldermann, A.; Domberger, G.; Schwab, C.; Dietzel, M. Green inhibitors reduce unwanted calcium carbonate precipitation: Implications for technical settings. *Water Res.* **2022**, *208*, 117850. [CrossRef]
35. Gauthier, G.; Chao, Y.; Horner, O.; Alos-Ramos, O.; Hui, F.; Lédion, J.; Perrot, H. Application of the Fast Controlled Precipitation method to assess the scale-forming ability of raw river waters. *Desalination* **2012**, *299*, 89–95. [CrossRef]
36. Lédion, J.; François, B.; Vienne, J. Characterization of the scaling properties of water by fast controlled precipitation test. *Eur. J. Water Qual.* **1997**, *28*, 15–35. [CrossRef]
37. Rosset, R.; Sok, P.; Poindessous, G.; Amor, M.B.; Walha, K. Caractérisation de la compacité des dépôts de carbonate de calcium d'eaux géothermales du Sud tunisien par impédancemétrie. *Comptes Rendus L'académie Sci. Ser. IIC Chem.* **1998**, *1*, 751–759. [CrossRef]
38. Amor, M.B.; Zgolli, D.; Tlili, M.M.; Manzola, A.S. Influence of water hardness, substrate nature and temperature on heterogeneous calcium carbonate nucleation. *Desalination* **2004**, *166*, 79–84. [CrossRef]
39. Fathi, A.; Mohamed, T.; Claude, G.; Maurin, G.; Mohamed, B.A. Effect of a magnetic water treatment on homogeneous and heterogeneous precipitation of calcium carbonate. *Water Res.* **2006**, *40*, 1941–1950. [CrossRef]
40. Waly, T.; Kennedy, M.D.; Witkamp, G.-J.; Amy, G.; Schippers, J.C. The role of inorganic ions in the calcium carbonate scaling of seawater reverse osmosis systems. *Desalination* **2012**, *284*, 279–287. [CrossRef]
41. Mullin, J.W. *Crystallization*; Butterworth-Heinemann: Oxford, UK, 2001.
42. Söhnel, O.; Garside, J.H. *Precipitation: Basic Principles and Industrial Applications*; Butterworth-Heinemann: Oxford, UK, 1992.
43. Chen, T.; Neville, A.; Yuan, M. Influence of on formation—Bulk precipitation and surface deposition. *Chem. Eng. Sci.* **2006**, *61*, 5318–5327. [CrossRef]
44. Fang, Y.; Zhang, F.; Farfan, G.A.; Xu, H. Low-Temperature Synthesis of Disordered Dolomite and High-Magnesium Calcite in Ethanol–Water Solutions: The Solvation Effect and Implications. *ACS Omega* **2022**, *7*, 281–292. [CrossRef]
45. Tlili, M.M.; Amor, M.B.; Gabrielli, C.; Joiret, S.; Maurin, G.; Rousseau, P. Study of Electrochemical Deposition of $CaCO_3$ by In Situ Raman Spectroscopy: II. Influence of the Solution Composition. *J. Electrochem. Soc.* **2003**, *150*, C485–C493. [CrossRef]
46. Zahid, A. *Mineral Scale Formation and Inhibition*; Plenum Press: New York, NY, USA, 1995.
47. Compton, R.G.; Brown, C.A. The Inhibition of Calcite Dissolution/Precipitation: $Mg^{2+}$ Cations. *J. Colloid Interface Sci.* **1994**, *165*, 445–449. [CrossRef]
48. Sheng, K.; Ge, H.; Huang, X.; Zhang, Y.; Song, Y.; Ge, F.; Zhao, Y.; Meng, X. Formation and Inhibition of Calcium Carbonate Crystals under Cathodic Polarization Conditions. *Crystals* **2020**, *10*, 275. [CrossRef]

**Disclaimer/Publisher's Note:** The statements, opinions and data contained in all publications are solely those of the individual author(s) and contributor(s) and not of MDPI and/or the editor(s). MDPI and/or the editor(s) disclaim responsibility for any injury to people or property resulting from any ideas, methods, instructions or products referred to in the content.

Article

# Optimizing Struvite Crystallization at High Stirring Rates

Atef Korchef [1,*], Salwa Abouda [2,3] and Imen Souid [1]

1. College of Sciences, King Khalid University (KKU), P.O. Box 960, Abha 61421, Saudi Arabia
2. LVMU, Centre National de Recherches en Sciences des Matériaux, Technopole de Borj-Cédria, BP 73, Soliman 8027, Tunisia
3. Centre de Recherches et Technologies des Eaux, Technopole Borj-Cédria, BP 273, Route Touristique de Soliman, Nabeul 8020, Tunisia
* Correspondence: akorchef@kku.edu.sa

**Abstract:** Phosphorus and ammonium can both be recovered in the presence of magnesium through struvite ($MgNH_4PO_4 \cdot 6H_2O$) crystallization. The present work aimed to optimize struvite crystallization at turbulent solution flow. Struvite was crystallized by magnetic stirring at different initial phosphorus concentrations between 200 and 800 mg·L$^{-1}$ and high stirring rates between 100 and 700 rpm. The crystals obtained were analyzed by powder X-ray diffraction, Fourier-transform infrared spectroscopy, and scanning electron microscopy. For all experiments, the only phase detected was struvite. It was shown that for an initial phosphorus concentration of 200 mg·L$^{-1}$, increasing the stirring rate to 500 rpm accelerated the precipitation of struvite, improved the phosphorus removal efficiency, and obtained larger struvite crystals. A decrease in the phosphorus removal efficiency and smaller struvite crystals were obtained at higher stirring rates. This was attributed to the solution turbulence. The limiting effect of turbulence could be overcome by enhancing the initial phosphorus concentration or by lowering the stirring rate. The highest phosphorus removal efficiency (~99%) through large struvite crystals (~400 μm in size) was obtained for an initial phosphorus concentration of 800 mg·L$^{-1}$ and a stirring rate of 100 rpm.

**Keywords:** struvite; fertilizer; phosphorus; ammonium; wastewater; stirring; turbulence

## 1. Introduction

The world's surface is covered by water to a great extent, most of it in the oceans. Only 3% of this water is fresh, of which only ~0.5% is available [1]. The rest (2.5%) is locked up in glaciers and soil or is too deep below the earth's surface to be easily extracted at an affordable cost. Unfortunately, freshwater can be polluted by industrial effluents, animal discharges, and chemical fertilizers. Phosphorus and nitrogen are two of the main nutrients found in water. Their presence in high concentrations pollutes the water and caused eutrophication. For this reason, their recycling has gained importance, especially since the European legislation against water pollution has imposed a phosphorus concentration below 2 mg·L$^{-1}$ [2]. Phosphorus and nitrogen can be simultaneously recycled, in a basic medium and in the presence of magnesium, through the crystallization of a sparingly soluble salt, struvite ($MgNH_4PO_4 \cdot 6H_2O$). This prevents eutrophication and varies phosphate resources. Struvite is recognized as a valuable fertilizer [3]. Recently, it was shown that struvite could be used effectively as a fire-extinguishing agent [4]. For these reasons, struvite crystallization from synthetic solutions and real wastewater has been intensively studied [5–8].

The control of both the nucleation and growth of struvite crystals is difficult because they depend on various physical and chemical parameters such as ion transfer between the liquid and solid phases, reaction kinetics, temperature, supersaturation, pH of the solution, concentrations of struvite constituent ions, foreign ions ($Ca^{2+}$, $Fe^{2+}$, $Cu^{2+}$, and others), and flow turbulence. The optimal temperature reported in the literature for struvite crystallization was in the range of 15–30 °C [9,10]. Indeed, higher temperatures decreased the struvite solubility and affected both its morphology and purity [11]. The precipitation pH used was in the range of 8–11 [9,10,12]. That is, alkaline solutions promoted struvite crystallization.

At a given phosphate concentration, enhancing the magnesium concentration in the solution lowered the struvite crystallization pH and considerably increased the effectiveness of phosphate and ammonium removal [13]. Depending on the added concentration, the addition of phosphate to the solution at a fixed magnesium concentration can either delay or entirely inhibit struvite crystallization [13]. An increase in magnesium or phosphate concentration affected the struvite crystals' purity, shape, and size. However, adding ammonium in excess to the solution favored struvite crystallization without affecting the purity of the obtained struvite crystals [13]. Jaffer et al. [14] showed that struvite crystallization occurred at a sewage treatment plant when the molar ratio of Mg:P was at least 1.05:1 and for lower ratios, phosphorus removal decreased but not exclusively as struvite. Kruk et al. [15] studied the crystallization of struvite from the supernatant of fermented waste-activated sludge using magnesium sacrificial anode at N:P molar ratios between 1.98 and 2.05. They found that up to 98% of soluble phosphorus was recovered through struvite crystallization. Korchef et al. [13] demonstrated that there was no optimal value for the Mg:P:N ratio for struvite crystallization that could be considered independently of the initial constituent ions concentrations, but it should be adjusted with respect to the initial phosphate, ammonium, and magnesium concentrations and operating parameters.

The presence of foreign ions in the solution such as the ions of calcium [16], copper [16,17], zinc [17], iron [18], aluminum [18,19], cadmium [19], and nickel [20], among other ions, affected the ammonium and phosphate recovery and limited struvite crystallization. Indeed, these ions can be incorporated with struvite, adsorbed on struvite surfaces, or precipitated as separated phases. Unlike heavy metal ions, the presence of phenolic organics enhanced the struvite crystallization rate [21].

Different techniques, such as stirring [22,23], $CO_2$ repelling [24,25], electrochemical deposition [26], and ion exchange [27], have been used to precipitate struvite from synthetic solutions and real wastewater. In this work, struvite was precipitated by magnetic stirring. It was shown that stirring strongly affected struvite crystallization. Indeed, increasing the stirring rate affected the mass transfer between the solution and the struvite crystals and enhanced the struvite crystal growth rate [28]. Capdevielle et al. [22] found that more than 90% of phosphorus was recovered through large struvite crystals at a high N:P ratio of 3:1 and moderate stirring rates between 45 and 90 rpm. Perera et al. [29] studied the precipitation of struvite in a stainless steel reactor at pH 9, and a N:P:Mg molar ratio equal to 1:1.2:1.2. They found that stirring at 50 rpm did not allow sufficient mixing, thus affecting the struvite growth. They found that the removal efficiencies of nitrogen and phosphate were 97% and 71%, respectively, for a stirring rate of 500 rpm. Xu et al. [30] investigated the effect of stirring on laboratory-scale recovery of phosphorus and potassium from urine. They showed that for stirring rates between 100 and 200 rpm, 68% of the phosphorus was recovered in the form of struvite and 76% of the potassium in the form of K-struvite ($MgKPO_4 \cdot 6H_2O$). Zhang et al. [31] investigated the effect of stirring and experiment time on struvite crystallization from swine wastewater as pretreatment to anaerobic digestion. They found that 38% and 44% of ammonium were recovered after 10 min at stirring rates of 160 and 400 rpm, respectively, and with increasing the reaction time no remarkable changes in the recovered amounts were observed. The optimal operating conditions for struvite crystallization were a P:Mg:N molar ratio of 1:1:1.2, pH = 10, and initially stirring at 400 rpm for 10 min and then stirring for 30 min at 160 rpm. From an experimental point of view, the precipitation of struvite under magnetic stirring is one of the easiest methods to implement on a laboratory scale. It requires space-saving equipment whose price remains reasonable. It saves consumables and time on the preparation and progress of manipulations.

The present work aimed to investigate struvite crystallization at turbulent solution flow caused by magnetic stirring. It is not unfounded to expect that the effect of turbulence due to high stirring rates on struvite crystallization can be overcome by controlling the solution volume or the concentrations of the struvite constituent ions. For this reason, the effect of stirring at a fixed volume and initial phosphorus concentration was first investigated. Then, the effect of the solution volume on the crystallization of struvite at a fixed initial phosphorus concentration and stirring rate was investigated. Finally, the effect of the initial phosphorus concentration on struvite crystallization was studied at a fixed stirring rate and solution volume. Phosphorus concentration and solution pH were monitored over time. Struvite crystals were examined by powder X-ray diffraction (XRD), and Fourier-transform infrared (FTIR) spectroscopy. Their morphology was observed by scanning electron microscopy (SEM).

## 2. Materials and Methods

The crystallization of struvite ($MgPO_4NH_4 \cdot 6H_2O$) was carried out by magnetic stirring (with an HI 190 stirrer, Hanna Instruments, Woonsocket, RI, USA) in a cylindrical 2 L Pyrex cell. The working solutions were prepared by mixing two solutions of magnesium chloride hexahydrate ($MgCl_2 \cdot 6H_2O$) and ammonium dihydrogen phosphate ($NH_4H_2PO_4$) in distilled water. The solutions of $NH_4H_2PO_4$ and $MgCl_2 \cdot 6H_2O$ were prepared by dissolving the corresponding solids $NH_4H_2PO_4$ (purity > 99%, Sigma Aldrich, Burlington, MA, USA) and $MgCl_2 \cdot 6H_2O$ (purity > 99%, Fluka, Charlotte, NC, USA) in distilled water. Tablets of NaOH (>99% purity, Sigma Aldrich) were dissolved in distilled water to obtain 1 M NaOH solution. This alkaline solution was used to adjust the initial pH of the working solution to the desired value. For all experiments, the molar ratio of Mg:P:N was set at 1:1:1 corresponding to struvite ($MgNH_4PO_4 \cdot 6H_2O$) solid. The solution temperature was fixed at 25 °C using a thermostat with circulating water. The solution pH was monitored using a pH meter (pH 213, Hanna Instruments, USA) with a combination glass/Ag/AgCl electrode after calibration with commercially available standard buffer solutions from Biopharm at pH 4, 7, and 10. Each experiment was performed at least four times for reproducibility, and the mean values are reported in the present work. The error on the reported values was less than 5%.

To study the effect of stirring on struvite precipitation, four series of experiments were performed. All the experiments performed in the present work aimed to optimize the crystallization of struvite at high stirring rates. In the first series of experiments, the stirring rate varied between 100 and 700 rpm (100, 300, 500, and 700 rpm), the initial solution pH was adjusted to 8, the initial phosphorus concentration was 200 mg·L$^{-1}$ (6.45 mM) and the solution volumes were 600 and 1200 mL. In the second series of experiments, the initial phosphorus concentration was assessed at 600 mg·L$^{-1}$ (19.35 mM) and the initial pH, the stirring rate, and the solution volume were assessed as those of the first series of experiments. The third series of experiments was performed at different solution volumes between 600 and 1200 mL, a stirring rate of 700 rpm, an initial solution pH of 8, and an initial phosphorus concentration of 200 mg·L$^{-1}$. The increase in the solution volume was expected to lower the flow turbulence caused by stirring and favor struvite precipitation. Finally, a series of experiments was set at initial phosphorus concentrations ranging from 200 to 800 mg·L$^{-1}$. The increase in the constituent ion concentrations may overcome the effect of turbulence on struvite crystallization. For this series of experiments, the stirring rate was set at 700 rpm, the initial pH at 8, and the volume of the solution at 600 mL.

During each experiment performed in this work, periodic samples of 1 mL were withdrawn from the solution and then filtered through a 0.45 μm membrane filter. Next, the concentration of phosphorus was determined photochemically by a HACH DR/4000 spectrophotometer. At the end of each experiment, the solution was filtered using a 0.45 μm membrane filter. The recovered precipitates were dried at room temperature and analyzed by XRD, SEM, and FTIR spectroscopy. XRD was carried out at room temperature using a Philips X'PERT PRO diffractometer in step scanning mode using Cu-Kα radiation. The

XRD patterns were recorded in the scanning range 2θ = 10–70° using a small angular step of 2θ = 0.017° and a fixed counting time of 4 s. The positions of the XRD reflections were determined using 'X-Pert HighScore Plus' software, Version 2.1. The XRD patterns of the collected precipitates were compared with the Joint Committee on Powder Diffraction Standards data. The FTIR spectra were recorded between 400 and 4000 cm$^{-1}$ (Shimadzu IRAffinity-1) using pressed powder samples in KBr medium. The size of particles was measured using a Mastersizer 2000 (Version 6.01, Malvern Instruments, Malvern, UK) combined with a Hydro 2000 MU.

## 3. Results

### 3.1. Effect of Stirring

For all the experiments carried out in the present work, the precipitated phase observed was struvite. A typical X-ray diffractogram and FTIR spectrum of the precipitates obtained are given in Figure 1a,b, respectively. The XRD pattern (Figure 1a) showed that the characteristic reflections of the obtained precipitates were comparable to those of the pattern for the struvite standard (JCPDS 15-0762) [32]. The FTIR spectrum (Figure 1b) presented the characteristic band of $PO_4^{3-}$ located at 1005 cm$^{-1}$ and the characteristic band of $NH_4^+$ located at 2933 cm$^{-1}$. Comparable struvite spectra were reported by Korchef et al. [13] and Zhang et al. [33]. The following absorption bands were also observed: a large band between 3600 and 2270 cm$^{-1}$ corresponding to O-H and N-H stretching vibrations, a low-intensity band located at 1660 cm$^{-1}$ attributed to H-O-H bending modes, which indicates water of crystallization, and a band at 1432 cm$^{-1}$ attributed to the asymmetric bending vibration of N-H in $NH_4^+$. The bands detected between 1005 and 456 cm$^{-1}$ are characteristics of the ion $PO_4^{3-}$, where the wide low-intensity band at 1005 cm$^{-1}$ was associated with the asymmetric bending vibration of $PO_4^{3-}$, the weak band at 870 cm$^{-1}$ was attributed to the vibration of coordinated water$^-$, the band at 566 cm$^{-1}$ was associated with the asymmetric bending modes of $PO_4^{3-}$, and the band at 456 cm$^{-1}$ was attributed to the symmetric bending vibration of $PO_4^3$ units. SEM analysis confirmed the results obtained by XRD and FTIR spectroscopy and showed precipitates with the typical prismatic pattern characteristic of struvite crystals (Figure 2).

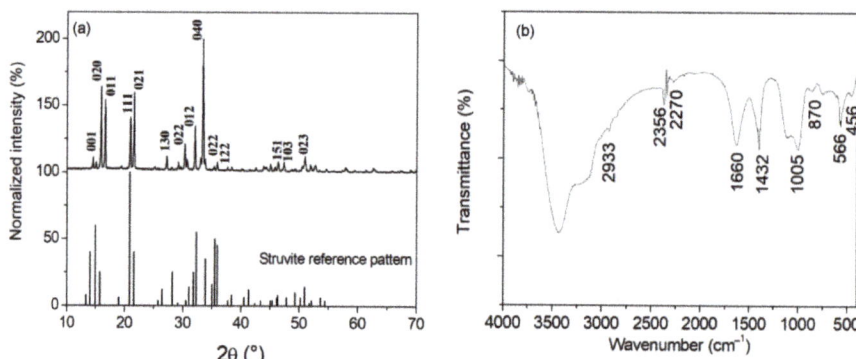

**Figure 1.** Typical examples of the (**a**) XRD pattern and (**b**) FTIR spectrum of the obtained precipitates.

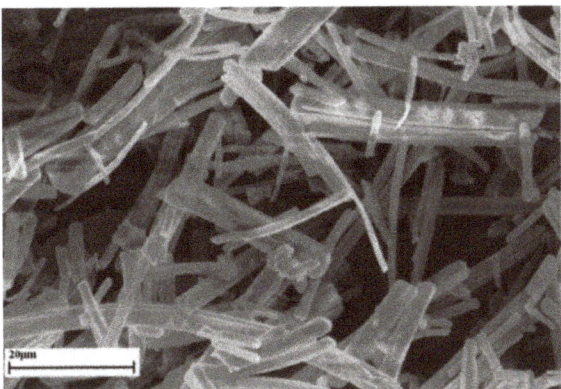

**Figure 2.** SEM micrograph of the precipitates obtained with the typical prismatic pattern characteristic of struvite crystals.

Figure 3 illustrates the changes in pH and phosphorus concentration over time at varied stirring rates between 100 and 700 rpm for an initial phosphorus concentration of 200 mg·L$^{-1}$ and solution volumes of 600 and 1200 mL. For all experiments performed, the pH decreased over time, and then from a certain time (depicted herein as the end of precipitation time), it remained practically constant (Figure 3a). Concomitantly, a decrease in phosphorus concentration was observed, followed by a plateau (Figure 3b). The higher the stirring rate was, the lower the concentration obtained at the end of the precipitation. The decrease in pH and phosphorus concentration over time was due to struvite precipitation according to the following reaction [34]:

$$Mg^{2+} + NH_4^+ + H_nPO_4^{n-3} + 6H_2O \rightarrow MgNH_4PO_4 \cdot 6H_2O + nH^+ \quad (1)$$

where $n$ = 0, 1, and 2 depending on the solution pH and the initial phosphorus concentration.

The constant values of pH and phosphorus concentration for advanced reaction times indicated the end of precipitation. The precipitation of struvite occurred as soon as the solution pH was adjusted to 8, and it was manifested by a significant decrease in phosphorus concentration and pH in the first minutes. Comparable results were found when struvite precipitated from pig manure in an anaerobic digester [35]. At the end of precipitation, the solution pH reached values between 7.2 and 7.4 for all experiments. This agreed with the results of Stratful et al. [36], who observed that at a pH equal to 7, no precipitation of struvite occurred, and at a slightly higher pH of 7.5, only a small amount was recovered through very small crystals. Several studies on the effect of pH on struvite crystallization were reported in the literature, and it was depicted that pH affected the solubility of struvite. Indeed, it was found that for pH values between 8 and 10 the solubility of struvite significantly decreased, and the struvite crystallization rate increased [23,37–39].

The pH and phosphorus concentration values obtained over time for the stirring rates 100 and 300 rpm and a solution volume of 600 mL were comparable but slightly higher than those obtained for 500 and 700 rpm (Figure 3a,b). When the volume increased to 1200 mL, the curves showing the changes in pH (Figure 3c) and phosphorus concentration (Figure 3d) over time presented the same evolution trends as those for a volume of 600 mL.

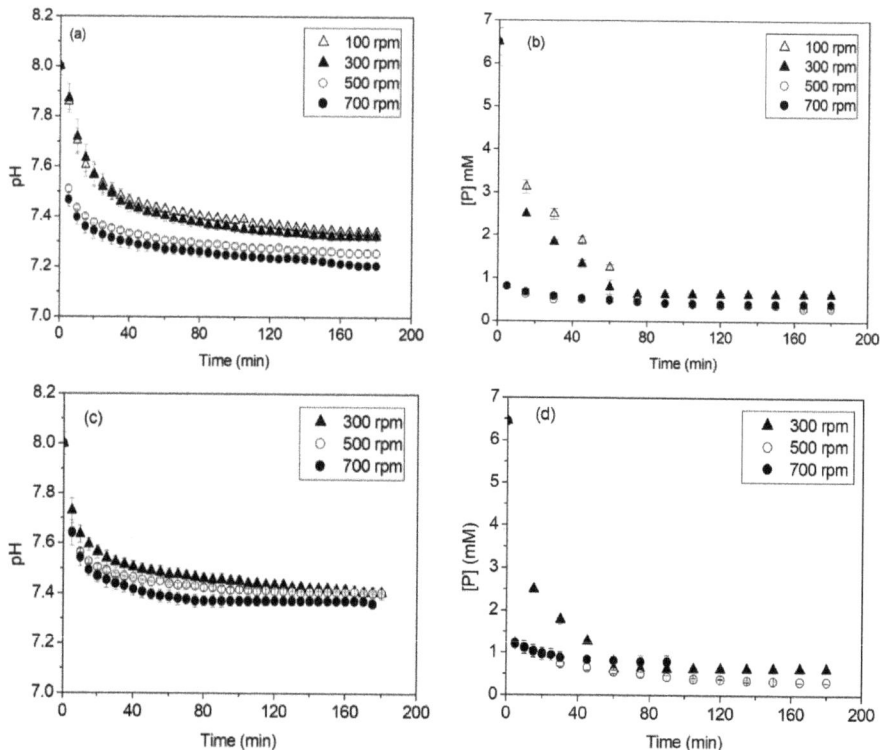

**Figure 3.** Variations of pH and phosphorus concentration over time for different stirring rates, an initial phosphorus concentration of 200 mg·L$^{-1}$ and solution volumes of (**a**,**b**) 600 mL and (**c**,**d**) 1200 mL.

Table 1 gives the precipitation end time ($t_f$), the initial precipitation rate ($V_i$), and the phosphorus removal efficiency, $R$ (%), obtained at different stirring rates and solution volumes of 600 and 1200 mL. The precipitation end time ($t_f$) corresponded to the first point of stabilization of the phosphorus concentration after precipitation. The initial precipitation rate ($V_i$) was determined from the slope of the linear part of the time curve of phosphorus concentration for times less than $t_f$. The phosphorus removal efficiency, $R$ (%), was calculated using the following equation:

$$R(\%) = \frac{(C_i - C_f)}{C_i} \times 100 \qquad (2)$$

where ($C_i$) and ($C_f$) are the initial and final phosphorus concentrations, respectively.

As the stirring rate increased, the precipitation end time ($t_f$) decreased for both volumes used. For example, $t_f$ decreased from 75 to 45 min when the stirring rate increased from 100 to 700 rpm, respectively, for a solution volume of 600 mL (Table 1). For a given stirring rate, the precipitation end time obtained for a solution volume of 1200 mL was greater than that obtained for a solution volume of 600 mL. The highest precipitation end time (90 min) was obtained for a stirring rate of 500 rpm and a solution volume of 1200 mL (Table 1) where, in addition to precipitation in the bulk solution, a significant amount of struvite precipitated on the cell walls. The change in the precipitation nature (from precipitation in the bulk solution to precipitation on the cell walls) could be the cause of the increase in the precipitation end time at the stirring rate of 500 rpm and the solution volume of 1200 mL. By increasing the stirring rate to 700 rpm, precipitation occurred mainly in the bulk solution and the precipitation end time decreased to 45 min for both volumes used.

**Table 1.** End of precipitation time ($t_f$), initial precipitation rate ($V_i$), and phosphorus removal efficiency ($R\%$) for different stirring rates, solution volumes of 600 and 1200 mL, and an initial phosphorus concentration of 200 mg·L$^{-1}$.

| Solution Volume | Stirring Rate (rpm) | $V_i$ (mmol·L$^{-1}$·min$^{-1}$) | $t_f$ (min) | $R$ (%) |
|---|---|---|---|---|
| 600 mL | 100 | 0.22 | 75 | 90.31 |
|  | 300 | 0.26 | 75 | 90.31 |
|  | 500 | 0.39 | 45 | 95.48 |
|  | 700 | 1.13 | 45 | 94.01 |
| 1200 mL | 100 | 0.22 | 75 | 90.31 |
|  | 300 | 0.26 | 60 | 90.31 |
|  | 500 | 0.36 | 90 | 95.04 |
|  | 700 | 1.04 | 45 | 88.71 |

The initial precipitation rate ($V_i$) increased with the increase in the stirring rate for both solution volumes of 600 and 1200 mL. At 100 rpm and a solution volume of 600 mL, the initial precipitation rate $V_i$ was equal to 0.22 mmol·L$^{-1}$·min$^{-1}$. The increase in the stirring rate to 500 rpm resulted in a slight increase in $V_i$ to 0.36 mmol·L$^{-1}$·min$^{-1}$ (Table 1). A more pronounced increase in the initial precipitation rate was observed when the stirring rate increased to 700 rpm., i.e., it became approximately five times higher ($V_i$ =1.13 mmol·L$^{-1}$·min$^{-1}$). Comparable results were found for a solution volume of 1200 mL. However, the acceleration of struvite precipitation was accompanied by a slight decrease in the phosphorus removal efficiency. Indeed, increasing the stirring rate from 100 to 500 rpm significantly improved the phosphorus removal efficiency, and comparable values were obtained for both solution volumes. For example, the phosphorus removal efficiency, at a solution volume of 600 mL, increased from ~90% to ~95% when the stirring rate increased from 100 to 500 rpm, respectively, and it decreased slightly to ~94% at a higher stirring rate of 700 rpm (Table 1). This decrease was more pronounced for a solution volume of 1200 mL where the removal efficiency of phosphorus reached ~89% at a stirring rate of 700 rpm. A detailed study of the effect of the solution volume on struvite precipitation is given herein.

To conclude, for an initial phosphorus concentration of 200 mg·L$^{-1}$, the optimal stirring rate (which gave the highest efficiency) was 500 rpm for both solution volumes of 600 and 1200 mL. Comparable results were depicted by Perera et al. [29], who found that magnetic stirring at 500 rpm gave the highest phosphorus removal through struvite crystallization from swine waste biogas digester effluent where the molar ratio of N:Mg:P was 1:1.2:1.2, the working volume was 3 L, and pH was equal to 9.

Figure 4 shows the variation in phosphorus concentration over time at stirring rates between 100 and 700 rpm, an initial phosphorus concentration of 600 mg·L$^{-1}$, and a solution volume of 600 mL. As soon as the initial pH was adjusted to 8, the concentration of phosphorus in the solution decreased. This decrease became more pronounced as the stirring rate increased. In addition, the pH attained at the end of precipitation reached ~6.9 for a stirring rate of 100 rpm and ~6.7 for stirring rates of 500 and 700 rpm. The initial rate of struvite precipitation increased with the stirring rate, and the end of precipitation time decreased (except at 700 rpm where a slight increase in $t_f$ was observed). In addition, the phosphorus removal efficiency increased with the stirring rate. Indeed, it increased from ~97.5% for a stirring rate of 100 rpm to ~99.4% for a stirring rate of 700 rpm (Table 2). Therefore, at the high initial phosphorus concentration of 600 mg·L$^{-1}$ and a solution volume of 600 mL, increasing the stirring rate accelerated the precipitation of struvite, and higher struvite amounts were obtained at relatively shorter times. Comparing the values of $t_f$, $V_i$, and $R$ (%) obtained for the initial phosphorus concentrations of 200 and 600 mg·L$^{-1}$

and the solution volume of 600 mL (Tables 1 and 2) indicated that the values obtained for 600 mg·L$^{-1}$ were significantly higher than those obtained for 200 mg·L$^{-1}$. This can be explained by the flow turbulence of the solution. Indeed, when the turbulence of the solution became stronger (by increasing the stirring rate at a fixed solution volume), the transition of the constituent ions of struvite from the liquid phase to the solid phase became more difficult, and consequently, the reaction with the struvite crystal surface was inhibited. This resulted in a decrease in struvite crystal growth and fewer amounts were obtained. The limiting effect of turbulence became less consequent to the precipitation of struvite when the phosphorus concentration increased to 600 mg·L$^{-1}$ at a fixed solution volume of 600 mL.

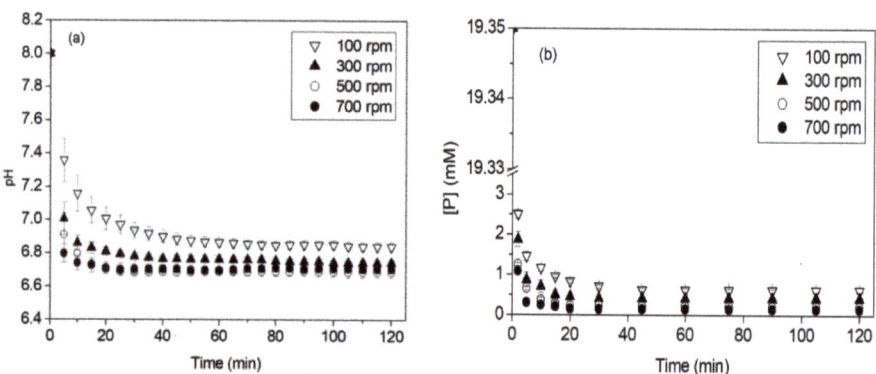

**Figure 4.** Variations of (**a**) pH and (**b**) phosphorus concentration over time for different stirring rates, an initial phosphorus concentration of 600 mg·L$^{-1}$, and a solution volume of 600 mL.

**Table 2.** End of precipitation time ($t_f$), initial precipitation rate ($V_i$), and phosphorus removal efficiency ($R$%) for different stirring rates, an initial phosphorus concentration of 600 mg·L$^{-1}$, and a solution volume of 600 mL.

| Stirring Rate (rpm) | $V_i$ (mmol·L$^{-1}$·min$^{-1}$) | $t_f$ (min) | R (%) |
|---|---|---|---|
| 100 | 8.42 | 45 | 97.51 |
| 300 | 8.74 | 30 | 97.92 |
| 500 | 9.05 | 20 | 98.73 |
| 700 | 9.13 | 30 | 99.42 |

Changes in pH and phosphorus concentration over time for different solution volumes between 600 and 1400 mL, an initial phosphorus concentration of 200 mg·L$^{-1}$, and a stirring rate of 700 rpm are shown in Figure 5a,b. The obtained curves showed the same evolution trends as those obtained for the different stirring rates presented in Figure 3. The increase in the solution volume resulted in an increase in the phosphorus concentration obtained at the end of precipitation. That is, fewer amounts of struvite were obtained when the solution volume increased. This explains the slight increase in the solution pH obtained at the end of struvite precipitation. Indeed, struvite precipitation is accompanied by the release of protons (see Equation (1)). The initial precipitation rate $V_i$, the end of precipitation time $t_f$, and the phosphorus removal efficiency $R$ (%) for different solution volumes, an initial phosphorus concentration of 200 mg·L$^{-1}$, and a stirring rate of 700 rpm are given in Figure 5c,d. The initial precipitation rate decreased from ~1.13 to ~1 mmol·L$^{-1}$·min$^{-1}$ when the volume of the solution increased from 600 to 800 mL, respectively. For higher volumes, it remained practically constant. The end of precipitation time remained practically constant up to a volume of 800 mL, then it increased significantly (Figure 5c). The phosphorus removal efficiency decreased as the volume of the solution increased. For example, it decreased from ~94% to ~85% as the volume increased from 600 to 1400 mL, respectively

(Figure 5d). At a constant stirring rate and initial phosphorus concentration, the increase in the solution volume affected the transfer of struvite constituent ions from the liquid phase to the solid phase and inhibited their reaction with the crystal surfaces, which limited the rate of struvite crystal growth. Consequently, the amount of precipitated struvite decreased. This was confirmed by the decrease in the initial precipitation rate and the increase in the end of precipitation time. This agreed with the fact that increasing the stirring rate up to 500 rpm at a constant volume of 600 mL increased the phosphorus removal efficiency by increasing the transfer rate of struvite constituent ions. At 700 rpm, the solution became strongly turbulent, and the opposite effect (limitation of the precipitation reaction) occurred. In conclusion, under the experimental conditions of the present work, to obtain a high phosphorus removal efficiency through struvite precipitation, the stirring rate should be set at 500 rpm when working at constant volume. For higher stirring rates, the volume of the solution should be reduced.

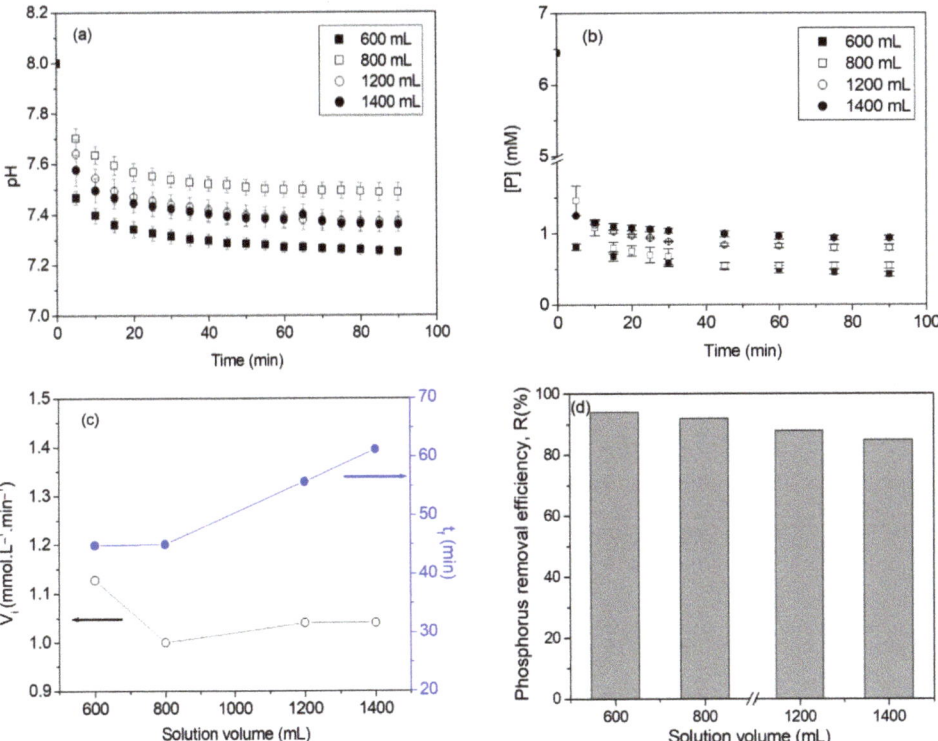

**Figure 5.** Variations of (**a**) pH, and (**b**) phosphorus concentration over time for different solution volumes, and variations of the (**c**) end of precipitation time $t_f$ and initial precipitation rate $V_i$, and (**d**) phosphorus removal efficiency with the solution volume for an initial phosphorus concentration of 200 mg·L$^{-1}$ and a stirring rate of 700 rpm.

*3.2. Effect of Phosphorus Concentration*

To study the effect of the initial phosphorus concentration on struvite crystallization, we performed a series of experiments at a constant stirring rate of 100 rpm, a solution volume of 600 mL, and an initial phosphorus concentration ranging from 200 to 800 mg·L$^{-1}$ (6.45 to 25.8 mM, respectively). The FTIR spectra of the precipitates obtained at different initial phosphorus concentrations are given in Figure 6. These spectra were comparable to the struvite spectrum. Therefore, the variation in the initial phosphorus concentration up to 800 mg·L$^{-1}$, at a constant stirring rate of 100 rpm and a solution volume of 600 mL, did not affect the nature of the precipitated phase.

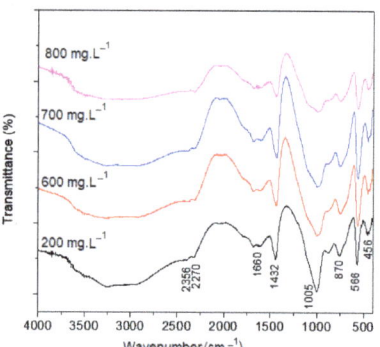

**Figure 6.** FTIR spectra of the precipitates obtained for different initial phosphorus concentrations, a stirring rate of 100 rpm, and a solution volume of 600 mL.

Changes in pH and phosphorus concentration over time for different initial phosphorus concentrations, a stirring rate of 100 rpm, and a solution volume of 600 mL (not presented herein) showed the same evolution trends as those obtained for different stirring rates, an initial phosphorus concentration of 600 mg·L$^{-1}$, and a solution volume of 600 mL presented in Figure 4. That is, the pH and the phosphorus concentration decreased over time, and then from a certain time remained practically constant, which indicated the end of struvite precipitation. The end of precipitation time ($t_f$), the initial precipitation rate ($V_i$), and the phosphorus removal efficiency $R$ (%) obtained for the different initial phosphorus concentrations are given in Table 3. The end of precipitation time decreased with increasing the initial phosphorus concentration, while the initial precipitation rate increased. For example, $t_f$ decreased from 50 to 20 min and $V_i$ increased from 2.33 to 12.55 mmol·L$^{-1}$·min$^{-1}$ when the phosphorus concentration increased from 300 to 800 mg·L$^{-1}$, respectively (Table 3). The phosphorus removal efficiency increased from ~87% to ~99% when the initial phosphorus concentration increased from 300 to 800 mg·L$^{-1}$, respectively. Therefore, the increase in the initial phosphorus concentration, at a fixed stirring rate of 100 rpm and a solution volume of 600 mL, improved phosphorus recovery, and more struvite crystals were precipitated at a shorter experiment time.

**Table 3.** End of precipitation time ($t_f$), the initial precipitation rate ($V_i$), and the phosphorus removal efficiency ($R$%) obtained for different initial phosphorus concentrations ([P]$_0$), a stirring rate of 100 rpm and a solution volume of 600 mL.

| [P]$_0$ (mg·L$^{-1}$) | $V_i$ (mmol·L$^{-1}$·min$^{-1}$) | $t_f$ (min) | $R$ (%) |
|---|---|---|---|
| 300 | 2.33 | 50 | 87.32 |
| 400 | 4.43 | 40 | 91.74 |
| 500 | 7.03 | 25 | 95.77 |
| 600 | 8.42 | 22 | 96.15 |
| 700 | 10.79 | 20 | 98.21 |
| 800 | 12.55 | 20 | 98.94 |

### 3.3. Particle Size

For an initial phosphorus concentration of 200 mg·L$^{-1}$, a solution volume of 600 mL and varied stirring rates, the particle size distribution curves showed a polymodal distribution, i.e., two maxima were observed for a stirring rate of 100 rpm, and three maxima were observed for a stirring rate of 700 rpm (Figure 7a). For both stirring rates, the first maximum was located at ~18 μm and the second one was located at ~100 μm. The third maximum observed at a stirring rate of 700 rpm appeared at ~875 μm. Therefore, the increase in the stirring rate led to the formation of large struvite crystals. This can be explained by a greater transfer of the struvite constituent ions from the liquid phase (solution) to the solid phase (the germs formed) for higher stirring. At a stirring rate of 700 rpm and an initial phosphorus concentration of 200 mg·L$^{-1}$, when the volume of the solution increased, the particle size distribution curves showed a polymodal distribution and a decrease in struvite particle size was observed (Figure 7b). For example, for a volume of 800 mL, the size distribution was trimodal and the maxima observed were located at ~2, ~18, and ~100 μm. For higher volumes of 1200 and 1400 mL, the size distribution became bimodal, and the maxima were located at ~18 and ~100 μm.

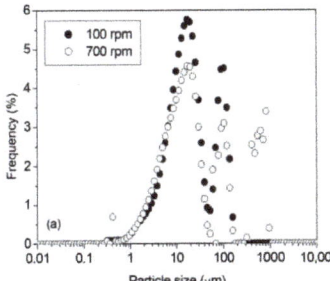
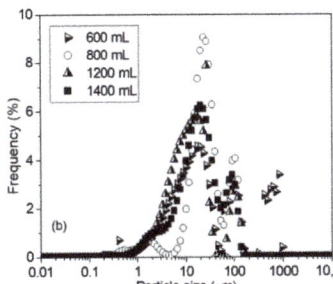
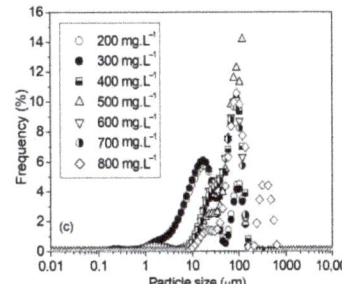

**Figure 7.** Particle size distributions of struvite crystals for (**a**) selected stirring rates at an initial phosphorus concentration of 200 mg·L$^{-1}$ and a solution volume of 600 mL, (**b**) different solution volumes at a stirring rate of 700 rpm, and (**c**) different initial phosphorus concentrations at a stirring rate of 100 rpm and a solution volume of 600 mL.

The particle size distributions of struvite crystals for phosphorus concentrations ranged from 200 to 700 mg·L$^{-1}$, a stirring rate of 100 rpm and a solution volume of 600 mL showed a bimodal distribution (Figure 7c). For a phosphorus concentration of 800 mg·L$^{-1}$, the size distribution became trimodal. For 200 and 300 mg·L$^{-1}$, the first maximum was observed at ~10 μm and the second one at ~100 μm. For initial phosphorus concentrations between 400 and 700 mg·L$^{-1}$, the first maximum was observed at ~30 μm and the second one at ~100 μm. For a phosphorus concentration of 800 mg·L$^{-1}$, besides the two maxima at 30 and 100 μm, a third maximum was observed at ~400 μm. Therefore, the increase in initial phosphorus concentration led to an increase in struvite particle size.

In summary, at an initial phosphorus concentration of 200 mg·L$^{-1}$ and a constant volume of 600 mL, increasing the stirring rate up to 500 rpm accelerated the precipitation of struvite, improved the phosphorus removal efficiency, and obtained larger struvite crystals. According to the results of Natsi et al. [7] and considering the dependence of phosphorus recovery through struvite crystallization on the solution turbulence depicted in the present work, the predominant mechanism of struvite crystallization can be surface-diffusion controlled. Indeed, the increase in the stirring rate from 100 to 500 rpm led to significant turbulence of the solution, which accelerated the diffusion of the constituent ions to the germs formed in the solution, and therefore an acceleration of struvite crystalline growth occurred. As a result, the particle size increased. This is of utmost importance from an experimental point of view since large struvite crystals formed in the solution facilitate its separation from the liquid phase. In addition, struvite shows slow-release properties [40] and a large surface area to volume ratio accelerates the release of ammonium and phosphorus from struvite used as fertilizer [41]. Increasing the stirring rate from 500 to 700 rpm to maintain more precipitates in suspension for a larger effective precipitation area resulted, however, in a decrease in phosphorus removal, and smaller struvite crystals were obtained. Indeed, when the stirring rate significantly increased, the mixing energy increased as well, the flow became turbulent, and the liquid shear stress was higher. This affected the transfer of struvite constituent ions from the liquid phase to the solid phase and inhibited their reaction with the crystal surfaces, limiting struvite crystal growth. By increasing the volume of the solution at a high stirring rate of 700 rpm, the turbulence of the solution decreased, and therefore, the diffusion of the constituent ions became less important, which led to smaller particles. Increasing the initial phosphorus concentration overcame the limiting effect of turbulence and resulted in an increase in struvite particle size.

In real wastewater, the concentration of total nitrogen can vary from ~50 mg·L$^{-1}$, i.e., in municipal wastewater with a low C/N ratio [42], to ~2700 mg·L$^{-1}$ in swine wastewater, for example [16]. The concentration of phosphorus and magnesium can vary from a few tens of mg·L$^{-1}$ to concentrations near 900 mg·L$^{-1}$ [14]. The constituent ion concentrations used in the present work fitted well with these ranges of concentrations. Real wastewater may also contain calcium (~20–60 mg·L$^{-1}$ [13,14,16]), chlorine (~90 mg·L$^{-1}$ [16]), and low concentrations of heavy metals such as Cu (~2 mg·L$^{-1}$), Zn (~6 mg·L$^{-1}$), Cd (~0.5 mg·L$^{-1}$), Pb (~0.8 mg·L$^{-1}$), Mn (~1 mg·L$^{-1}$) [16], and Fe (~5 mg·L$^{-1}$) [43]. Note that the concentrations are given as an indication. These foreign ions can adsorb on struvite surfaces, incorporate in the struvite lattice, or precipitate as separate phases that disturb struvite growth in all cases. For example, calcium precipitated in treated wastewater as calcium phosphate salts [44,45], which affected struvite crystallization. Copper and iron precipitated in the form of hydroxides during struvite crystallization [43], and zinc precipitated as Zn-PO$_4$ at low Zn concentration and Zn-OH at high Zn concentration [46]. Depending on the added concentration, lead can be adsorbed on struvite surfaces ([Pb] < 1 mg·L$^{-1}$) or precipitated as Pb hydroxide and Pb phosphate ([Pb] between 10 and 100 mg·L$^{-1}$) [47]. Like lead, the presence of nickel in the wastewater competed with struvite crystallization, i.e., at low concentrations (<1 mg·L$^{-1}$), nickel formed Ni-OH and Ni-PO$_4$ on the struvite surface and precipitated separately as amorphous Ni-struvite, Ni hydroxide, and Ni phosphate at higher concentrations between 10 and 100 mg·L$^{-1}$ [20]. Additionally, we showed in the present work that under a constant stirring rate, in the range between 100 and 500 rpm, the increase in the solution volume from 600 to 1200 mL did not significantly affect the struvite crystal growth, and for a higher stirring rate of 700 rpm, the volume of the solution should be reduced to obtain high amounts of struvite. However, experiments in the present work were conducted on free foreign ion solutions with relatively low volumes that did not exceed 1.4 L. Obviously, the volumes of wastewater treated in industrial processes are significantly larger than those used in the present work. For those reasons, further work is needed to investigate struvite crystallization from real wastewater at laboratory and industrial scales.

## 4. Conclusions

Phosphorus and ammonium can be recovered in the presence of magnesium through struvite (MgNH$_4$PO$_4$·6H$_2$O) crystallization. Struvite is recognized as an effective fertilizer. In the present work, the crystallization of struvite by magnetic stirring was investigated at different initial phosphorus concentrations, solution volumes, and stirring rates. The crystals obtained were characterized by XRD, FTIR spectroscopy, and SEM. For all experiments, struvite was the only solid observed. It was shown that:

- For an initial phosphorus concentration of 200 mg·L$^{-1}$, as the stirring rate increased from 100 to 500 rpm, the initial precipitation rate increased, the precipitation end time decreased, the phosphorus removal efficiency increased from 90% to 95%, respectively, and struvite particle size increased.
- A decrease in the removal efficiency and the struvite particle size was found at a higher stirring rate of 700 rpm. This was attributed to the solution turbulence, which caused the struvite precipitation reaction to be limited.
- At a fixed stirring rate and initial phosphorus concentration, the increase in the solution volume decreased the initial precipitation rate, increased the precipitation end time, and decreased the phosphorus removal efficiency.

- An increase in the phosphorus concentration overcame the limiting effect of turbulence. Indeed, it accelerated the precipitation of struvite and increased the phosphorus recovery (~99%) through large struvite crystals (~400 μm in size).
- Large struvite crystals formed in the solution facilitate the separation of struvite from the liquid phase and accelerate the release of ammonium and phosphate ions in soils when struvite is used as a fertilizer.
- Real wastewater contains ions of calcium, sulfate, and heavy metals, among others. The presence of these foreign ions influences the nature, shape, and size of the precipitated phase. Therefore, further study on the recycling of phosphorus and nitrogen through struvite crystallization from real wastewater is needed since it will allow us to identify these phases and to determine, in the case of insertion or adsorption, the rate of removal of these ions by the crystallization of struvite.

**Author Contributions:** A.K. was involved in the methodology, experimental investigation, supervision, and writing and editing of the work. S.A. was involved in the experimental investigation. I.S. was involved in ensuring resources, data curation, and formal analysis of the work. All authors have read and agreed to the published version of the manuscript.

**Funding:** This research was funded by the Deanship of Scientific Research at King Khalid University, Saudi Arabia, grant number GRP/87/44.

**Institutional Review Board Statement:** Not applicable.

**Informed Consent Statement:** Not applicable.

**Data Availability Statement:** Not available.

**Acknowledgments:** The authors extend their appreciation to the Deanship of Scientific Research at King Khalid University, Saudi Arabia, for funding this work through General Research Project under grant number GRP/87/44.

**Conflicts of Interest:** The authors declare no conflict of interest.

## References

1. US Bureau of Reclamation. Water Facts—Worldwide Water Supply. 2022. Available online: https://www.usbr.gov/mp/arwec/water-facts-ww-water-sup.html (accessed on 25 March 2023).
2. EU. Drinking Water Directive; Council Directive 76/464/EEC of 4 May 1976 on Pollution Caused by Certain Dangerous Substances Discharged into the Aquatic Environment of the Community. 1976. Available online: https://eur-lex.europa.eu/legal-content/EN/TXT/PDF/?uri=CELEX:31976L0464&from=EN (accessed on 25 March 2023).
3. Wang, L.; Ye, C.; Gao, B.; Wang, X.; Li, Y.; Ding, K.; Li, H.; Ren, K.; Chen, S.; Wang, W.; et al. Applying struvite as a N-fertilizer to mitigate N2O emissions in agriculture: Feasibility and mechanism. *J. Environ. Manag.* **2023**, *330*, 117143. [CrossRef] [PubMed]
4. Liang, Z.; Liu, J.; Wan, Y.; Feng, Z.; Zhang, P.; Wang, M.; Zhang, H. Preparation and fire extinguishing mechanism of novel fire extinguishing powder based on recyclable struvite. *Mater. Today Comm.* **2023**, *34*, 105410. [CrossRef]
5. Giulio, G.; Ferretti, G.; Rosinger, C.; Huber, S.; Medoro, V.; Mentler, A.; Díaz-Pinés, E.; Gorfer, M.; Faccini, B.; Keiblinger, K.M. Recycling nitrogen from liquid digestate via novel reactive struvite and zeolite minerals to mitigate agricultural pollution. *Chemosphere* **2023**, *317*, 137881. [CrossRef]
6. Ha, T.-H.; Mahasti, N.N.; Lu, M.C.; Huang, Y.H. Ammonium-nitrogen recovery as struvite from swine wastewater using various magnesium sources. *Sep. Purif. Technol.* **2023**, *308*, 122870. [CrossRef]
7. Natsi, P.D.; Goudas, K.-A.; Koutsoukos, P.G. Phosphorus recovery from municipal wastewater: Brucite from MgO hydrothermal treatment as magnesium source. *Crystals* **2023**, *13*, 208. [CrossRef]
8. Sultana, R.; Kékedy-Nagy, L.; Daneshpour, R.; Greenlee, L. Electrochemical recovery of phosphate from synthetic wastewater with enhanced salinity. *Electrochim. Acta* **2022**, *426*, 140848. [CrossRef]
9. Song, L.; Li, Z.; Wang, G.; Tian, Y.; Yang, C. Supersaturation control of struvite growth by operating pH. *J. Mol. Liq.* **2021**, *336*, 116293. [CrossRef]
10. Abbona, F.; Madsen, H.L.; Boistelle, R. Crystallization of two magnesium phosphates, struvite and newberyite: Effect of pH and concentration. *J. Cryst. Growth* **1982**, *57*, 6–14. [CrossRef]
11. Aage, H.; Andersen, B.; Blom, A.; Jensen, I. The solubility of struvite. *J. Radioanal. Nucl. Chem.* **1997**, *223*, 213–215. [CrossRef]
12. Li, M.; Zhang, H.; Sun, H.; Mohammed, A.; Liu, Y.; Lu, Q. Effect of phosphate and ammonium concentrations, total suspended solids and alkalinity on lignin-induced struvite precipitation. *Sci. Rep.* **2022**, *12*, 2901. [CrossRef]

13. Korchef, A.; Saidou, H.; Ben Amor, M. Phosphate recovery through struvite precipitation by $CO_2$ removal: Effect of magnesium, phosphate and ammonium concentrations. *J. Hazard. Mater.* **2011**, *186*, 602–613. [CrossRef]
14. Jaffer, Y.; Clark, T.A.; Pearce, P.; Parsons, S.A. Potential phosphorus recovery by struvite formation. *Water Res.* **2002**, *36*, 1834–1842. [CrossRef]
15. Kruk, D.J.; Elektorowicz, M.; Oleszkiewicz, J.A. Struvite precipitation and phosphorus removal using magnesium sacrificial anode. *Chemosphere* **2014**, *101*, 28–33. [CrossRef]
16. Wang, Y.; Da, J.; Deng, Y.; Wang, R.; Liu, X.; Chang, J. Competitive adsorption of heavy metals between Ca-P and Mg-P products from wastewater during struvite crystallization. *J. Environ. Manag.* **2023**, *335*, 117552. [CrossRef]
17. Wang, Y.; Deng, Y.; Liu, X.; Chang, J. Adsorption behaviors and reduction strategies of heavy metals in struvite recovered from swine wastewater. *J. Chem. Eng.* **2022**, *437*, 135288. [CrossRef]
18. Hutnik, N.; Stanclik, A.; Piotrowski, K.; Matynia, A. Kinetic conditions of struvite continuous reaction crystallisation from wastewater in presence of aluminium(III) and iron(III) ions. *Int. J. Environ. Pollut.* **2019**, *64*, 358–374. [CrossRef]
19. Saidou, H.; Korchef, A.; Moussa, S.B.; Amor, M.B. Study of $Cd^{2+}$, $Al^{3+}$, and $SO_4^{2-}$ ions influence on struvite precipitation from synthetic water by dissolved $CO_2$ degasification technique. *Open J. Inorg. Chem.* **2015**, *5*, 41. [CrossRef]
20. Lu, X.; Xu, W.; Zeng, Q.; Liu, W.; Wang, F. Quantitative, morphological, and structural analysis of Ni incorporated with struvite during precipitation. *Sci. Total Environ.* **2022**, *817*, 152976. [CrossRef]
21. Rabinovich, A.; Rouff, A.A. Effect of phenolic organics on the precipitation of struvite from simulated dairy wastewater. *ACS EST Water* **2021**, *1*, 910–918. [CrossRef]
22. Capdevielle, A.; Sýkorová, E.; Biscans, B.; Béline, F.; Daumer, M.L. Optimization of struvite precipitation in synthetic biologically treated swine wastewater-Determination of the optimal process parameters. *J. Hazard. Mater.* **2013**, *244*, 357–369. [CrossRef]
23. Bouropoulos, N.C.; Koutsoukos, P.G. Spontaneous precipitation of struvite from aqueous solutions. *J. Cryst. Growth* **2000**, *213*, 381–388. [CrossRef]
24. Fattah, K.P.; Zhang, Y.; Mavinic, D.S.; Koch, F.A. Application of carbon dioxide stripping for struvite crystallization-I: Development of a carbon dioxide stripper model to predict $CO_2$ removal and pH changes. *J. Environ. Eng. Sci.* **2008**, *7*, 345–356. [CrossRef]
25. Saidou, H.; Korchef, A.; Moussa, S.B.; Amor, M.B. Struvite precipitation by the dissolved $CO_2$ degasification technique: Impact of the airflow rate and pH. *Chemosphere* **2009**, *74*, 338–343. [CrossRef] [PubMed]
26. Wang, L.; Gu, K.; Zhang, Y.; Sun, J.; Gu, Z.; Zhao, B.; Hu, C. Enhanced struvite generation and separation by magnesium anode electrolysis coupled with cathode electrodeposition. *Sci. Total Environ.* **2022**, *804*, 150101. [CrossRef] [PubMed]
27. Chen, H.; Shashvatt, U.; Amurrio, F.; Stewart, K.; Blaney, L. Sustainable nutrient recovery from synthetic urine by Donnan dialysis with tubular ion-exchange membranes. *J. Chem. Eng.* **2023**, *460*, 141625. [CrossRef]
28. Perwitasari, D.S.; Santi, S.S.; Yahya, A. The effect of stirrer rotation on crystallization of struvite that can be used as fertilizer. *Int. J. Mech. Eng.* **2022**, *7*, 2969–2972.
29. Perera, P.W.A.; Wu, W.-X.; Chen, Y.-X.; Han, Z.-Y. Struvite Recovery from Swine Waste Biogas Digester Effluent through a Stainless Steel Device under Constant pH Conditions. *Biomed. Environ. Sci.* **2009**, *22*, 201–209. [CrossRef]
30. Xu, K.; Wang, C.; Wang, X.; Qian, Y. Laboratory experiments on simultaneous removal of K and P from synthetic and real urine for nutrient recycle by crystallization of magnesium–potassium–phosphate-hexahydrate in a draft tube and baffle reactor. *Chemosphere* **2012**, *88*, 219–223. [CrossRef]
31. Zhang, D.; Chen, Y.; Jilani, G.; Wu, W.; Liu, W.; Han, Z. Optimization of struvite crystallization protocol for pretreating the swine wastewater and its impact on subsequent anaerobic biodegradation of pollutants. *Bioresour. Technol.* **2012**, *116*, 386–395. [CrossRef]
32. International Centre for Diffraction Data, Ammonium Magnesium Phosphate Hydrate (Standard #15-0762), A Computer Database. 1996. Available online: https://www.icdd.com (accessed on 25 March 2023).
33. Zhang, T.; Ding, L.; Ren, H. Pretreatment of ammonium removal from landfill leachate by chemical precipitation. *J. Hazard. Mater.* **2009**, *166*, 911–915. [CrossRef]
34. Korchef, A.; Naffouti, S.; Souid, I. Recovery of high concentrations of phosphorus and ammonium through struvite crystallization by $CO_2$ repelling. *Cryst. Res. Technol.* **2022**, *57*, 2200123. [CrossRef]
35. Nelson, N.O.; Mikkelsen, R.L.; Hesterberg, D.L. Struvite precipitation in anaerobic swine lagoon liquid: Effect of pH and Mg:P ratio and determination of rate constant. *Bioresour. Technol.* **2003**, *89*, 229–236. [CrossRef]
36. Stratful, I.; Scrimshaw, M.D.; Lester, J.N. Conditions influencing the precipitation of magnesium ammonium phosphate. *Water Res.* **2001**, *35*, 4191–4199. [CrossRef]
37. Muys, M.; Cámara, S.J.G.; Derese, S.; Spiller, M.; Verliefde, A.; Vlaeminck, S.E. Dissolution rate and growth performance reveal struvite as a sustainable nutrient source to produce a diverse set of microbial protein. *Sci. Total Environ.* **2023**, *866*, 161172. [CrossRef]
38. González-Morales, C.; Fernández, B.; Molina, F.J.; Naranjo-Fernández, D.; Matamoros-Veloza, A.; Camargo-Valero, M.A. Influence of pH and temperature on struvite purity and recovery from anaerobic digestate. *Sustainability* **2021**, *13*, 10730. [CrossRef]
39. Buchanan, J.R.; Mote, C.R.; Robinson, R.B. Thermodynamics of struvite formation. *Trans. ASAE.* **1994**, *37*, 617–621. [CrossRef]
40. Mancho, C.; Diez-Pascual, S.; Alonso, J.; Gil-Díaz, M.; Lobo, M.C. Assessment of recovered struvite as a safe and sustainable phosphorous fertilizer. *Environments* **2023**, *10*, 22. [CrossRef]

41. Li, B.; Boiarkina, I.; Young, B.; Yu, W. Quantification and mitigation of the negative impact of calcium on struvite purity. *Adv. Powder Technol.* **2016**, *27*, 2354–2362. [CrossRef]
42. Yang, Y.; Peng, Y.; Cheng, J.; Zhang, S.; Liu, C.; Zhang, L. A novel two-stage aerobic granular sludge system for simultaneous nutrient removal from municipal wastewater with low C/N ratios. *Chem. Eng. J.* **2023**, *462*, 142318. [CrossRef]
43. Guan, Q.; Zeng, G.; Gong, B.; Li, Y.; Ji, H.; Zhang, J.; Song, J.; Liu, C.; Wang, Z.; Deng, C. Phosphorus recovery and iron, copper precipitation from swine wastewater via struvite crystallization using various magnesium compounds. *J. Clean. Prod.* **2021**, *328*, 129588. [CrossRef]
44. Pastor, L.; Mangin, D.; Barat, R.; Seco, A. A pilot-scale study of struvite precipitation in a stirred tank reactor: Conditions influencing the process. *Bioressour. Technol.* **2008**, *99*, 6285–6291. [CrossRef] [PubMed]
45. Le Corre, K.S.; Valsami-Jones, E.; Hobbs, P.; Parsons, S.A. Impact of calcium on struvite crystal size, shape and purity. *J. Cryst. Growth.* **2005**, *283*, 514–522. [CrossRef]
46. Rouff, A.A.; Juarez, K.M. Zinc interaction with struvite during and after mineral formation. *Environ. Sci. Technol.* **2014**, *48*, 6342. [CrossRef] [PubMed]
47. Lu, X.; Zhong, R.; Liu, Y.; Li, Z.; Yang, J.; Wang, F. The incorporation of $Pb^{2+}$ during struvite precipitation: Quantitative, morphological and structural analysis. *J. Environ. Manag.* **2020**, *276*, 111359. [CrossRef]

**Disclaimer/Publisher's Note:** The statements, opinions and data contained in all publications are solely those of the individual author(s) and contributor(s) and not of MDPI and/or the editor(s). MDPI and/or the editor(s) disclaim responsibility for any injury to people or property resulting from any ideas, methods, instructions or products referred to in the content.

Article

# The Influence of Hydrothermal Temperature on Alumina Hydrate and Ammonioalunite Synthesis by Reaction Crystallization

Junkai Wang, Laishi Li *, Yusheng Wu * and Yuzheng Wang

School of Material Science and Engineering, Shenyang University of Technology, Shenyang 110870, China
* Correspondence: lilaishi@sut.edu.cn (L.L.); wuyus@sut.edu.cn (Y.W.)

**Abstract:** With the rapid development of the alumina industry and the shortage of bauxite, high-alumina coal fly ash (HACFA) has attracted more and more attention as a potential alternative alumina resource. In order to extract alumina from HACFA with newly developed technology, the investigation of the crucial step, the reaction between $NH_4Al(SO_4)_2 \cdot 12H_2O$ and $NH_3 \cdot H_2O$, is necessary and valuable. Thermodynamic analyses have shown that four kinds of alumina hydrate (boehmite, diaspore, gibbsite, and bayerite) might be formed at 120–200 °C, and ammonioalunite might be formed at temperatures over 180 °C. A hydrothermal reaction crystallization method was applied to this reaction. The experimental results showed that boehmite (AlOOH) could be formed at 150 °C and 200 °C after 12 h and $NH_4Al_3(SO_4)_2(OH)_6$, an unstable intermediate, is formed during the initial stage and transformed into boehmite, eventually. The higher temperature (200 °C) was more energetically favorable for the formation of $NH_4Al_3(SO_4)_2(OH)_6$, and the crystallinity of the products was better. More importantly, the sheet-like structure of boehmite (AlOOH) could be formed at 150 °C after 24 h of reaction time. The SEM results proved that the sheet-like structures evolutionary process of boehmite.

**Keywords:** alumina hydrate; boehmite; ammonioalunite; evolutionary process; reaction crystallization

Citation: Wang, J.; Li, L.; Wu, Y.; Wang, Y. The Influence of Hydrothermal Temperature on Alumina Hydrate and Ammonioalunite Synthesis by Reaction Crystallization. *Crystals* **2023**, *13*, 763. https://doi.org/10.3390/cryst13050763

Academic Editor: James L. Smialek

Received: 10 April 2023
Revised: 25 April 2023
Accepted: 29 April 2023
Published: 4 May 2023

**Copyright:** © 2023 by the authors. Licensee MDPI, Basel, Switzerland. This article is an open access article distributed under the terms and conditions of the Creative Commons Attribution (CC BY) license (https://creativecommons.org/licenses/by/4.0/).

## 1. Introduction

Alumina ($Al_2O_3$), one of the most important powder materials obtained from calcined alumina hydrates, has been applied in various fields such as in the development of heat-resistant, abrasion-resistant, and corrosion-resistant materials [1,2]. The microscopic morphology of alumina powder is the same as one of its precursors, alumina hydrate [3]. Therefore, it is also necessary to research the synthesis of alumina hydrate. As an important alumina hydrate, boehmite is not only a precursor of alumina, but it also has wide applications in the context of catalysis, surfactants, and adsorbents due to its high surface area mesoporous structure [4,5].

It is well known that alumina is mainly extracted from bauxite; however, the shortage of bauxite is a growing problem in China. In recent years, the extraction of alumina from high-alumina coal fly ash (HACFA), which is the solid waste contained in aluminum resources, has stimulated great research interest [6–8]. A new process for alumina extraction was developed using HACFA as follows: (a) aluminum ammonium sulfate solutions containing impurities are firstly acquired via leaching HACFA and by using ammonium hydrogen sulfate solutions; (b) relatively pure solid aluminum ammonium sulfate is obtained via impurity removal by cooling crystallization; and (c) the alumina hydrate is prepared via the reaction of solid aluminum ammonium sulfate with ammonia ($NH_3 \cdot H_2O$). For aluminum resource extraction by leaching HACFA and ammonium hydrogen sulfate solutions, the preparation of alumina hydrate by reactive crystallization is a crucial step for controlling the morphology of alumina [9,10].

Hydrothermal methods can generate high-temperature and high-pressure environments, which can help insoluble or poorly soluble powders to achieve dissolution and recrystallization. The phase transformation and achievement of various microscopic morphologies can also be obtained using hydrothermal methods. Peng et al. [11] discovered that the different phase structures of $MnO_2$ could be obtained using hydrothermal methods. As it is an important chemical synthesis method, reaction crystallization has been used widely in industrial fields. Moreover, hydrothermal reaction crystallization is a method through which crystallization takes place in hydrothermal reactors, and it allows one to obtain supersaturation via a coordination reaction [12]. Huang et al. [13] found that hierarchically boehmite and urchin-like hollow alumina microspheres with interconnected needle-like building blocks can be synthesized using hydrothermal methods. Therefore, the application of hydrothermal reaction crystallization to solid $NH_4Al(SO_4)_2 \cdot 12H_2O$ and $NH_3 \cdot H_2O$ as a new process for extracting alumina from HACFA could lead to new findings.

Regarding the crystalline structure of alumina hydrate, there are two major categories: $Al(OH)_3$ (gibbsite and bayerite) and AlOOH (diaspore and boehmite) [14]. The preparation of alumina hydrate plays a decisive role in determining the quality and properties of $Al_2O_3$. The microscopic morphology of the alumina powder inherits the morphology of its precursor, aluminum hydroxide powder, due to the topological conversion between aluminum hydroxide and alumina. For the reaction crystallization between solid $NH_4Al(SO_4)_2 \cdot 12H_2O$ and $NH_3 \cdot H_2O$, Ji et al. [15] revealed that pseudoboehmite was obtained through the reaction crystallization of a solution of ammonium aluminum sulfate [$NH_4Al(SO_4)_2$ (aq)] and ammonia water [$NH_3 \cdot H_2O$ (aq)]. Compared with liquid–liquid reactions, solid–liquid reactions represent a more energy-efficient option. In the liquid–liquid reaction, the solid $NH_4Al(SO_4)_2 \cdot 12H_2O$ was first dissolved into a solution system and then reacted with ammonia to produce alumina hydrate.

In this paper, we focused on the reaction crystallization process between the solid $NH_4Al(SO_4)_2 \cdot 12H_2O$ and $NH_3 \cdot H_2O$. A thermodynamic analysis was used to assess the feasibility of the reaction product at 120–200 °C. The samples were prepared at 150 °C and 200 °C with different reaction times. The phases of the samples were prepared at 150 °C and 200 °C, and the processes that led to the transformation to alumina hydrate and ammonioalunite products were investigated. Additionally, the morphology of the samples obtained at different reaction times was studied, and the evolution process that led to the creation of sheet-like boehmite structures was also investigated.

## 2. Experimental Section

### 2.1. Chemicals and Materials

$NH_4Al(SO_4)_2 \cdot 12H_2O$ and $NH_3 \cdot H_2O$ were purchased from Tianjin Zhiyuan Chemical Reagent Co. LTD. (Tianjin, China)

### 2.2. Sample Preparation

In the hydrothermal (HT) synthesis process, 40 mL of 2.5% $NH_3 \cdot H_2O$ solution and 5 g of solid $NH_4Al(SO_4)_2 \cdot 12H_2O$ were added to a Teflon-lined stainless steel autoclave and kept at different temperatures (150 °C and 200 °C) for a certain time. The samples were collected by centrifugation and washed with distilled water several times. The samples were named as "HT- temperature-time" (e.g., HT-150-24 h means that the sample was kept at 150 °C for 24 h).

### 2.3. Sample Characterization

X-ray diffraction (XRD) patterns were performed using a Rigaku SmartLab SE diffractometer equipped with a Cu Kα X-ray source and a 2θ range of 10°–80° with 5°/min. The morphology of the sample was monitored using a scanning electron microscope (SEM, TESCAN MIRA LMS). The surface areas and pore volumes of the samples were evaluated by nitrogen physisorption using a surface area and pore size analyzer (Gold APP Instruments V-Sorb 2008P, Beijing, China). The surface areas were estimated using the BET

(Brunauer−Emmett−Teller) method and the pore volumes were determined by the BJH (Barrett−Joyner−Halenda) method.

## 3. Thermodynamic Analysis

In the standard state, the Van't Hoff formula used was as follows [16]:

$$\Delta G = \Delta G^0 + RT \ln J \qquad (1)$$

where $\Delta G^0$ is the Gibbs free energy change at the same temperature and standard pressure, R is the gas constant (8.3145 J·K$^{-1}$·mol$^{-1}$), T is the thermodynamic temperature (K), and J is the entropy in either reaction state.

When the reaction is in equilibrium, i.e., $\Delta G^0 = 0$, J is the equilibrium constant K (m/M) [17]:

$$\Delta G_T^0 = -RT \ln K \qquad (2)$$

The Gibbs equation of the reaction in the standard state is as follows [18]:

$$\Delta G = \Delta H - T\Delta S \qquad (3)$$

where $\Delta H$ is the enthalpy change in the standard state and $\Delta S$ is the entropy change in the standard state.

The reactions between the $NH_4Al(SO_4)_2 \cdot 12H_2O$ and $NH_3 \cdot H_2O$ solutions that may have occurred are as follows:

$$NH_4Al(SO_4)_2 + 3NH_3 \cdot H_2O = AlOOH \text{ (boehmite)} + 2(NH_4)_2SO_4 + H_2O \qquad (4)$$

$$NH_4Al(SO_4)_2 + 3NH_3 \cdot H_2O = AlOOH \text{ (diaspore)} + 2(NH_4)_2SO_4 + H_2O \qquad (5)$$

$$NH_4Al(SO_4)_2 + 3NH_3 \cdot H_2O = Al(OH)_3 \text{ (gibbsite)} + (NH_4)_2SO4 \qquad (6)$$

$$NH_4Al(SO_4)_2 + 3NH_3 \cdot H_2O = Al(OH)_3 \text{ (bayerite)} + (NH_4)_2SO4 \qquad (7)$$

In the reaction crystallization between $NH_4Al(SO_4)_2 \cdot 12H_2O$ and $NH_3 \cdot H_2O$, the $NH_4Al(SO_4)_2 \cdot 12H_2O$ solid is first dissolved into a solution system. Next, hydroxide is obtained by the $Al^{3+}$ combining with the $OH^-$. Due to the reaction crystallization that took place in the alkaline environment, the $OH^-$ could be effectively disassociated from the $NH_3 \cdot H_2O$ solution. Neither $NH_4^+$ nor $SO_4^{2-}$ is involved in the main chemical reaction, and the reaction of Equations (4)–(7) can be written as follows:

$$Al^{3+} + 3OH^- = AlOOH \text{(boehmite)} + H_2O \qquad (8)$$

$$Al^{3+} + 3OH^- = AlOOH \text{(diaspore)} + H_2O \qquad (9)$$

$$Al^{3+} + 3OH^- = Al(OH)_3 \text{ (gibbsite)} \qquad (10)$$

$$Al^{3+} + 3OH^- = Al(OH)_3 \text{ (bayerite)} \qquad (11)$$

The following products may be produced during reaction crystallization: boehmite, diaspore, gibbsite, and bayerite. However, it is noted that alumina hydrate cannot be formed at a high temperature as reaction (12) may occur and ammonioalunite $(NH_4Al_3(SO_4)_2(OH)_6)$ is obtained when $NH_4^+$, $Al^{3+}$, and $SO_4^{2-}$ exist simultaneously in the solution [19,20]. However, ammonioalunite is not a common compound. It has rarely been reported and little is known about its chemical and physical characteristics. According to our knowledge, there are only two papers about ammonioalunite. Wang et al. [20] revealed that an ammonioalunite with a hexagonal rose-like morphology was synthesized via a hydrothermal method at 165 °C, which used $Al_2(SO_4)_3 \cdot 18H_2O$ and urea as the raw materials and cetyltrimethylammonium bromide (CTAB) as the additive. Yang et al. [19] found that ammonioalunite can be obtained, via the hydrothermal synthesis method, in an ammonium aluminum sulfate solution without other raw materials at temperatures of over 180 °C. Additionally, the morphology of ammonioalunite is constituted by irregular particles. If reaction (12) happens in a reaction crystallization system between $NH_4Al(SO_4)_2 \cdot 12H_2O$ and $NH_3 \cdot H_2O$, it will be the first time that ammonioalunite has been synthesized in the alkaline initial environment.

$$NH_4^+ + 3Al^{3+} + 2SO_4^{2-} + 6OH^- = NH_4Al_3(SO_4)_2(OH)_6 \qquad (12)$$

In summary, it can be seen that there is a complexity to the reaction crystallization of alumina hydrate. In order to judge the reaction between $NH_4Al(SO_4)_2 \cdot 12H_2O$ and $NH_3 \cdot H_2O$, thermodynamic analysis can be used as the first effective method; furthermore, the important thermodynamic parameters should be calculated, including $\Delta G$, $\Delta H$, and $\Delta S$. These parameter values for the formation of alumina hydrate from $NH_4Al(SO_4)_2 \cdot 12H_2O$ and $NH_3 \cdot H_2O$ at 120 °C, 150 °C, 180 °C, and 200 °C were calculated and are summarized in Table 1. It could be summarized that $\Delta G_T^0$ in reactions (8)–(11) was negative at 120–200 °C; this indicates that these reactions were thermodynamically feasible and spontaneous in nature and that the values did not differ much at the same temperature [21]. In addition, the value of $\Delta G_T^0$ for reactions (8)–(11) decreased with an increase in temperature. Additionally, the increased temperature could contribute to the appropriate formation of alumina hydrate. At 120 °C, the order of the $\Delta G_T^0$ value was as follows: reaction (11) > reaction (8) > reaction (10) > reaction (9). At 200 °C, the order of the $\Delta G_T^0$ value became as follows: reaction (11) > reaction (10) > reaction (8) > reaction (9).

**Table 1.** Thermodynamic calculations of reactions (8)–(11).

|  | T/°C | ΔH /(KJ/mol) | ΔS/(J/K) | ΔG /(KJ/mol) | K | Log(K) |
|---|---|---|---|---|---|---|
| Reaction (8) | 120 | 3.188 | 151.956 | −56.554 | $2.757 \times 10^{31}$ | 31.441 |
|  | 150 | 8.558 | 165.101 | −61.305 | $4.628 \times 10^{31}$ | 31.665 |
|  | 180 | 15.431 | 180.77 | −66.485 | $1.169 \times 10^{32}$ | 32.068 |
|  | 200 | 21.146 | 193.103 | −70.221 | $2.742 \times 10^{32}$ | 32.438 |
| Reaction (9) | 120 | −1.001 | 148.645 | −59.444 | $1.116 \times 10^{33}$ | 33.048 |
|  | 150 | 4.314 | 161.664 | −64.094 | $1.277 \times 10^{33}$ | 33.106 |
|  | 180 | 11.132 | 177.207 | −69.170 | $2.305 \times 10^{33}$ | 33.363 |
|  | 200 | 16.81 | 189.461 | −72.834 | $4.414 \times 10^{33}$ | 33.645 |
| Reaction (10) | 120 | 1.246 | 138.819 | −53.330 | $4.452 \times 10^{29}$ | 29.649 |
|  | 150 | 6.464 | 151.589 | −57.681 | $6.221 \times 10^{29}$ | 29.794 |
|  | 180 | 13.218 | 166.987 | −62.452 | $1.326 \times 10^{29}$ | 30.123 |
|  | 200 | 18.87 | 179.185 | −65.911 | $2.799 \times 10^{29}$ | 30.447 |
| Reaction (11) | 120 | −2.841 | 138.507 | −57.295 | $7.123 \times 10^{31}$ | 31.853 |
|  | 150 | 2.361 | 151.24 | −61.636 | $6.864 \times 10^{31}$ | 31.837 |
|  | 180 | 9.096 | 166.592 | −66.396 | $1.058 \times 10^{32}$ | 32.025 |
|  | 200 | 14.733 | 178.757 | −69.846 | $1.841 \times 10^{32}$ | 32.265 |

For the synthesis of $NH_4Al_3(SO_4)_2(OH)_6$ in the $H_2O$ system, containing $NH_4^+$, $Al^{3+}$, and $SO_4^{2-}$, the related thermodynamic calculations were conducted by Yang [19], and the relationships between $\Delta G_T^0$ and T in reaction (13) were explored. The results showed that a hydrothermal reaction condition of over 177 °C was required to obtain $NH_4Al_3(SO_4)_2(OH)_6$. Reaction (12) was similar to reaction (13), while the only different reactant was $OH^-$ in reaction (12) when compared with $H_2O$ in reaction (13). Therefore, the values of $\Delta G_T^0$ are close together and similar.

$$NH_4^+ + 3Al^{3+} + 2SO_4^{2-} + 6H_2O = NH_4Al_3(SO_4)_2(OH)_6 + 6H^+ \qquad (13)$$

The negative $\Delta H$ for reactions (9) and (11) at 120–200 °C indicated that the reactions were exothermic, while the rest of the $\Delta H$ values were positive, which indicated that the reactions were endothermic [22]. The positive $\Delta S$ values suggested a degree of disorder in the system and $\Delta S$ increased with temperature; thus, the degree of disorder in the reaction system increased [23]. Additionally, all the theoretical possibilities and the feasibility of these reactions were illustrated by thermodynamic calculations, but the actual products could not be determined by thermodynamic data, requiring further investigation.

Previous work by our group indicated that boehmite would be produced by reaction crystallization between $NH_4Al(SO_4)_2 \cdot 12H_2O$ and $NH_3 \cdot H_2O$ in a low-temperature and atmospheric pressure environment [24]. As one of the effective methods for controlling product morphology and phases, the change in hydrothermal temperature should be focused on. In this study, 150 °C and 200 °C were selected as the research temperatures, considering the reasons for forming the condition of $NH_4Al_3(SO_4)_2(OH)_6$.

## 4. Results and Discussion

### 4.1. XRD Analysis

To determine the phase structure of the reactant and products, XRD analysis was used, and the patterns are presented in Figures 1–3. As is shown in Figure 1, the XRD patterns of the $NH_4Al(SO_4)_2 \cdot 12H_2O$ (starting solid) reactant corresponded with the standard XRD pattern (PDF No. 71-2203).

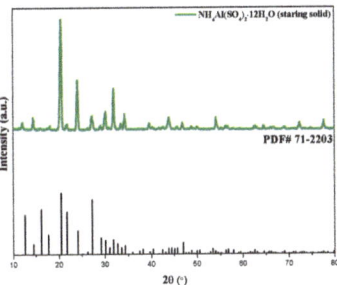

**Figure 1.** XRD patterns of $NH_4Al(SO_4)_2 \cdot 12H_2O$ (starting solid).

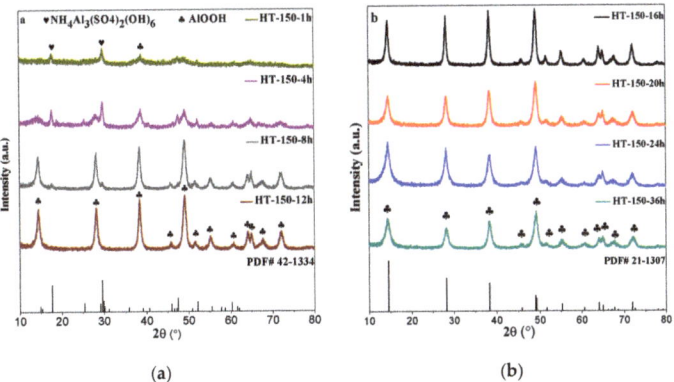

**Figure 2.** XRD patterns of samples at 150 °C. (**a**) HT-150-1 h, HT-150-4 h, HT-150-8 h, and HT-150-12 h samples, and (**b**) HT-150-16 h, HT-150-20 h, HT-150-24 h, and HT-150-36 h samples.

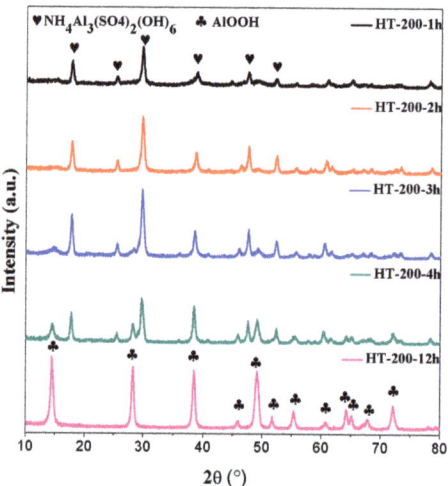

**Figure 3.** XRD patterns of samples at 200 °C.

The XRD patterns of the hydrothermal samples at 150 °C with different reaction times (1 h, 4 h, 8 h, 12 h, 16 h, 20 h, 24 h, and 36 h) are shown in Figure 2. For the HT-150-1 h sample, only three small peaks at 17.6°, 29.5°, and 38.8° could be observed. The peaks at 17.6° and 29.5° were associated with $NH_4Al_3(SO_4)_2(OH)_6$ (PDF No. 42-1334) [19], and the peak at 38.8° was attributed to the characteristic peaks of boehmite (AlOOH) [6]. This indicated that the crystallized product was formed after 1 h with the fundamental amorphous phase. Additionally, there were no obvious diffraction peaks that could be observed; thus, the phase of the HT-150-1 h sample could not be accurately determined. For the HT-150-4 h sample, diffraction peaks at 17.6° and 29.5° could also be observed, and other diffraction peaks with smaller intensities and broader widths could be observed at 14.5°, 28.1°, 38.3°, 48.8°, 51.5°, 55.1°, 60.5°, 63.8°, 64.8°, 67.5°, and 72.1°, which were associated with the crystal structure of boehmite. This indicated that the phase of boehmite formed gradually in the HT-150-4 h sample. When the reaction time reached eight hours, diffraction peaks corresponding to $NH_4Al_3(SO_4)_2(OH)_6$ were barely observed, except for the small peaks at 17.6° and 29.5°. Additionally, the remaining peaks belonged to boehmite. With the reaction time extended to twelve hours, the peaks of $NH_4Al_3(SO_4)_2(OH)_6$ completely disappeared, and the intensity of the peaks corresponding to boehmite were enhanced, which indicated that good boehmite crystallinity was formed in the HT-150-12 h sample. Between the HT-150-16 h, HT-150-20 h, HT-150-24 h, and HT-150-36 h samples, the XRD patterns showed similar diffraction peaks, as is shown in Figure 2b. All the diffraction peaks corresponding to boehmite (PDF No. 21-1307) were found in the hydrothermal product. Noticeably, diffraction peaks assigned to $NH_4Al_3(SO_4)_2(OH)_6$ were observed in the HT-150-1 h and HT-150-4 h samples at 17.6° and 29.5°, which indicated that a decent amount of $NH_4Al_3(SO_4)_2(OH)_6$ could be formed at 150 °C.

Figure 3 shows the XRD patterns of the hydrothermal samples at 200 °C with different reaction times (1 h, 2 h, 3 h, 4 h, and 12 h). For the HT-200-1 h sample, diffraction peaks around 17.6°, 25.2°, 29.5°, 38.6°, 47.3°, and 51.9° were ascribed to the characteristic reflections of $NH_4Al_3(SO_4)_2(OH)_6$ [19]. This indicated that when the hydrothermal temperature reached 200 °C, the reaction product was not alumina hydrate but $NH_4Al_3(SO_4)_2(OH)_6$. This was also the first time that the synthesis of $NH_4Al_3(SO_4)_2(OH)_6$ occurred under initial alkaline conditions. Noticeably, with prolonged reaction times, the transformation of phases for the hydrothermal sample at 200 °C took place. When the reaction time reached 12 h, the crystallized product was completely transformed into boehmite at a 150 °C reaction temperature. Additionally, the same condition occurred at a 200 °C hydrothermal temperature, where no $NH_4Al_3(SO_4)_2(OH)_6$ peak could be observed in the HT-200-12 h sample and diffraction peaks of boehmite with a good crystallinity were observed.

The XRD data indicated that the reaction product was boehmite, transformed from $NH_4Al(SO_4)_2(OH)_6$ with the increase in reaction time, which occurred regardless of whether the hydrothermal temperature was 150 °C or 200 °C. Therefore, $NH_4Al_3(SO_4)_2(OH)_6$ is an unstable intermediate. Combined with the results of Figures 1–3, the solid phase transition that took place in this reaction system was as follows: $NH_4Al(SO_4)_2 \cdot 12H_2O \rightarrow NH_4Al_3(SO_4)_2(OH)_6 \rightarrow AlOOH$. In addition, the generation of $NH_4Al(SO_4)_2 \cdot 12H_2O$ in the initial phase when the temperature increased from 150 °C to 200 °C was favored. Moreover, the crystallinity of the product (for both $NH_4Al(SO_4)_2 \cdot 12H_2O$ and AlOOH) was also beneficial.

For this solution system—which contained $NH_4^+$, $Al^{3+}$, $SO_4^{2-}$, and $OH^-$ simultaneously—the formation of alumina hydrate or $NH_4Al_3(SO_4)_2(OH)_6$ has been previously reported; however, the transformations of these compounds have yet to be reported. It has been reported that rose-like ammonioalunite produced from $Al_2(SO_4)_3 \cdot 18H_2O$ and urea, with cetyltrimethylammonium bromide (CTAB) as the additive, is produced via the hydrothermal method at 165 °C and 4 h [20]. Nevertheless, boehmite (γ-AlOOH) hollow microspheres were synthesized from $Al_2(SO_4)_3 \cdot 18H_2O$ and urea, and the production of the amphiphilic block copolymer of poly-styrene-block-poly-hydroxyl-ethyl acrylate (PS-b-PHEA), as a structure-directing reagent, via hydrothermal synthesis at 150 °C and 24 h was also achieved [25]. In the two studies mentioned above, different products—$(NH_4Al_3(SO_4)_2(OH)_6$ and γ-AlOOH)—could be obtained from the same reactants ($Al_2(SO_4)_3 \cdot 18H_2O$ and urea) at a similar temperature. It is well known that the phase of the product is not affected by the template. Thus, this reflects the complicated nature of a solution system that contains $NH_4^+$, $Al^{3+}$, $SO_4^{2-}$, and $OH^-$ simultaneously. Significantly, it could be found that the synthesis time for $NH_4Al(SO_4)_2(OH)_6$ is lower than alumina hydrate, and that the synthesis time is relatively long for alumina hydrate when using the hydrothermal method. This was also consistent with our experimental results, although aluminum ammonium sulfate ($NH_4Al(SO_4)_2 \cdot 12H_2O$) was used as an aluminum source in our experiment. In addition, we have reason to believe that alumina hydrate with a specific morphology can be obtained as the reaction is extended.

$NH_4Al_3(SO_4)_2(OH)_6$ has potential application value in the fields of artificial gemstones, catalyst carriers, sewage treatment, corundum products, etc. [20,26]. However, alumina hydrate from CFA is a better choice for recycled aluminum resources because it is more energy efficient when compared to ammonioalunite. It is the most widespread and in demand application, as is the preparation of aluminum (Al) through the electrolytic reduction of alumina ($Al_2O_3$) and as $Al_2O_3$ is from the calcination of alumina hydrate. More synthetic paths would provide a more certain basis for its application, and the synthesis of it from CFA would provide a much greater variety.

*4.2. SEM Analysis*

Figure 4 shows the morphology of the HT-150-24 h sample. A large amount of the sheet-like structure, with distinguishable boundaries of boehmite (AlOOH), was observed in the HT-150-24 h sample. Additionally, the rose-like structures were composed of nanosheet building blocks, which were also unstable. It was reported that boehmite crystals prefer to grow into nano-flakes under weak, basic hydrothermal conditions, owing to their distinctly $AlO_6$-octahedral-layered structure, which possesses a great deal of surface-located OH-groups [27,28]. The boehmite samples in the literature exhibit mainly flake-like structures that have a length of approximately 600 nm and a width of around 250 nm [28]. In our research, the boehmite sample exhibited mainly sheet-like morphologies with a length of around 1.6 μm and a width of approximately 0.5 μm. In order to investigate the evolutionary process of the sheet-like structure of the AlOOH crystals, SEM was performed on the whole samples at a hydrothermal temperature of 150 °C. Several representative pictures of this process are shown in Figure 5.

**Figure 4.** SEM images of the HT-150-24 h sample.

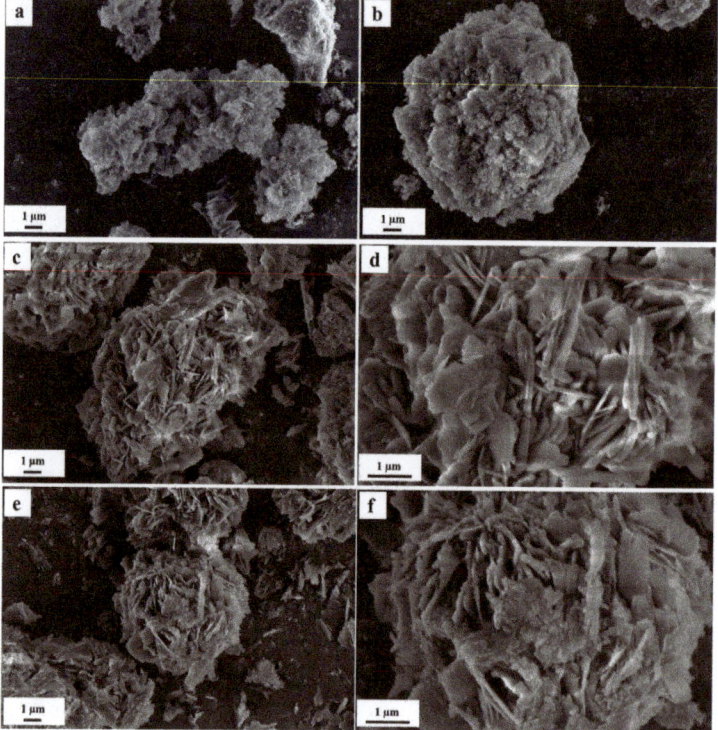

**Figure 5.** SEM images of the different samples. (**a**): HT-150-1 h; (**b**): HT-150-4 h; (**c**,**d**): HT-150-8 h; and (**e**,**f**): HT-150-20 h.

As is shown in Figure 5, structures of irregular morphology with coarse and loose surfaces appeared in the HT-150-1 h and HT-150-4 h samples. These samples had a certain pore-like structure, and the integrated granules could hardly be observed in the HT-150-1 h sample. In the beginning of the reactive crystallization process, along with the gradual hydrolysis of $NH_4Al(SO_4)_2 \cdot 12H_2O$, $Al^{3+}$ and $SO_4^{2-}$ began to increase substantially, and $NH_4Al_3(SO_4)_2(OH)_6$ might thus be formed. Combined with the XRD results, certain $NH_4Al_3(SO_4)_2(OH)_6$ crystals were formed in the HT-150-1 h and HT-150-4 h samples. According to the reported literature, amorphous phase clusters form in the initial stage of the hydrothermal reaction; furthermore, the amorphous phase is unstable due to its higher solubility [29]. Additionally, its dissolved–recrystallized process and slow-phase transformation could have occurred in the amorphous phase as a more thermodynamically stable product could be obtained step by step. It can be inferred that boehmite is a more thermodynamically stable crystal when compared with $NH_4Al_3(SO_4)_2(OH)_6$. In the HT-150-8 h sample, particles with lamellar- and strip-like structures can be simultaneously observed. Additionally, as the reaction progressed in the HT-150-20 h sample, the strip-like structure disappeared, and the particle with an extensive lamellar-like structure was formed. On the basis of the XRD results, boehmite was completely formed in the HT-150-12 h sample without $NH_4Al_3(SO_4)_2(OH)_6$. Therefore, it could be suggested that the lamellar-like structure of the particle shown in Figure 6e is boehmite. When the reaction time reached 24 h, it was presumed that the sheet-like structure of boehmite appeared through a partial dissolution of particles with a lamellar-like structure. In conclusion, the morphology's evolutionary process of boehmite was as follows: amorphous phase clusters → particles with lamellar-like and strip-like structures → particles with more lamellar-like structures → sheet-like structures. Additionally, as the reaction time continued to extend, no regular and special morphologies could be observed in the HT-150-36 h sample.

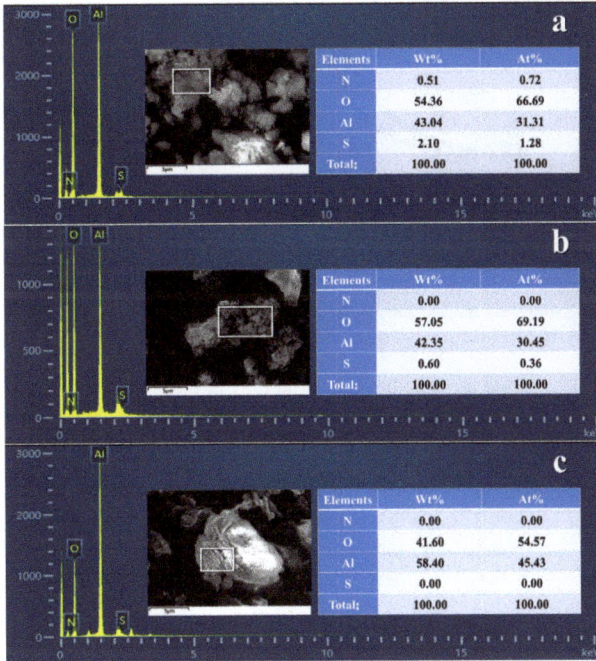

**Figure 6.** The energy spectra of the samples. (**a**): HT-150-1 h; (**b**): HT-150-4 h; and (**c**): HT-150-16 h.

*4.3. Energy Spectrum Analysis*

The energy spectra of the HT-150-1 h, HT-150-4 h, and HT-150-16 h samples are presented in Figure 6. For the HT-150-1 h sample, the element wt% values of N, O, Al, and S were 0.51%, 54.36%, 43.04%, and 2.10%, respectively. It was confirmed that the minor quantities of the S and N elements were present in the HT-150-1 h sample, which further validated the little $NH_4Al_3(SO_4)_2(OH)_6$ crystals that formed in the HT-150-1 h sample. As the reaction progressed, the element wt% values of the N and S in the HT-150-16 h sample decreased gradually until they reached 0. Furthermore, the element wt% values of the O and Al were 41.60% and 58.4%, respectively. These findings were fully consistent with the XRD results.

### 4.4. BET Analysis

Figure 7 shows the $N_2$ absorption–desorption isotherm and the pore-size distribution of the HT-150-16 h, HT-150-20 h, and HT-150-24 h samples. All the samples showed type IV isotherms, according to the BDDT classification, which implies that all the three samples belong to mesoporous materials. The isotherms of the three samples exhibited a typical H3 hysteresis loop as the closure points existed at a low $P/P_0$ ($P/P_0$ = 0.45). The BJH desorption pore-size distributions that were evaluated from the $N_2$ desorption isotherm are presented in Figure 7b. It was evident that the desorption pore-size distributions of the three catalysts gathered at 10–100 nm. The HT-150-16 h and HT-150-20 h samples showed a wider distribution peak, which indicated that the HT-150-16 h and HT-150-20 h samples had a more abundant mesoporous structure when compared with the HT-150-24 h sample.

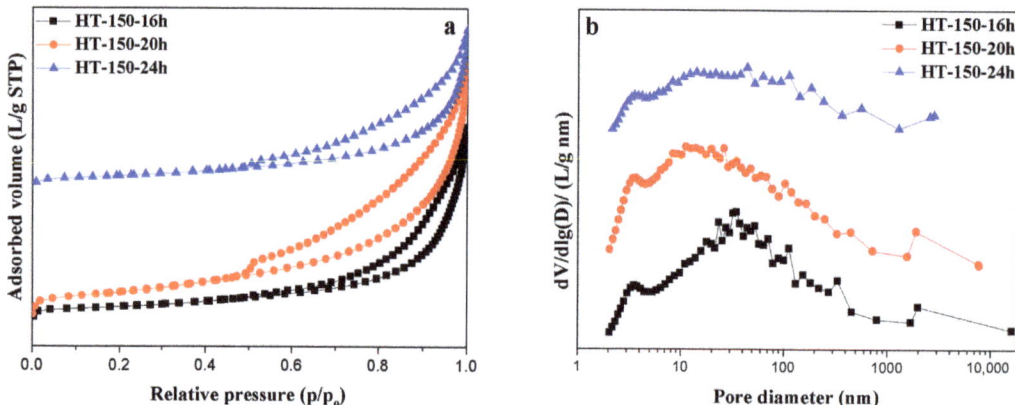

**Figure 7.** $N_2$ absorption–desorption results of the HT-150-16 h, HT-150-20 h, and HT-150-24 h samples. (**a**) $N_2$ absorption–desorption isotherm; and (**b**) pore-size distributions.

Table 2 lists the BET surface areas, pore volumes, and average pore size of the HT-150-16 h, HT-150-20 h, and HT-150-24 h samples. It can be seen that the HT-150-20 h sample possesses the largest specific surface area (88.317 m$^2$/g) and pore volume (0.344 cm$^3$/g). The specific surface area and pore volume were greater when compared with the HT-150-16 h sample. However, the BET surface areas, pore volume, and average pore size of the HT-150-24 h sample were 41.573 m$^2$/g, 0.212 cm$^3$/g, and 15.257 nm, respectively. This represented the smallest value among the three samples. Combined with the results of the SEM analysis, when the reaction time of 16 h was increased to 20 h, the specific surface area and pore volume were increased due to more lamellar-like structures appearing and the strip-like structure disappearing. Additionally, when the reaction time increased to 24 h, the particles with a great number of lamellar-like structures dissolved into sheet-like structures. The specific surface area decreased to 46.744 m$^2$/g and the pore volume decreased to 0.132 cm$^3$/g. The BET results provided supporting evidence for the process of the sheet-like-structured boehmite.

**Table 2.** Structural parameters of the samples.

| Samples | $S_{BET}$ (m$^2$/g) | Pore Volume (cm$^3$/g) | Average Pore Size (nm) |
|---|---|---|---|
| HT-150-16 h | 45.110 | 0.258 | 17.939 |
| HT-150-20 h | 88.317 | 0.344 | 12.309 |
| HT-150-24 h | 41.573 | 0.212 | 15.257 |

## 5. Conclusions

For the hydrothermal reaction crystallization process between a $NH_4Al(SO_4)_2 \cdot 12H_2O$ solid and $NH_3 \cdot H_2O$, the thermodynamic analysis indicated that all four kinds of alumina hydrate (boehmite, diaspore, gibbsite, and bayerite) might be formed. The experimental and characterization results showed that boehmite (AlOOH) was formed at 150 °C and 200 °C after a 12 h reaction. Additionally, $NH_4Al_3(SO_4)_2(OH)_6$ was found to be an unstable intermediate; however, it did eventually transform into boehmite. This study represents the first time that $NH_4Al_3(SO_4)_2(OH)_6$ has been prepared under initial alkaline conditions. The higher temperature (200 °C) was more energetically favorable for the formation of $NH_4Al_3(SO_4)_2(OH)_6$, and the crystallinity of the products (both $NH_4Al_3(SO_4)_2(OH)_6$ and AlOOH) could be improved. Most importantly, however, it was found that the sheet-like structure of boehmite (AlOOH) can be formed at 150 °C and in 24 h. The SEM results prove the evolutionary process of boehmite with a sheet-like-structure.

**Author Contributions:** Conceptualization, J.W. and L.L.; methodology, Y.W. (Yusheng Wu); formal analysis, J.W.; investigation, J.W.; resources, L.L. and Y.W. (Yusheng Wu); data curation, Y.W. (Yuzheng Wang); writing—original draft preparation, J.W.; writing—review and editing, L.L.; visualization, Y.W. (Yuzheng Wang); project administration, Y.W. (Yusheng Wu). All authors have read and agreed to the published version of the manuscript.

**Funding:** This work was supported by the National Natural Science Foundation of China (Grant No. 51974188) and the LiaoNing Revitalization Talents Program (No. XLYC2008014).

**Institutional Review Board Statement:** "Not applicable" for studies not involving humans or animals.

**Data Availability Statement:** The data underlying the results presented in this paper are not publicly available at this time but may be obtained from the authors upon reasonable request.

**Conflicts of Interest:** We have no relevant financial interests to disclose. None of the authors have any financial or scientific conflicts of interest with respect to the research described in this manuscript.

## References

1. Zhang, W.; Wang, S.; Zhao, L.; Ran, J.; Kang, W.; Feng, C.; Zhu, J. Investigation of Low-Calcium Circulating Fluidized Bed Fly Ash on the Mechanical Strength and Microstructure of Cement-Based Material. *Crystals* **2022**, *12*, 400. [CrossRef]
2. Luțcanu, M.; Cimpoeșu, R.; Abrudeanu, M.; Munteanu, C.; Moga, S.G.; Coteata, M.; Zegan, G.; Benchea, M.; Cimpoeșu, N.; Murariu, A.M. Mechanical Properties and Thermal Shock Behavior of $Al_2O_3$-YSZ Ceramic Layers Obtained by Atmospheric Plasma Spraying. *Crystals* **2023**, *13*, 614. [CrossRef]
3. Frazer, I.H.; Leggatt, G.R.; Mattarollo, S.R. Prevention and treatment of papillomavirus-related cancers through immunization. *Annu. Rev. Immunol.* **2011**, *29*, 111–138. [CrossRef] [PubMed]
4. Chen, S.; Yang, S.; Cheung, C.F.; Liu, T.; Duan, D.; Ho, L.-T.; Jiang, Z. Study on the Surface Generation Mechanism during Ultra-Precision Parallel Grinding of SiC Ceramics. *Crystals* **2023**, *13*, 646. [CrossRef]
5. Vo, D.D.; Alsarraf, J.; Moradikazerouni, A.; Afrand, M.; Salehipour, H.; Qi, C. Numerical investigation of γ-AlOOH nano-fluid convection performance in a wavy channel considering various shapes of nanoadditives. *Powder Technol.* **2019**, *345*, 649–657. [CrossRef]
6. Wu, Y.S.; Xu, P.; Li, L.S. Synthesis of alumina with coarse particle by precipitating aluminum ammonium sulfate solution with ammonia. *Adv. Powder Technol.* **2016**, *27*, 124–129. [CrossRef]
7. Alteary, S.S.; Marei, N.H. Fly ash properties, characterization, and applications: A review. *J. King Saud Univ. Sci.* **2021**, *33*, 101536. [CrossRef]
8. Ma, Y.; Stopic, S.; Xakalashe, B.; Ndlovu, S.; Forsberg, K.; Friedrich, B. A cleaner approach for recovering Al and Ti from coal fly ash via microwave assisted baking, leaching and precipitation. *Hydrometallurgy* **2021**, *206*, 105754. [CrossRef]

9. Yang, X.; Wu, Y.; Li, L.; Wang, Y.; Li, M. Crystallization mechanism of ammonium aluminum sulfate during cooling process. *J. Cryst. Growth* **2021**, *560–561*, 126064. [CrossRef]
10. Wu, Y.; Yang, X.; Li, L.; Wang, Y.; Li, M. Kinetics of extracting alumina by leaching coal fly ash with ammonium hydrogen sulfate solution. *Chem. Pap.* **2019**, *73*, 2289–2295. [CrossRef]
11. Peng, Y.; Chang, H.; Dai, Y.; Li, J. Structural and surface effect of $MnO_2$ for low temperature selective catalytic reduction of NO with $NH_3$. *Procedia Environ. Sci.* **2013**, *18*, 384–390. [CrossRef]
12. Wang, F.; Zhu, L.; Wei, Q.; Wang, Y. Research on the effects of hydrothermal synthesis conditions on the crystal habit of MIL-121. *R. Soc. Open Sci.* **2020**, *7*, 201212. [CrossRef] [PubMed]
13. Huang, H.; Wang, L.; Cai, Y.; Zhou, C.; Yuan, Y.; Zhang, X.; Wan, H.; Guan, G. Facile fabrication of urchin-like hollow boehmite and alumina microspheres with a hierarchical structure via Triton X-100 assisted hydrothermal synthesis. *CrystEngComm* **2015**, *17*, 1318–1325. [CrossRef]
14. Kloprogge, J.T.; Duong, L.V.; Wood, B.J.; Frost, R.L. Xps study of the major minerals in bauxite: Gibbsite, bayerite and (pseudo-)boehmite. *J. Colloid Interface* **2006**, *296*, 572–576. [CrossRef]
15. Ji, Y.; Wu, Y.; Li, L. Synthesis and characterization of pseudoboehmite by neutralization method. *Ceram. Int.* **2021**, *47*, 15923–15930. [CrossRef]
16. Ahsendorf, T.; Wong, F.; Eils, R.; Gunawardena, J. A framework for modelling gene regulation which accommodates non-equilibrium mechanisms. *BMC Biol.* **2014**, *12*, 102. [CrossRef]
17. Teng, Y.; Liu, Z.; Yao, K.; Song, W.; Sun, Y.; Wang, H.; Xu, Y. Preparation of Attapulgite/$CoFe_2O_4$ magnetic composites for efficient adsorption of tannic acid from aqueous solution. *Int. J. Environ. Res. Public Health* **2019**, *16*, 2187. [CrossRef]
18. Xie, F.; Zhang, J.; Chen, J.; Wang, J.; Wu, L. Research on enrichment of $P_2O_5$ from low-grade carbonaceous phosphate ore via organic acid solution. *J. Anal. Method Chem.* **2019**, *3*, 1–7. [CrossRef]
19. Yang, X.; Wu, Y.; Li, L.; Wang, Y.; Li, M. Thermodynamics of ammonioalunite precipitation in ammonium aluminum sulfate solution. *Hydrometallurgy* **2020**, *195*, 105393. [CrossRef]
20. Wang, F.; Zhu, J.; Liu, H. Hydrothermal Synthesis of Ammonioalunite with Hexagonal Rose-like Morphology. *Mater. Lett.* **2018**, *216*, 269–272. [CrossRef]
21. Ueoka, N.; Oku, T. Stability Characterization of $PbI_2$-Added $CH_3NH_3PbI_{3-x}Cl_x$ Photovoltaic Devices. *ACS Appl. Mater. Interfaces* **2018**, *51*, 44443–44451. [CrossRef] [PubMed]
22. Lee, J.; Hwang, T.; Cho, H.B.; Kim, J.; Choa, Y.H. Near theoretical ultra-high magnetic performance of rare-earth nanomagnets via the synergetic combination of calcium-reduction and chemoselective dissolution. *Sci. Rep.* **2018**, *8*, 15656. [CrossRef] [PubMed]
23. Liu, J.; Huang, L.; Xie, W.; Kuo, J.; Buyukada, M.; Evrendilek, F. Bioresource Technology Characterizing and optimizing (co-)pyrolysis as a function of different feedstocks, atmospheres, blend ratios, and heating rates. *Bioresour. Technol.* **2019**, *277*, 104–116. [CrossRef] [PubMed]
24. Ma, X.; Wu, Y.; Li, L.; Wang, Y. Effect of SDBS on crystallization behavior of pseudoboehmite. *J. Phys. Chem. C* **2021**, *125*, 26039–26048. [CrossRef]
25. Wu, X.; Wang, D.; Hu, Z.; Gu, G. Synthesis of γ-AlOOH (γ-$Al_2O_3$) self-encapsulated and hollow architectures. *Mater. Chem. Phys.* **2008**, *109*, 560–564. [CrossRef]
26. Akar, S.T.; Akar, T.; San, E. Chitosan-alunite composite: An effective dye remover with high sorption, regeneration and application potential. *Carbohydr. Polym.* **2016**, *143*, 318–326. [CrossRef]
27. Zhang, L.; Lu, W.; Cui, R.; Shen, S. One-pot template-free synthesis of mesoporous boehmite core–shell and hollow spheres by a simple solvothermal route. *Mster Res. Bull.* **2010**, *45*, 429–436. [CrossRef]
28. Cai, W.; Yu, J.; Gu, S.; Jaroniec, M. Facile hydrothermal synthesis of hierarchical boehmite: Sulfate-mediated transformation from nanoflakes to hollow microspheres. *Cryst. Growth Des.* **2010**, *10*, 3977–3982. [CrossRef]
29. Wu, X.; Zhang, B.; Wang, D.; Hu, Z. Morphology evolution studies of boehmite hollow microspheres synthesized under hydrothermal conditions. *Mater. Lett.* **2012**, *70*, 128–131. [CrossRef]

**Disclaimer/Publisher's Note:** The statements, opinions and data contained in all publications are solely those of the individual author(s) and contributor(s) and not of MDPI and/or the editor(s). MDPI and/or the editor(s) disclaim responsibility for any injury to people or property resulting from any ideas, methods, instructions or products referred to in the content.

Article

# Sn-Doped Hydrated V₂O₅ Cathode Material with Enhanced Rate and Cycling Properties for Zinc-Ion Batteries

Kai Guo [1,2], Wenchong Cheng [1], Haiyuan Liu [1], Wenhao She [1], Yinpeng Wan [1], Heng Wang [1], Hanbin Li [1], Zidan Li [1], Xing Zhong [1], Jinbo Ouyang [1,*] and Neng Yu [1,*]

[1] Jiangxi Province Engineering Research Center of New Energy Technology and Equipment, School of Chemistry, Biology and Materials Science, East China University of Technology, Nanchang 330013, China
[2] State Key Laboratory of Materials Processing and Die & Mould Technology, Huazhong University of Science and Technology, Wuhan 430074, China
* Correspondence: oyjb1001@163.com (J.O.); neng063126@ecut.edu.cn (N.Y.)

**Abstract:** Water molecules and cations with mono, binary, and triple valences have been intercalated into $V_2O_5$ to significantly improve its electrochemical properties as a cathode material of zinc-ion batteries. Sn as a tetravalent element is supposed to interact aggressively with the $V_2O_5$ layer and have a significant impact on the electrochemical performance of $V_2O_5$. However, it has been rarely investigated as a pre-intercalated ion in previous works. Hence, it is intriguing and beneficial to develop water molecules and Sn co-doped $V_2O_5$ for zinc-ion batteries. Herein, Sn-doped hydrated $V_2O_5$ nanosheets were prepared by a one-step hydrothermal synthesis, and they demonstrated that they had a high specific capacity of 374 mAh/g at 100 mA/g. Meanwhile, they also showed an exceptional rate capability with 301 mAh/g even at a large current density of 10 A/g, while it was only 40 mAh/g for the pristine hydrated $V_2O_5$, and an excellent cycling life (87.2% after 2500 cycles at 5 A/g), which was far more than the 25% of the pure hydrated $V_2O_5$. The dramatic improvement of the rate and cycling performance is mainly attributed to the faster charge transfer kinetics and the enhanced crystalline framework. The remarkable electrochemical performance makes the Sn-doped hydrate $V_2O_5$ a potential cathode material for zinc-ion batteries.

**Keywords:** vanadium pentoxide; cathode; doping; rate performance; stability

## 1. Introduction

Aqueous zinc-ion batteries (ZIBs), as one of the candidates for next-generation rechargeable batteries, have attracted tremendous interest because their zinc metal anodes have some unique features, including a high theoretical capacity (819 mA h/g), a low redox potential (−0.76 V vs. SHE), a small radius (0.74 Å), and the two-electron reaction of $Zn/Zn^{2+}$ [1–5]. Unfortunately, the relatively large radius (4.3 Å) of the hydrated $Zn^{2+}$ ion in an aqueous electrolyte and a strong electrostatic interaction with the cathode host both add to a high energy barrier for its intercalation/deintercalation in the cathode materials, resulting in sluggish electrochemical kinetics, serious electrochemical polarization, as well as unsatisfied cycling and rate performances [6–10]. Therefore, it is crucial to design and develop suitable cathode materials for constructing high-performance ZIBs [3,11,12].

A variety of cathode materials have been investigated, such as manganese-based oxides, Prussian blue analogs, conducting polymers, and vanadium-based oxides, over the past few years [13–17]. Among these cathode materials, the vanadium-based oxides have been widely studied for ZIBs because of their multivalence, open skeleton structure, and high theoretical capacities [13,14,18–21]. Vanadium pentoxide ($V_2O_5$) is one of the promising materials due to its high theoretical capacity and layered structure with it having a large interspace [1,3,22,23]. However, $V_2O_5$ usually displays a low conductivity, a poor ion diffusion coefficient, a long activation process, and an unsatisfying cyclic stability [1,24–26].

Generally, the nanostructures with high specific surface areas and short ion diffusion paths are conducive to good $Zn^{2+}$ diffusion rates and rapid electrochemical kinetics [27]. Additionally, recent research suggests that the pre-insertion of water molecules or foreign metal ions (e.g., $Na^+$ and $Ca^{2+}$) into $V_2O_5$ not only strongly modifies its crystal structure, but it also plays an important role in its electrochemical kinetics [28–30]. The water molecules intercalated in the $V_2O_5$ interlayers pillar its layered structure and effectively function like a "lubricant" to facilitate fast $Zn^{2+}$ transport, significantly improving the rate and cycle performance of $V_2O_5$ [31].

As another choice, metal ions incorporated between the $V_2O_5$ layers may covert the crystal structure of $V_2O_5$ to a more stable tunnel framework or enlarge the interlayer spacing and strongly bond to the apical oxygens of the $V_2O_5$ layers to maintain the structural stability of $V_2O_5$, depending on radiuses and charges of the metal ions [7,32]. Cations with mono, binary, and triple valences, such as Na, K, Mg, Ca, Zn, Mn, and Fe, have already been studied [10,11,28,33]. Generally, multivalent metal ions with a higher charge density and stronger electrostatic interaction than those of the monovalent cations are beneficial to build a stronger bond with the vanadium oxide layers, resulting in a better structural stability and cycling performance [5]. Meanwhile, the strong electrostatic interaction between the $V_2O_5$ host and the foreign cations shields the interaction between the oxygen atom and $Zn^{2+}$ and thus, it reduces the energy barrier of the $Zn^{2+}$ diffusion inside $V_2O_5$, which is conducive to a better rate performance [27]. However, the charge numbers of the doped metal ions are no more than three. What will happen to the zinc-ion storage capability if alien ions with a charge number of more than three are hybridized with $V_2O_5$ is unsure. The element tin, which is commonly in a tetravalent state with ion $Sn^{4+}$ and with a charge number of four, is believed to interact more strongly with the $V_2O_5$ layer than other previously reported elements do, and it will have a significant impact on the electrochemical performance of $V_2O_5$ if it is doping $V_2O_5$. However, it has been rarely investigated as a pre-intercalated ion in previous works. Hence, it is interesting and worthwhile to develop Sn-doped $V_2O_5$ cathodes for ZIBs and to clarify the role of the doped Sn element on the zinc-ion storage capability.

Herein, Sn-doped hydrated $V_2O_5$ was synthesized in a one-step hydrothermal method to realize a cathode material with a superior Zn-storage performance by a hydrothermal reaction. Compared with the pristine $V_2O_5$, the obtained SnVO displays larger interlayer spacing, a greatly improved rate performance, and a superior cycling stability. SnVO delivers a high reversible specific capacity of 374 mAh/g at a current density of 100 mA/g, retains 320 mAh/g at 5000 mA/g, and maintains 87.5% of its initial capacity after cycling at 2 A/g for 2000 times. The facile synthesis route and its significant electrochemical performance enhancement suggest that Sn doping is an effective strategy and Sn-doped hydrated $V_2O_5$ is a prospective cathode materials for zinc-ion batteries.

## 2. Experimental Section

### 2.1. Preparation of Sn Doped $V_2O_5 \cdot nH_2O$

Two mmol of $V_2O_5$ (Sinopharm Chemical, Shanghai, China) and 0.2 mmol of $SnCl_4 \cdot 5H_2O$ (Adamas-beta, Shanghai, China) were dissolved in a mixture of 20 mL of deionized water and 10 mL $H_2O_2$ (30 wt%, Sinopharm Chemical, Shanghai, China) by magnetic stirring at room temperature. Then, the solution was poured into the autoclave reactor and heated at 200 °C for 48 h. The resulting reactants were vacuum filtered, thoroughly washed with deionized water, and then, they were dried at 80 °C for 6 h to obtain a dark red, dry gel, which was labeled as SnVOH. For the comparison, a sample was synthesized following the same procedure without the addition of $SnCl_4 \cdot 5H_2O$, and it was labeled as VOH.

### 2.2. Materials Characterizations

The crystalline structure of the samples was identified using an X'Pert 3 diffractometer (PANalytical, Almelo, The Netherlands) at the range from 5 to 80°. The morphologies and structure of the samples were investigated using a field emission scanning electron microscope (FESEM, HITACHI SU8200, Tokyo, Japan) equipped with an energy-dispersive X-ray (EDX) detector and transmission electron microscopy (TEM, FEI Talos S-FEG). X-ray photoelectron spectroscopy (XPS, PHI QUANTERA-II SXM) was used to identify the valence state of the samples.

*2.3. Electrochemical Measurements*

The obtained SnVOH or VOH were mixed with a carbon nanotube (CNT) solution (10 mg/mL, XFNANO, Nanjing, China) and then, it was filtered using a cellulose film (0.22 μm, Shanghai Xinya, Shanghai, China) to form freestanding composite films, for which the mass ratios of SnVOH or VOH were 70%. Additionally, the as-prepared composite films were cut into small pieces and used directly as the cathode to assemble the CR2025 button cells in the air with metallic zinc foil (100 μm thick) as the anode, 3 M zinc trifluoromethane sulfonate ($Zn(CF_3SO_3)_2$) solution as the electrolyte, and glass fiber as the separator. Cycle voltammetry (CV) was performed on the button cells using an electrochemical workstation (CHI 660E, CH Instruments, Inc., Bee Cave, TX, USA) within the potential range of 0.2–1.7 V vs. $Zn/Zn^{2+}$. The electrochemical impedance spectra (EIS) were acquired with the same electrochemical workstation over the frequency range of $0.01$–$10^5$ Hz at a voltage of 5 mV. Additionally, galvanostatic charge–discharge (GCD) tests and galvanostatic intermittent titration technique (GITT) tests were carried out using a NEWARE 4000 system (Neware Technology Limited, Shenzhen, China) at room temperature. For the GITT test, the coin cell was charging/discharging for 20 min at 0.1 A/g with a 120 min relaxation duration. The solid diffusion coefficient was calculated according to the equation below:

$$D = \frac{4L^2}{\pi\tau}\left(\frac{\Delta E_s}{\Delta E_t}\right)^2 \quad (1)$$

where $t$, $\tau$, and $\Delta E_s$ represent the duration of the current pulse (s), the relaxation time (s), and the steady-state voltage change (V) that was induced by the current pulse, respectively [34]. $\Delta E_t$ is the voltage change (V) during the galvanostatic current pulse after eliminating the IR drop. $L$ is the ion diffusion length (cm) of the electrode, which was equal to the thickness of the composite film electrode (40.2 and 41.7 μm for the VOH and SnVOH composite film, respectively). The specific capacity, the energy density, and the power density were calculated based on the mass of active materials from the cathode.

## 3. Results and Discussion

The VOH and SnVOH were both synthesized via a facile hydrothermal reaction, as shown in Figure S1. The compositions of the as-prepared samples were characterized by XRD and XPS. No peaks related to tin metal or tin oxides were detected, and there are only six discrete peaks in the XRD pattern for both VOH and SnVOH, corresponding to the (00n) planes of a hydrated $V_2O_5$ phase (JCPDS NO. 40–1297) (Figure 1a), which is similar to previous research [35]. Additionally, there are slight blue shifts which occurred at the characteristic peaks of VOH after doping the Sn element, suggesting that the (00n) interlayer spacing of VOH was expanded through the Sn doping. The expanded interlayer spacing is commonly induced by the volume change after introducing doping ions to the interlayers of the V-O bilayers in the hydrated $V_2O_5$ [1,3]. According to the Bragg Equation in the Supporting Information file, the calculated interplanar distances of the (001) plane are $d_{001}$ = 14.91 and 14.81 Å for SnVOH and VOH, respectively. The small differences in the interplanar distance of two samples could be ascribed to a combination of two factors. On one hand, the intercalated $Sn^{4+}$ can lead to an expanded interlayer spacing. On the other hand, the strong attraction between the intercalated $Sn^{4+}$ ions and the adjacent V-O bilayers tends to bring the V-O bilayers close to or narrow to the interlayer spacing. These two factors counteract each other and they induce a small interplanar distance change [7,32]. The schematic crystalline structure of SnVOH is illustrated in Figure S2. In the XPS spectrum of SnVOH (Figure 1b), two pairs of characteristic peaks can be found in the V 2p spectra; the stronger peaks at 516.9 and 524.4 eV are ascribed to $V^{5+}$, and the weaker peaks at 516.0 and 522.8 eV are from $V^{4+}$, respectively [36]. The weak peaks from $V^{4+}$ indicate a very small amount of $V^{4+}$, which may be caused by the doping of $Sn^{4+}$. The two O1s XPS peaks are deconvoluted into (V-O) lattice oxygen at 530.4 eV and a hydroxyl (V-OH) of defective oxygen at 532.1 eV, respectively. The peaks that can be observed in the Sn 3d spectrum belong to $Sn^{4+}$ (Figure 1c), confirming the successful doping of the Sn element in the SnVOH [37].

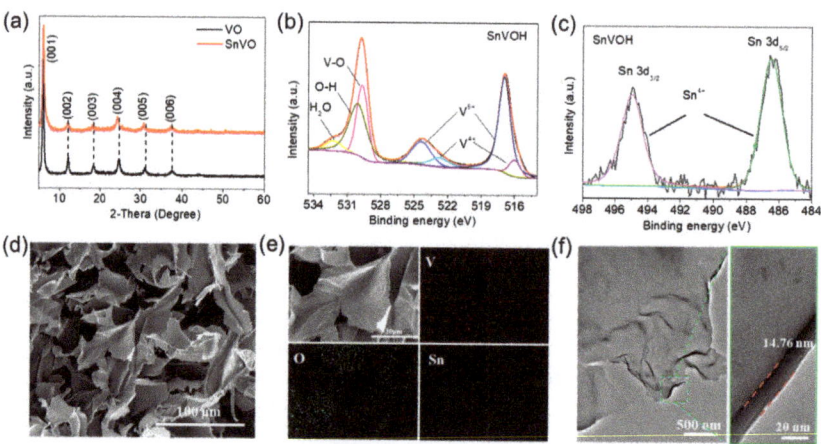

**Figure 1.** Characterization of SnVOH. (**a**) XRD patterns of SnVOH sample; (**b**) XPS spectra of (**b**) O1s and V 2p and (**c**) Sn 3d; (**d**) SEM of SnVOH sample; (**e**) EDS mapping of SnVOH sample; (**f**) TEM of SnVOH sample.

SEM and TEM were performed to observe the morphology and structure of SnVOH. The SEM image (Figure 1d) shows that the SnVOH sample is composed of nanosheets with lateral sizes of tens of micrometers, and the VOH sample has a similar morphology with an even element distribution (Figure S3). The diagram in Figure 1e shows the uniform distribution of the V, O, and Sn elements. Additionally, the EDS elemental analysis result demonstrates that the molar ratio of the Sn element in the SnVOH accounts for 3.85%, as shown in Table S1. The TEM image of SnVOH in Figure 1f confirms the nanosheet morphology and demonstrates that the thickness of the nanosheets is in the range of 10–20 nanometers. The distribution of the elements Sn, V, and O in the SnVOH nanosheet was characterized by the EDS. The above analysis results suggest the Sn-doped $V_2O_5$ has been successfully synthesized.

To evaluate the zinc-ion storage performance, VOH and SnVOH were mixed with the CNT solution and then, they filtered to form free-standing composite films, which were cut into small pieces and used directly as the cathode. The VOH and SnVOH nanosheets were both evenly mixed with the CNTs (Figure S4a,c), and the composite films were layered at a thickness of 40.2 and 41.7 μm, respectively (Figure S4b,d). The coin cells were assembled using the VOH and SnVOH-based cathodes and tested in the voltage range of 0.2–1.7 V vs. $Zn/Zn^{2+}$. There are two pairs of redox peaks in the CV profile of the VOH-based cathode, as shown in Figure S5a. Similarly, the CV profile of SnVOH in Figure 2a shows two oxidation peaks at 0.75 and 1.11 V along with two reduction peaks at 0.37 and 0.75 V, which correspond to the multi-step extraction/insertion of the $Zn^{2+}$ ion in the framework of SnVOH, respectively [37,38]. The redox reactions are consistent with the plateaus of the initial three galvanostatic charge and the discharge curves of SnVOH at 0.1 A/g in Figure 2b. The GCD profiles of SnVOH in Figure 2c show the maximum discharge capacity is 387 mAh/g at a current density of 0.1 A/g. Additionally, its capacity retention is impressive, at 77.8%, when the current density increases from 0.1 to 10 A/g. The pristine VOH shows an average discharge capacity of 395 mAh/g at

0.1 A/g (Figure 2d), which is close to that of SnVOH. However, its rate performance is unsatisfactory by possessing a dramatically lower retention of 12.6% when the current density increased to 10 A/g. The average discharge capacities of 387, 365, 349, 348, 345, 330, 321, 309, and 301 mAh/g were recorded for SnVOH at the current densities of 0.1, 0.3, 0.5, 0.8, 1, 3, 5, 8, and 10 A/g, respectively (Figure 2d). SnVOH had higher capacities than VOH did at the current densities over 0.5 A/g, and the capacity advantage is larger at a higher current density. The significant improvement in the rate performance of SnVOH can be accounted for by the highly facilitated $Zn^{2+}$ diffusion in the VOH framework after incorporating the Sn element. Not only is there a dramatically enhanced rate performance, but SnVOH also demonstrates an improved electrochemical stability at both the small and large current densities, as shown in Figure 2e,f, respectively. During the cycling test at 10 A/g for 500 cycles, the capacity of SnVOH goes through a slow increasing process during the initial 100 cycles and then, it remains stable in the following cycles, while the capacity of VOH decays rapidly in the first 10 cycles, then, it slowly increases in the following 200 cycles, and stays steady in the last 300 cycles. The capacity retentions for SnVOH and VOH are 135.4% and 76.1% after the cycling at 10 A/g for 500 times, respectively (Figure 2e). During the cycling at 5 A/g for 2500 cycles, the VOH cathode undergoes fast capacity fading, leading to a final capacity retention of 24%, which is much lower than the 71% of the SnVOH cathode in the same condition (Figure 2f). In addition, the rate and cycling performance of the Sn-doped hydrated $V_2O_5$ is superior to the results from the recent research on hydrated $V_2O_5$, as shown in Table S2. The rapid capacity decay of VOH may be mainly attributed to the unstable layered structure and the dissolution of VOH during repeated insertion/extraction of zinc ions. Additionally, the enhanced cycling performance of SnVOH reveals the outstanding structural stabilizer function of the doped Sn ions.

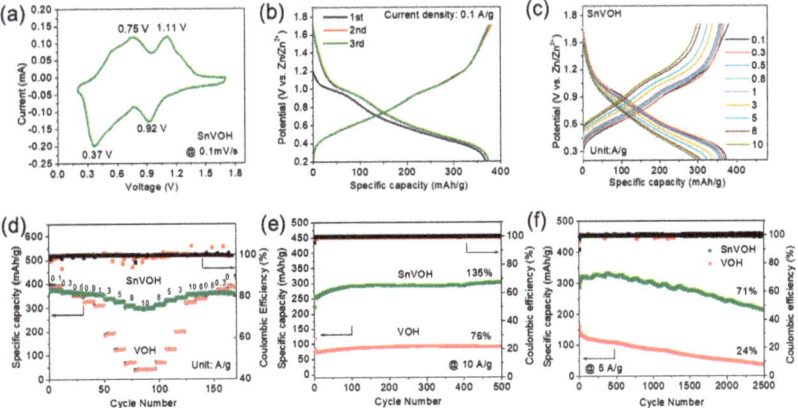

**Figure 2.** Electrochemical performance of VOH and SnVOH. (**a**) C–V curves at 0.1 mV/s in the voltage range of 0.2–1.7 V vs. $Zn^{2+}$/Zn, (**b**) corresponding GCD curves, (**c**) representative galvanostatic charge–discharge curves at different current densities, (**d**) rate capacities at current densities between 0.1 to 10 A/g, (**e**) cycling performance at 10 A/g and (**f**) cycling performance at 5 A/g for the VOH and SnVOH sample.

The high-rate performance and long-term stability of the SnVOH cathode are substantially controlled by the electrochemical kinetics, which were analyzed through the C–V curves at different scanning rates and during the GITT test, as shown in Figure 3. The C–V curves of SnVOH in Figure 3a maintain a similar shape with two pairs of charge and discharge peaks, while the reduction peaks and oxidation peaks shift to lower and higher voltages at increased scan rates, respectively. The peak current (*i*) and scan rate (*v*) have a relationship that is defined by the following equations [39]:

$$I = av^b \quad (2)$$

which can be transformed to

$$\log(i) = b\log(v) + \log(a) \quad (3)$$

where $a$ and $b$ are the variables. The b value ranges from 0.5 to 1.0, which indicates a different mechanism. A $b$ value of one is indicative of the dominated contribution of the surface capacitance to the total capacity, and a $b$ value of 0.5 exhibits diffusion-controlled charge storage [27]. In addition, the slope of the log (*i*) versus the log (*v*) plot can be used to estimate the b value, as shown in Figure 3b,c. The $b$ values of the peaks 1–4 for the SnVOH electrode are 0.97, 0.81, 0.93, and 0.89, (Figure 3b), which implies the SnVOH has considerable kinetics, and its charge storage mainly comes from surface capacitance, and it is slightly influenced by the diffusion process. By fitting the C–V curves at different scan rates in Figure S5b, the $b$ values of the peaks 1–4 for the VOH electrode are 0.85, 0.86, 0.72, and 0.96 (Figure S5c), suggesting the capacity of the VOH cathode is also influenced by both the capacitive and diffusion processes, and it is dominated by the surface capacitive capacity. Furthermore, the capacity is divided as a capacitive-controlled part ($k_1v$) and diffusion-induced part ($k_2v^{1/2}$) which are described by the following equations [40]:

$$i = k_1 v + k_2 v^{1/2} \quad (4)$$

or

$$i/v^{1/2} = k_1 v^{1/2} + k_2 \quad (5)$$

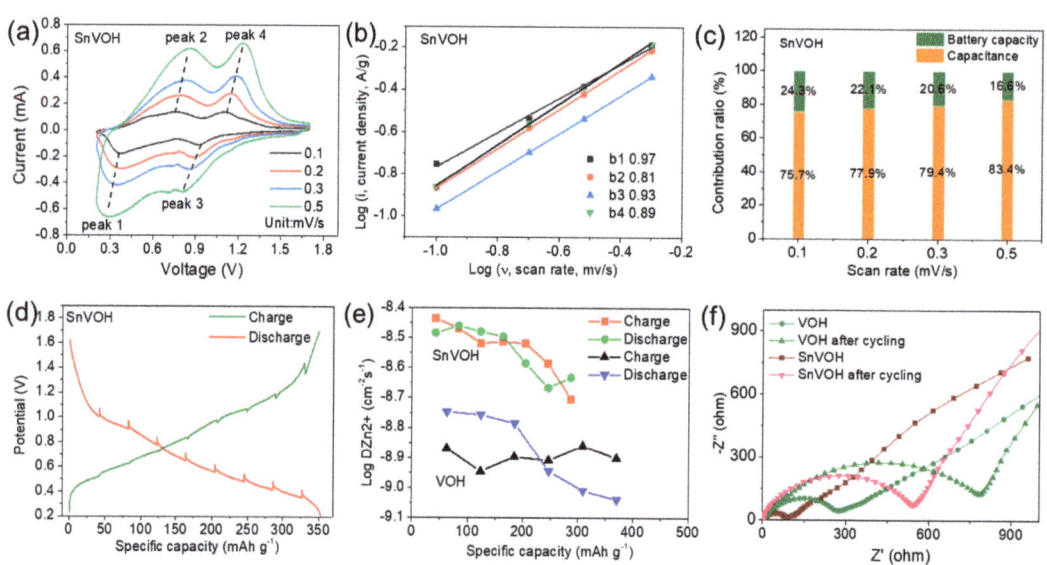

**Figure 3.** Kinetics study of VOH and SnVOH. (**a**) C–V curves of SnVOH electrode at different scan rates, (**b**) log (current) vs. log (scan rate) plots of four peaks in C–V curves during the cycles, and (**c**) capacity contribution ratios of battery type capacity and capacitance, (**d**) the discharge/charge curves in GITT measurement of SnVOH, (**e**) corresponding diffusivity coefficient of $Zn^{2+}$ in discharge and charge processes of SnVOH samples, and (**f**) electrochemical impedance profiles recorded for the VOH and SnVOH electrodes before electrochemical tests and after cycling at 5 A/g for 2500 times.

The ratios of the surface-controlled capacitive and the diffusion-induced parts of SnVOH at various scan rates are displayed in Figure 3c. The surface-controlled capacitive contribution ratio increases from 75.7% (0.1 mV/s) to 83.4% (0.5 mV/s), indicating that the batteries possess fast charge-transfer kinetics. For VOH, the surface-controlled capacitive contribution ratio is 21.5% (0.1 mV/s), and this increases to 36.3% (0.5 mV/s), as presented in Figure S5d. The much higher surface-controlled capacitive contribution ratio for SnVOH implies that it has much faster kinetics when one is doping the Sn element, which may be ascribed to the effect of $Sn^{4+}$ shielding the interaction between the oxygen atom and $Zn^{2+}$ [41].

Then, the constant current intermittent titration technique (GITT) was used to determine the ion diffusion coefficient of $Zn^{2+}$ in the VOH and SnVOH cathodes (Figures S5e and S3d). As demonstrated in Figure 3e, The diffusion coefficients of SnVOH are from $10^{-8.9}$ to $10^{-8.6}/(cm^2\ s)$ during the charge and discharge processes, respectively, which are higher than those of the VOH samples ($10^{-9.0}$ to $10^{-8.8}/(cm^2\ s)$, in Figure S5f). The results confirm that $Zn^{2+}$ migration in SnVOH is faster when it is compared to that in VOH. Thus, the much higher rate performance of SnVOH can be ascribed to a reduced interfacial impedance and an enhanced ion diffusion. Moreover, doping the Sn element largely reduces the charge transfer resistance from 300 to 100 $\Omega$ (Figure 3f), which is indicative of a reduced ion diffusion impedance between the interface of the cathode and electrolyte. After cycling at 5 A/g for 2500 times, the charge transfer resistance of SnVOH-based cell rises to about 580 $\Omega$, which is still smaller than 880 $\Omega$ of VOH based cell. The results confirm that faster electrochemical kinetics can be achieved by doping the Sn element.

## 4. Conclusions

In summary, the Sn-doped hydrated $V_2O_5$ ZIB cathode materials were prepared by simple one-step hydrothermal synthesis. Compared to the undoped sample, the Sn-doped hydrated $V_2O_5$ demonstrates a significant enhancement in its rate performance and cyclic stability. SnVOH shows a high initial reversible capacity of 387 mAh/g at 0.1 A/g, an excellent rate capability with 301 mAh/g even at a large current density of 10 A/g, and its retains 87.5% of its initial capacity after the cycling at 2 A/g for 2000 times. The rate and cycling performance of Sn-doped hydrated $V_2O_5$ are superior to the results from recent research on hydrated $V_2O_5$. These great improvements can be due the smaller charge transfer resistance and the higher zinc diffusion coefficient that occur after the Sn doping. Therefore, this work reveals that Sn doping is an effective strategy to improve the zinc storage performance of hydrated $V_2O_5$, and the Sn-doped hydrated $V_2O_5$ is a promising cathode material candidate to construct ZIBs of a high specific capacity, an excellent rate performance, and a high durability.

**Supplementary Materials:** The following supporting information can be downloaded at: https://www.mdpi.com/article/10.3390/cryst12111617/s1, Figure S1: Scheme of synthesis of SnVOH; Figure S2: The schematic crystalline structure of SnVOH; Figure S3: EDS characterization of VOH sample; Table S1: The molar ratio of different elements in SnVOH; Figure S4: Characterization of VOH and SnVOH composite cathodes; Figure S5: Electrochemical characterization of VOH-based cathode; Table S2. Electrochemical performance comparison of SnVOH with recent literature data on doped hydrated $V_2O_5$-based cathodes in ZIBs. Refs. [42–45] are cited in the Supplementary Materials file.

**Author Contributions:** Conceptualization, K.G. and N.Y.; investigation, W.C. and H.L. (Haiyuan Liu); validation, H.L. (Haiyuan Liu), W.S. and Y.W.; writing—original draft preparation, K.G.; writing—review and editing, N.Y. and J.O.; visualization, H.W., H.L. (Hanbin Li) and Z.L.; supervision, N.Y.; project administration, K.G.; funding acquisition, K.G., X.Z. and N.Y. All authors have read and agreed to the published version of the manuscript.

**Funding:** This research was funded by the National Natural Science Foundation of China (52102214, 51702048, and 22166003), the Jiangxi Provincial Natural Science Foundation (20202BAB213008, 20202BABL213003, and 20202BABL203026), the Opening Project of Jiangxi Province Engineering Research Center of New Energy Technology and Equipment (JXNE2021-04, JXNE2021-05), State Key Laboratory of Materials Processing and Die & Mould Technology, Huazhong University of Science and Technology (P2022-009), and the National (Jiangxi Province) College Students Innovation and Entrepreneurship Training Program 2020 (S202110405018, S202010405031 and S202010405018) supported this work.

**Institutional Review Board Statement:** Not applicable.

**Acknowledgments:** The authors would like to thank Xiaoming Chen from Shiyanjia Lab (www.shiyanjia.com) for the material characterization.

**Conflicts of Interest:** The authors declare no conflict of interest.

## References

1. Ming, F.; Liang, H.; Lei, Y.; Kandambeth, S.; Eddaoudi, M.; Alshareef, H.N. Layered $Mg_xV_2O_5 \cdot nH_2O$ as Cathode Material for High-Performance Aqueous Zinc Ion Batteries. *ACS Energy Lett.* **2018**, *3*, 2602–2609. [CrossRef]
2. Pan, Q.; Dong, R.; Lv, H.; Sun, X.; Song, Y.; Liu, X.-X. Fundamental understanding of the proton and zinc storage in vanadium oxide for aqueous zinc-ion batteries. *Chem. Eng. J.* **2021**, *419*, 129491. [CrossRef]
3. Pang, Q.; He, W.; Yu, X.; Yang, S.; Zhao, H.; Fu, Y.; Xing, M.; Tian, Y.; Luo, X.; Wei, Y. Aluminium pre-intercalated orthorhombic $V_2O_5$ as high-performance cathode material for aqueous zinc-ion batteries. *Appl. Surf. Sci.* **2021**, *538*, 148043. [CrossRef]
4. Qi, Z.; Xiong, T.; Chen, T.; Shi, W.; Zhang, M.; Ang, Z.W.J.; Fan, H.; Xiao, H.; Lee, W.S.V.; Xue, J. Harnessing oxygen vacancy in $V_2O_5$ as high performing aqueous zinc-ion battery cathode. *J. Alloy. Compd.* **2021**, *870*, 159403. [CrossRef]
5. Qiu, N.; Yang, Z.; Xue, R.; Wang, Y.; Zhu, Y.; Liu, W. Toward a High-Performance Aqueous Zinc Ion Battery: Potassium Vanadate Nanobelts and Carbon Enhanced Zinc Foil. *Nano Lett.* **2021**, *21*, 2738–2744. [CrossRef]
6. Lübke, M.; Ning, D.; Armer, C.F.; Howard, D.; Brett, D.J.L.; Liu, Z.; Darr, J.A. Evaluating the Potential Benefits of Metal Ion Doping in $SnO_2$ Negative Electrodes for Lithium Ion Batteries. *Electrochim. Acta* **2017**, *242*, 400–407. [CrossRef]
7. Song, H.; Liu, C.; Zhang, C.; Cao, G. Self-doped $V^{4+}$–$V_2O_5$ nanoflake for 2 Li-ion intercalation with enhanced rate and cycling performance. *Nano Energy* **2016**, *22*, 1–10. [CrossRef]
8. Sun, R.; Qin, Z.; Liu, X.; Wang, C.; Lu, S.; Zhang, Y.; Fan, H. Intercalation Mechanism of the Ammonium Vanadate ($NH_4V_4O_{10}$) 3D Decussate Superstructure as the Cathode for High-Performance Aqueous Zinc-Ion Batteries. *ACS Sustain. Chem. Eng.* **2021**, *9*, 11769–11777. [CrossRef]
9. Suresh, R.; Giribabu, K.; Manigandan, R.; Munusamy, S.; Kumar, S.P.; Muthamizh, S.; Stephen, A.; Narayanan, V. Doping of Co into $V_2O_5$ nanoparticles enhances photodegradation of methylene blue. *J. Alloy. Compd.* **2014**, *598*, 151–160. [CrossRef]
10. Wang, H.; Ye, W.; Yang, Y.; Zhong, Y.; Hu, Y. Zn-ion hybrid supercapacitors: Achievements, challenges and future perspectives. *Nano Energy* **2021**, *85*, 105942. [CrossRef]
11. Song, M.; Tan, H.; Chao, D.; Fan, H.J. Recent Advances in Zn-Ion Batteries. *Adv. Funct. Mater.* **2018**, *28*, 1802564. [CrossRef]
12. Wang, H.; Jing, R.; Shi, J.; Zhang, M.; Jin, S.; Xiong, Z.; Guo, L.; Wang, Q. Mo-doped $NH_4V_4O_{10}$ with enhanced electrochemical performance in aqueous Zn-ion batteries. *J. Alloy. Compd.* **2021**, *858*, 158380. [CrossRef]
13. Fu, Q.; Wang, J.; Sarapulova, A.; Zhu, L.; Missyul, A.; Welter, E.; Luo, X.; Ding, Z.; Knapp, M.; Ehrenberg, H.; et al. Electrochemical performance and reaction mechanism investigation of $V_2O_5$ positive electrode material for aqueous rechargeable zinc batteries. *J. Mater. Chem. A* **2021**, *9*, 16776–16786. [CrossRef]
14. Geng, H.; Cheng, M.; Wang, B.; Yang, Y.; Zhang, Y.; Li, C.C. Electronic Structure Regulation of Layered Vanadium Oxide via Interlayer Doping Strategy toward Superior High-Rate and Low-Temperature Zinc-Ion Batteries. *Adv. Funct. Mater.* **2019**, *30*, 1907684. [CrossRef]
15. He, Z.; Jiang, Y.; Zhu, J.; Li, Y.; Jiang, Z.; Zhou, H.; Meng, W.; Wang, L.; Dai, L. Boosting the performance of $LiTi_2(PO_4)_3$/C anode for aqueous lithium ion battery by Sn doping on Ti sites. *J. Alloy. Compd.* **2018**, *731*, 32–38. [CrossRef]
16. Kuang, M.; Han, P.; Huang, L.; Cao, N.; Qian, L.; Zheng, G. Electronic Tuning of Co, Ni-Based Nanostructured (Hydr)oxides for Aqueous Electrocatalysis. *Adv. Funct. Mater.* **2018**, *28*, 1804886. [CrossRef]
17. Liu, R.; Liang, Z.; Gong, Z.; Yang, Y. Research Progress in Multielectron Reactions in Polyanionic Materials for Sodium-Ion Batteries. *Small Methods* **2018**, *3*, 1800221. [CrossRef]
18. Kim, J.; Lee, S.H.; Park, C.; Kim, H.S.; Park, J.H.; Chung, K.Y.; Ahn, H. Controlling Vanadate Nanofiber Interlayer via Intercalation with Conducting Polymers: Cathode Material Design for Rechargeable Aqueous Zinc Ion Batteries. *Adv. Funct. Mater.* **2021**, *31*, 2100005. [CrossRef]
19. Ni, S.; Liu, J.; Chao, D.; Mai, L. Vanadate-Based Materials for Li-Ion Batteries: The Search for Anodes for Practical Applications. *Adv. Energy Mater.* **2019**, *9*, 1803324. [CrossRef]
20. Venkatesan, R.; Bauri, R.; Mayuranathan, K.K. Zinc Vanadium Oxide Nanobelts as High-Performance Cathodes for Rechargeable Zinc-Ion Batteries. *Energy Fuels* **2022**, *36*, 7854–7864. [CrossRef]

21. Li, R.; Xing, F.; Li, T.; Zhang, H.; Yan, J.; Zheng, Q.; Li, X. Intercalated polyaniline in $V_2O_5$ as a unique vanadium oxide bronze cathode for highly stable aqueous zinc ion battery. *Energy Storage Mater.* **2021**, *38*, 590–598. [CrossRef]
22. Lv, T.-T.; Liu, Y.-Y.; Wang, H.; Yang, S.-Y.; Liu, C.-S.; Pang, H. Crystal water enlarging the interlayer spacing of ultrathin $V_2O_5 \cdot 4VO_2 \cdot 2.72H_2O$ nanobelts for high-performance aqueous zinc-ion battery. *Chem. Eng. J.* **2021**, *411*, 128533. [CrossRef]
23. Tang, B.; Zhou, J.; Fang, G.; Guo, S.; Guo, X.; Shan, L.; Tang, Y.; Liang, S. Structural Modification of $V_2O_5$ as High-Performance Aqueous Zinc-Ion Battery Cathode. *J. Electrochem. Soc.* **2019**, *166*, A480–A486. [CrossRef]
24. Moretti, A.; Giuli, G.; Trapananti, A.; Passerini, S. Electrochemical and structural investigation of transition metal doped $V_2O_5$ sono-aerogel cathodes for lithium metal batteries. *Solid State Ion.* **2018**, *319*, 46–52. [CrossRef]
25. Wu, Y.; Liu, S.; Hu, L.; Chen, S. Constructing electron pathways by graphene oxide for $V_2O_5$ nanoparticles in ultrahigh-performance and fast charging aqueous zinc ion batteries. *J. Alloy. Compd.* **2021**, *878*, 160324. [CrossRef]
26. Yu, H.; Rui, X.; Tan, H.; Chen, J.; Huang, X.; Xu, C.; Liu, W.; Yu, D.Y.; Hng, H.H.; Hoster, H.E.; et al. Cu doped $V_2O_5$ flowers as cathode material for high-performance lithium ion batteries. *Nanoscale* **2013**, *5*, 4937–4943. [CrossRef]
27. Yang, Y.; Tang, Y.; Fang, G.; Shan, L.; Guo, J.; Zhang, W.; Wang, C.; Wang, L.; Zhou, J.; Liang, S. Li+ intercalated $V_2O_5 \cdot nH_2O$ with enlarged layer spacing and fast ion diffusion as an aqueous zinc-ion battery cathode. *Energy Environ. Sci.* **2018**, *11*, 3157–3162. [CrossRef]
28. Sheng, X.; Li, Z.; Cheng, Y. Electronic and Thermoelectric Properties of $V_2O_5$, $MgV_2O_5$, and $CaV_2O_5$. *Coatings* **2020**, *10*, 453. [CrossRef]
29. Xu, X.; Xiong, F.; Meng, J.; Wang, X.; Niu, C.; An, Q.; Mai, L. Vanadium-Based Nanomaterials: A Promising Family for Emerging Metal-Ion Batteries. *Adv. Funct. Mater.* **2020**, *30*, 1904398. [CrossRef]
30. Zhang, X.; Dong, M.; Xiong, Y.; Hou, Z.; Ao, H.; Liu, M.; Zhu, Y.; Qian, Y. Aqueous Rechargeable Li(+) /Na(+) Hybrid Ion Battery with High Energy Density and Long Cycle Life. *Small* **2020**, *16*, e2003585. [CrossRef]
31. Zhu, K.; Wu, T.; van den Bergh, W.; Stefik, M.; Huang, K. Reversible Molecular and Ionic Storage Mechanisms in High-Performance $Zn_{0.1}V_2O_5 \cdot nH_2O$ Xerogel Cathode for Aqueous Zn-Ion Batteries. *ACS Nano* **2021**, *15*, 10678–10688. [CrossRef] [PubMed]
32. Shao, Y.; Sun, Z.; Tian, Z.; Li, S.; Wu, G.; Wang, M.; Tong, X.; Shen, F.; Xia, Z.; Tung, V.; et al. Regulating Oxygen Substituents with Optimized Redox Activity in Chemically Reduced Graphene Oxide for Aqueous Zn-Ion Hybrid Capacitor. *Adv. Funct. Mater.* **2020**, *31*, 2007843. [CrossRef]
33. Han, L.; Dong, S.; Wang, E. Transition-Metal (Co, Ni, and Fe)-Based Electrocatalysts for the Water Oxidation Reaction. *Adv. Mater.* **2016**, *28*, 9266–9291. [CrossRef] [PubMed]
34. Liao, M.; Wang, J.; Ye, L.; Sun, H.; Wen, Y.; Wang, C.; Sun, X.; Wang, B.; Peng, H. A Deep-Cycle Aqueous Zinc-Ion Battery Containing an Oxygen-Deficient Vanadium Oxide Cathode. *Angew. Chem. Int. Ed.* **2020**, *59*, 2273–2278. [CrossRef]
35. Wang, H.; Bi, X.; Bai, Y.; Wu, C.; Gu, S.; Chen, S.; Wu, F.; Amine, K.; Lu, J. Open-Structured $V_2O_5 \cdot nH_2O$ Nanoflakes as Highly Reversible Cathode Material for Monovalent and Multivalent Intercalation Batteries. *Adv. Energy Mater.* **2017**, *7*, 1602720. [CrossRef]
36. Du, Y.; Wang, X.; Sun, J. Tunable oxygen vacancy concentration in vanadium oxide as mass-produced cathode for aqueous zinc-ion batteries. *Nano Res.* **2020**, *14*, 754–761. [CrossRef]
37. Li, Y.; Yao, J.; Uchaker, E.; Zhang, M.; Tian, J.; Liu, X.; Cao, G. Sn-Doped $V_2O_5$ Film with Enhanced Lithium-Ion Storage Performance. *J. Phys. Chem. C* **2013**, *117*, 23507–23514. [CrossRef]
38. Wu, C.; Zhu, G.; Wang, Q.; Wu, M.; Zhang, H. Sn-based nanomaterials: From composition and structural design to their electrochemical performances for Li- and Na-ion batteries. *Energy Storage Mater.* **2021**, *43*, 430–462. [CrossRef]
39. Li, Z.; Zhang, C.; Liu, C.; Fu, H.; Nan, X.; Wang, K.; Li, X.; Ma, W.; Lu, X.; Cao, G. Enhanced Electrochemical Properties of Sn-doped $V_2O_5$ as a Cathode Material for Lithium Ion Batteries. *Electrochim. Acta* **2016**, *222*, 1831–1838. [CrossRef]
40. He, D.; Peng, Y.; Ding, Y.; Xu, X.; Huang, Y.; Li, Z.; Zhang, X.; Hu, L. Suppressing the skeleton decomposition in Ti-doped $NH_4V_4O_{10}$ for durable aqueous zinc ion battery. *J. Power Sources* **2021**, *484*, 229284. [CrossRef]
41. Byeon, Y.W.; Ahn, J.P.; Lee, J.C. Diffusion Along Dislocations Mitigates Self-Limiting Na Diffusion in Crystalline Sn. *Small* **2020**, *16*, e2004868. [CrossRef] [PubMed]
42. Wu, F.; Wang, Y.; Ruan, P.; Niu, X.; Zheng, D.; Xu, X.; Gao, X.; Cai, Y.; Liu, W.; Shi, W.; et al. Fe-doping enabled a stable vanadium oxide cathode with rapid Zn diffusion channel for aqueous zinc-ion batteries. *Mater. Today Energy* **2021**, *21*, 100842. [CrossRef]
43. Zhao, H.; Fu, Q.; Yang, D.; Sarapulova, A.; Pang, Q.; Meng, Y.; Wei, L.; Ehrenberg, H.; Wei, Y.; Wang, C.; et al. In Operando Synchrotron Studies of $NH_4^+$ Preintercalated $V_2O_5 \cdot nH_2O$ Nanobelts as the Cathode Material for Aqueous Rechargeable Zinc Batteries. *ACS Nano.* **2020**, *14*, 11809–11820. [CrossRef] [PubMed]
44. Xu, G.; Liu, X.; Huang, S.; Li, L.; Wei, X.; Cao, Z.; Yang, L.; Chu, P.K. Freestanding, Hierarchical, and Porous Bilayered $Na_xV_2O_5 \cdot nH_2O$/rGO/CNT Composites as High-Performance Cathode Materials for Nonaqueous K-Ion Batteries and Aqueous Zinc-Ion Batteries. *ACS Appl. Mater. Interfaces* **2020**, *12*, 706–716. [CrossRef] [PubMed]
45. Jiang, H.; Zhang, Y.; Xu, L.; Gao, Z.; Zheng, J.; Wang, Q.; Meng, C.; Wang, J. Fabrication of $(NH_4)_2V_3O_8$ nanoparticles encapsulated in amorphous carbon for high capacity electrodes in aqueous zinc ion batteries. *Chem. Eng. J.* **2020**, *382*, 122844. [CrossRef]

Article

# 2D Layer Structure in Two New Cu(II) Crystals: Structural Evolvement and Properties

Jia-Jing Luo [1], Xiang-Xin Cao [1], Qi-Wei Chen [1], Ying Qin [1], Zhen-Wei Zhang [1,*], Lian-Qiang Wei [2,*] and Qing Chen [1]

[1] College of Pharmacy, Guangxi Zhuang Yao Medicine Center of Engineering and Technology, Guangxi University of Chinese Medicine, Nanning 530200, China; 15777113797@163.com (J.-J.L.); caoxiangxin@zmc.top (X.-X.C.); cqw77914@163.com (Q.-W.C.); qinying12531@sohu.com (Y.Q.); qing0082@163.com (Q.C.)
[2] College of Chemistry and Bio-Engineering, Hechi University, Hechi 546300, China
* Correspondence: charliezh@163.com (Z.-W.Z.); wlq259@163.com (L.-Q.W.)

**Abstract:** Two new Cu(II) crystals, {[Cu(dtp)]·H$_2$O}$_n$ (**1**) and [Cu(Hdtp)(bdc)$_{0.5}$]$_n$ (**2**) (H$_2$dtp = 4′-(3,5-dicarboxyphenyl)-2,2′:6′,2′′′-terpyridine, H$_2$bdc = 1,4-benzenedicarboxylic acid) were synthesized under hydrothermal conditions. X-ray single-crystal structural analysis revealed that the 5-connective Cu(II) is in a distorted tetragonal-pyramidal coordination sphere for both compounds. Crystal **1** shows a "wave-shaped" 2D layer in the structure, while **2** bears a 1D coordination chain structure and a supermolecular 2D layer structure with a thickness of 7.9 Å via 1D chain stacking. PXRD and TGA measurements showed that **1** and **2** are air stable, with thermal stabilities near 300 °C.

**Keywords:** crystals; metal-organic frameworks; Cu(II) ion; 2D layer; structural evolvement

## 1. Introduction

In the past few decades, Metal-Organic Framework (MOF) material has been a hot topic in the field of chemistry and materials. It is a kind of hybrid material with a highly ordered network formed by the coordination bond connection between metal ions and organic ligands [1]. It has aroused great interest because of its structural diversity and its wide applications in catalysis [2,3], light-emitting sensors [4–8], gas adsorption/separation [9], magnetism [10], biomedicine [11] and other advanced materials [12]. MOF-based 2D nanosheet materials arouse great interest [13] due to the application of gas separation [14] and molecular sieving membranes [15]. Even though the "top-down" and the "bottom-up" methods have been developed to fabricate this 2D material [16], the understanding of the structure and the consequential tuning of the 2D MOF layer growth are still the key issues at this point.

The 4′-(3,5-dicarboxyphenyl)-2,2′:6′,2′′′-terpyridine (H$_2$dtp) ligand is a ditopic near-plane shape linker with *m*-dicarboxylic and tribipyridine groups [17–21], which is a good candidate for the construction of 2D MOFs structures [22]. In order to fulfill the purpose of the 2D layer configuration, it's important to further govern the coordination when a metal ion is coordinated to the tribipyridine group of H$_2$dtp, as one can expect that a lower coordination number of the metal ion will reduce the possibility of the 3D network extending. Besides six, the coordination number of Cu(II) can be varied to five [23] or four [24], which makes it a potential low-coordinative ion to serve as a node in the linking of the 2D coordination network.

In this paper, two new MOFs with different kinds of 2D structure are synthesized, namely {[Cu(dtp)]·H$_2$O}$_n$ (**1**) and [Cu(Hdtp)(bdc)$_{0.5}$]$_n$ (**2**) (H$_2$bdc=1,4-benzenedicarboxylic acid). The coordination number of the Cu(II) is five in both compounds. Besides this, **1** shows a "wave-shaped" 2D layer in the structure, as we expected. As for **2**, the introduction of bdc$^{2-}$ into the Cu-H$_2$dtp coordination system reduces the 2D network into

the 1D coordination chain; however, a unique supermolecular 2D layer with a thickness of 7.9 Å is also found, which is formed by the orderly array of the 1D chain. A detailed structural analysis shows how the network evolution comes about, and the PXRD and TGA measurements help in understanding the air and thermal stability of the compounds.

## 2. Materials and Methods

### 2.1. Materials and General Methods

The reagents and solvents were commercially available and used as received. All of the other starting materials were of analytical grade, and were used as received, without further purification. The powder X-ray diffraction (PXRD) data were recorded on a Rigaku Miniflex 600 diffractometer (Rigaku, Tokyo, Japan) using Cu K$\alpha$ radiation ($\lambda$ = 1.54056 Å), with a scan speed of $4°$ min$^{-1}$ in the $2\theta$ = 5–45° region. Thermogravimetric analyses (TGA) were carried out on a Netzsch STA449 F5 analyzer (Netzsch, Serb, Germany), with the heating of the crystalline samples from room temperature to 800 °C at a rate of 10 °C min$^{-1}$ in a nitrogen atmosphere.

### 2.2. Synthesis of the Complexes

Regarding the synthesis of $\{[Cu(H_2dtp)]\cdot H_2O\}_n$ (**1**), a mixture of 0.1 mmol Cu(NO$_3$)$_2$·3H$_2$O (0.024 g), 0.05 mmol H$_2$dtp (0.020 g), 1.3 mL H$_2$O and 3.6 mL DMA (N, N'-dimethylacetamide), together with 3.6 mL methanol, was sealed in a 15 mL capped vial. The vial was heated at 150 °C for 48 h under autogenous pressure, and then cooled slowly down to room temperature. Light-green crystals with a regular cuboid structure were obtained. The yield was 35% based on H$_2$dtp. The elemental analysis (%) was calculated for C23H15CuN3O5: C 57.92 H 3.17 N 8.81; it found: C 57.20 H 3.35 N 8.42. IR (cm$^{-1}$): 3420 (vs), 1618 (s), 1560 (s), 1443 (s), 1388 (s), 1227 (m), 1178 (m), 1002 (w), 890 (w), 763 (m), 710 (vw), 665 (vw).

Regarding the synthesis of $[Cu(H_2dtp)(bdc)_{0.5}]_n$ (**2**), the synthesis of compound **2** was similar to that of **1**. In addition to using 15 mL H$_2$O instead of the solvent condition, 0.05 mmol H$_2$bdc (8 mg) and 0.25 mmol NaOH (10 mg) were added into the system. The mixture was sealed in a 25 mL capped vial, heated at 140 °C for 48 h under autogenous pressure, and then cooled gradually down to room temperature. Similarly, dark-green, slender bulk crystals were obtained. The yield was 32% based on H$_2$dtp. Elemental analysis (%) was calculated for C27H16CuN3O6: C 59.83 H 2.98 N 7.75; it found: C 59.20 H 3.25 N 8.01. IR (cm$^{-1}$): 3444 (vs), 1620 (s), 1554 (s), 1444 (s), 1398 (s), 1250 (m), 1174 (m), 1002 (w), 878 (w), 796 (m), 770 (m), 658 (m).

### 2.3. Crystal Structure Determination

Single crystals of **1** and **2** with the proper dimensions were chosen under an optical microscope and coated with high vacuum grease (Dow Corning Corporation) before being mounted on a glass fiber for the data collection. X-ray crystallography data of **1** and **2** were gathered on a Bruker Apex Smart CCD diffractometer (Bruker, Bremen, Germany) at 293 K with graphite-monochromated Mo K$\alpha$ ($\lambda$ = 0.71073 Å) or Cu K$\alpha$ ($\lambda$ = 1.54184 Å) radiation by using the $\omega$-$2\theta$ scan mode. The intensity data were corrected for Lorentz and polarization effects (SAINT), and empirical absorption corrections based on equivalent reflections were applied (SADABS) [25]. The structures ware solved by direct methods, and were refined by the full-matrix least-squares method on $F^2$ with the SHELXTL program package [26]. All of the non-hydrogen atoms were refined with anisotropic displacement parameters. The hydrogen atoms were calculated and refined as a riding model. The hydrogen atoms of carboxylic groups and water molecules were located from difference maps. The disordered guest molecules were treated by a solvent mask with the Olex2 program [27]. Crystallographic data for **1** and **2** are given in Table 1. The hydrogen-bonding parameters and selected bond lengths and angles for **1** and **2** are listed in Tables S1–S4 (see Supplementary Materials).

**Table 1.** Crystal data and structure refinement for **1** and **2**.

| Compound | 1 | 2 |
|---|---|---|
| Empirical formula | $C_{23}H_{15}CuN_3O_5$ | $C_{27}H_{16}CuN_3O_6$ |
| Formula weight | 476.92 | 541.97 |
| Temperature/K | 293 | 293 |
| Crystal system | monoclinic | triclinic |
| Space group | $P2_1/n$ | $P\bar{1}$ |
| a/Å | 12.3459(3) | 8.68340(10) |
| b/Å | 11.0867(3) | 11.25970(10) |
| c/Å | 14.8560(4) | 11.57420(10) |
| α/° | 90 | 85.1240(10) |
| β/° | 101.181(2) | 88.0210(10) |
| γ/° | 90 | 85.7880(10) |
| Volume/Å³ | 1994.83(8) | 1124.042(19) |
| Z | 4 | 2 |
| $\rho_{calc}$ g/cm³ | 1.588 | 1.601 |
| μ/mm⁻¹ | 1.138 | 1.824 |
| F(000) | 972.0 | 552.0 |
| Radiation | MoKα (λ = 0.71073) | CuKα (λ = 1.54184) |
| 2θ range for data collection/° | 3.934 to 59.824 | 7.67 to 147.946 |
| Index ranges | −16 ≤ h ≤ 16, −14 ≤ k ≤ 15, −20 ≤ l ≤ 20 | −10 ≤ h ≤ 10, −13 ≤ k ≤ 14, −14 ≤ l ≤ 13 |
| Reflections collected | 28,754 | 23,945 |
| Independent reflections | 5115 [$R_{int}$ = 0.0398, $R_{sigma}$ = 0.0294] | 4345 [$R_{int}$ = 0.0223, $R_{sigma}$ = 0.0143] |
| Data/restraints/parameters | 5115/0/292 | 4345/0/335 |
| Goodness-of-fit on $F^2$ | 1.078 | 1.072 |
| Final R indexes [I ≥ 2σ (I)] | $R_1$ = 0.0321, $wR_2$ = 0.0834 | $R_1$ = 0.0277, $wR_2$ = 0.0789 |
| Final R indexes [all data] | $R_1$ = 0.0416, $wR_2$ = 0.0871 | $R_1$ = 0.0289, $wR_2$ = 0.0800 |
| Largest diff. peak/hole/e Å⁻³ | 0.33/−0.38 | 0.25/−0.33 |

$R_1 = \sum ||F_o| - |F_c||/\sum |F_o|, wR_2 = [\sum w(|F_o| - |F_c|)^2/\sum w|F_o|^2]^{1/2}$.

## 3. Results and Discussion

### 3.1. Structure Analysis

Uniform single crystals of **1** and **2** were harvested under hydrothermal conditions (Figure 1). Single-crystal X-ray diffraction (SCXRD) analysis revealed that **1** crystallizes in a monoclinic crystal system $P2_1/n$. The asymmetrical unit of **1** is composed of fully deprotonating dtp²⁻, one Cu(II) ion, and a guest molecule of $H_2O$ (Figure 2a). The Cu(II) ion is five-coordinated by nitrogen atoms from pyridine rings and two oxygen atoms from the carboxylic group of neighboring dtp²⁻, forming a distorted tetragonal-pyramidal coordination sphere with the Cu(II)−N and Cu(II)−O bond length in the scale of 1.9463(14)−2.0452(15) Å and 1.9399(12)−2.0950(13) Å, respectively. The coordination properties enable the asymmetry unit of **1** to behave as a four-connective node to link the neighboring units, further forming a "wave-shaped" 2D layer (Figure 2b). The void between the neighboring layers is dotted by the guest water molecules. The two hydrogen atoms of $H_2O$ form a double H-bond (O−H⋯O) interaction with the carboxylic groups of neighboring layers with the D-A length of 2.757(2) and 2.880(2) Å [28]. Besides this, π-π stacking interactions with a center–center distance of 3.71 and 4.25 Å [29] are also found between the pyridines of dtp²⁻ from the neighboring layers (Figure 3). Those intermolecular forces direct the 3D packing of the "wave-shaped" layers in the crystal b direction (Figure 2c).

**Figure 1.** The obtained single crystal photos for **1** (**a**) and **2** (**b**).

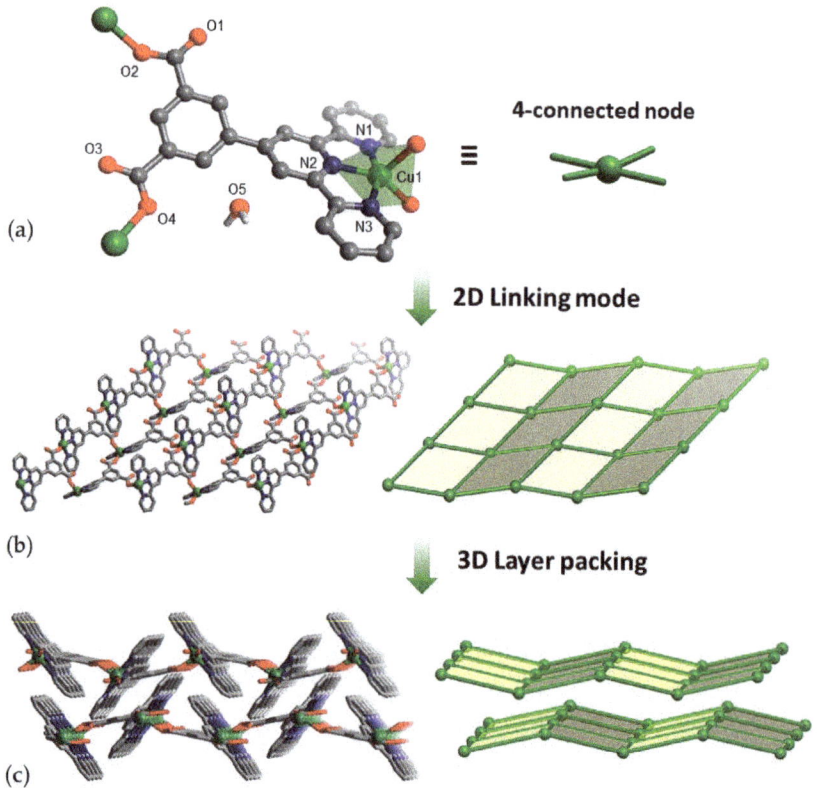

**Figure 2.** Asymmetric unit of **1** (**a**); the 2D "wave-shaped" layer structure as well as the linking mode of the nodes in **1** (**b**) and the 3D stacking mode of the layers in **1** (**c**) (some H atoms are omitted for clarity).

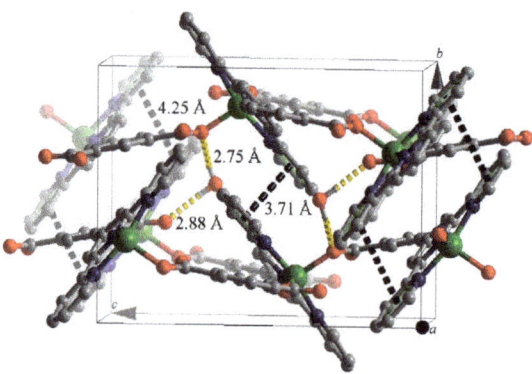

**Figure 3.** H-bond and π-π stacking as the interlayer force in **1**.

SCXRD analysis revealed that **2** crystallizes in monoclinic crystal system $P\bar{1}$. The asymmetry unit of **2** is composed of partially deprotonating Hdtp$^-$, one Cu(II) ion, and half of bdc$^{2-}$ (Figure 4a). The Cu(II) ion is five-coordinated by nitrogen atoms from pyridine rings and two oxygen atoms from carboxylic group of bdc$^-$ and neighboring Hdtp$^-$, forming a distorted tetragonal-pyramidal coordination sphere with Cu(II)-N and

Cu(II)-O bond lengths in the scale of 1.9330(13)–2.0217(14) Å and 1.9102(11)–2.2431(13) Å, respectively. The introduction of bdc⁻ into **2** results in a linkage reduction of the asymmetric unit; compared with **1**, the asymmetric unit of **2** behaves as a three-connective node to link the neighboring units, further forming a "ladder-shaped" 1D chain extending in the crystal *b* direction (Figure 4b). The 1D chain was found arrayed in the crystal *a* direction through the neighboring chain interaction of the H-bond (O−H···O) and π-π stacking interactions; the former exist between the carboxylic and carboxylate groups with a D-A distance 2.5770(16) Å [28], while the latter is formed between pyridine rings with a center–center distance of 4.38 Å [29]. The inter-chain interactions afford the orderly array of 1D chains, and further give birth to 2D supermolecular layers with a thickness of ca. 7.9 Å (Figure 4c). The final 3D structure of **2** is furnished by the stacking of the 2D supermolecular layers through the neighboring layer interaction of π-π stacking within the pyridine rings as well as the benzene rings; the center–center distances are 3.77 and 3.75 Å [29], respectively (Figure 4d).

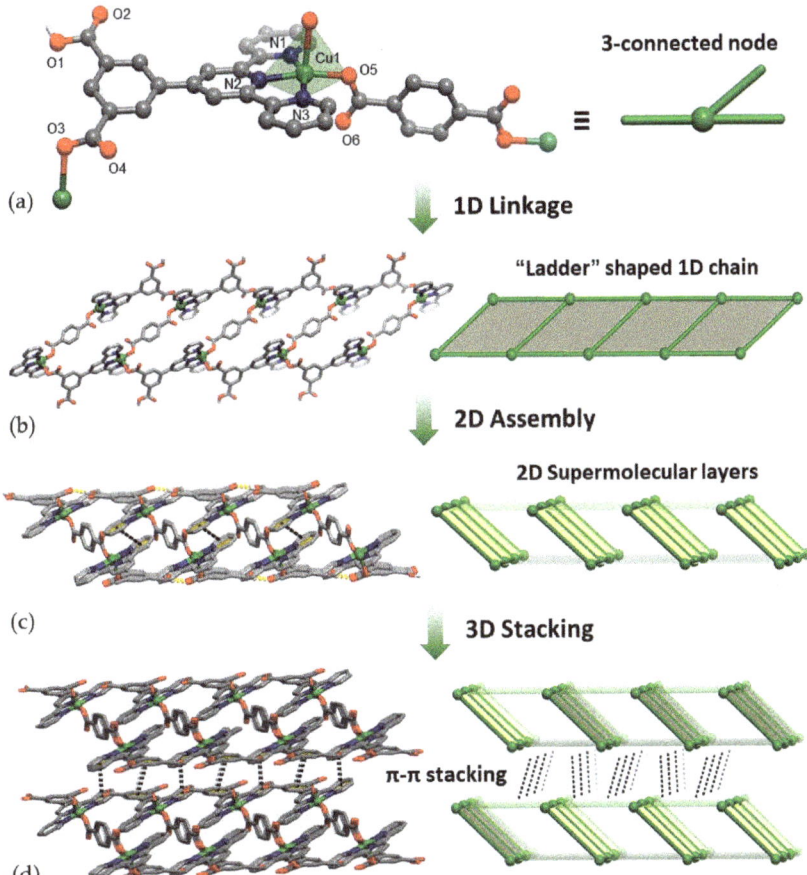

**Figure 4.** Asymmetric unit of **2** (**a**); the 1D "ladder-shaped" chain structure in **2**, as well as the linking mode of the nodes (**b**); a 2D supermolecular layer is formed by the orderly array of the 1D chain (**c**); the 3D structure of **2** (**d**). Some H atoms are omitted for clarity.

## 3.2. X-ray Diffraction Patterns

The PXRD (powder X-ray diffraction) patterns of complexes **1** and **2** were measured with crystalline samples at room temperature. As is shown in Figure 5, the experimentally determined PXRD patterns and the simulated ones from the SCXRD analyses are in accordance in general, suggesting their phase homogeneity.

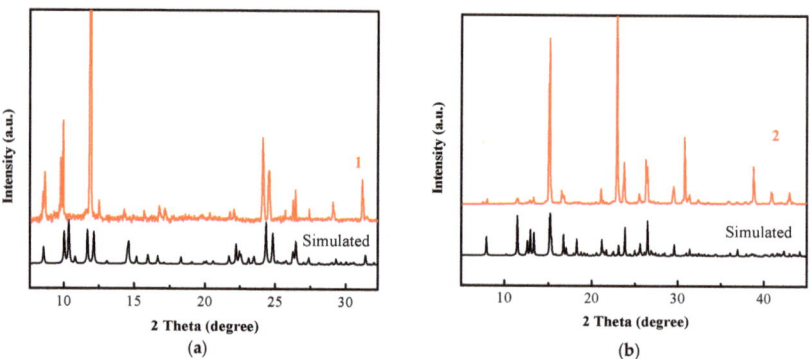

**Figure 5.** The experimental and simulated PXRD patterns for **1** (**a**) and **2** (**b**).

## 3.3. Thermal Analysis

In order to characterize the thermal stability of **1** and **2**, their thermal behaviors were investigated by TGA (Figure 6). For **1**, a weight loss of 9.8% was witnessed in the temperature scale of 25–60 °C, which corresponds to the removal of guest molecules of water. The TG curve goes into a platoon within 160–270 °C, which implies that the framework stability of **1** is up to 270 °C. After that, a weight loss step occurrs until 800 °C, which corresponds to the decomposition of organic ligands. (Figure 6a). As for **2**, the first weight loss is 3.6% at 200 °C. Due to the absence of guest molecules according to the SCXRD analysis, the first weight loss was attributed to the elimination of water molecules adhering to the sample's surface. The second weight loss begins at 300 °C; after that, a sharp weight loss occurred, which corresponds to the decomposition of organic ligands. The TGA result of **2** implies that the framework stability of **2** is up to 300 °C (Figure 6b).

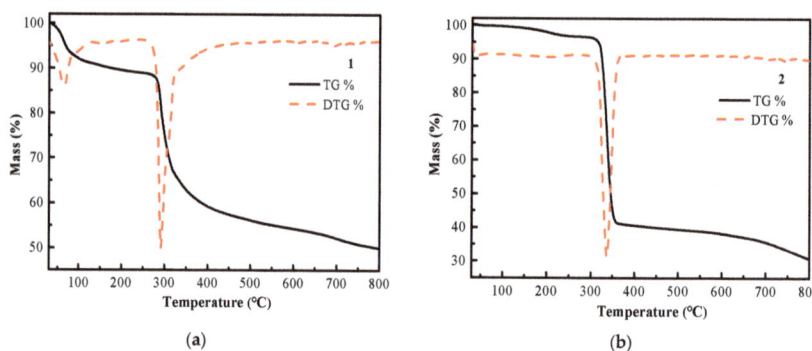

**Figure 6.** The TGA diagrams for **1** (**a**) and **2** (**b**).

## 4. Conclusions

In this work, we presented a structural study of a Cu(II)-H$_2$dtp-based MOF system, in which the rational assembly of Cu(II), dtp$^{2-}$ or Hdtp$^{-}$ as well as auxiliary bdc$^{2-}$ linker give birth to two new MOF materials with the formulae of {[Cu(dtp)](H$_2$O)$_2$}$_n$ (**1**) and [Cu(Hdtp)(bdc)$_{0.5}$]$_n$ (**2**). The harvest of 2D "wave-shaped" networks in **1** benefits from the five-coordinative Cu(II), which suggests a low metal ion coordination number in controlling the formation of the 2D Cu(II)-dtp$^{2-}$ network. The introduction of bdc$^{2-}$ into the Cu(II)-dtp$^{2-}$ system results in a 1D chain structure with a "ladder-shape" in **2**. A unique 2D supermolecular layer with a thickness of 7.9 Å is formed by the orderly array of the 1D chain, which may help to broaden the horizon for the assembly of 2D materials through the intermolecular interactions. The air stability of two MOFs is witnessed by PXRD, and the TGA indicates that the thermal stability of the two MOFs is as high as 300 °C. The unique structure and relatively good stability of these Cu(II)-MOFs inspired us to further isolate the 2D structures of **1** and **2** as 2D materials for further research.

**Supplementary Materials:** The following supporting information can be downloaded at: https://www.mdpi.com/article/10.3390/cryst12050585/s1, Table S1: Hydrogen-bonding parameters in **1**; Table S2: Hydrogen-bonding parameters in **2**; Table S3: Selected bond lengths (Å) and angles (o) for **1**; Table S4: Selected bond lengths (Å) and angles (o) for **2**. CCDC 2164393 (**1**) and 2164388 (**2**) contain the supplementary crystallographic data for this paper. These data can be obtained free of charge via www.ccdc.cam.ac.uk/data_request/cif (accessed on 5 April 2022), by emailing data_request@ccdc.cam.ac.uk, or by contacting The Cambridge Crystallographic Data Centre, 12 Union Road, Cambridge CB2 1EZ, U. K.; Fax: +44-1223-336033.

**Author Contributions:** Conceptualization, Z.-W.Z. and L.-Q.W.; methodology, Z.-W.Z., L.-Q.W. and J.-J.L.; validation, J.-J.L., X.-X.C., Q.-W.C. and Y.Q.; formal analysis, J.-J.L., X.-X.C., Q.-W.C. and Y.Q.; investigation, Z.-W.Z. and L.-Q.W.; resources, Z.-W.Z., L.-Q.W. and J.-J.L.; data curation, J.-J.L.; writing—original draft preparation, Z.-W.Z. and L.-Q.W.; writing—review and editing, Z.-W.Z., L.-Q.W. and Q.C.; visualization, Q.C.; supervision, Z.-W.Z. and L.-Q.W.; project administration, Z.-W.Z. and L.-Q.W.; funding acquisition, Z.-W.Z., L.-Q.W. and J.-J.L. All authors have read and agreed to the published version of the manuscript.

**Funding:** This research was funded by National Natural Science Foundation of China grant number 22161009, Thousands of Young and Middle-aged Backbone Teachers Training Project of Guangxi Colleges and Universities grant number Gui-Jiao 2020-58, University innovation foundation of Guangxi Medicine grant number Gui-Degree 2021 No. 6, Collaborative Innovation Center of Zhuang and Yao Ethnic Medicine grant number 2013 No. 20, Guangxi University of Chinese Medicine Key Project of First-class Discipline Construction grant number 2018XK050, Guangxi Natural Science Foundation grant number 2022GXNSFAA035477.

**Institutional Review Board Statement:** Not applicable.

**Informed Consent Statement:** Not applicable.

**Data Availability Statement:** Not applicable.

**Conflicts of Interest:** The authors declare no conflict of interest.

# References

1. Yin, Z.; Zhou, Y.-L.; Zeng, M.-H.; Kurmoo, M. The Concept of Mixed Organic Ligands in Metal-Organic Frameworks: Design, Tuning and Functions. *Dalton. Trans.* **2015**, *44*, 5258–5275. [CrossRef] [PubMed]
2. Wang, Q.; Astruc, D. State of the Art and Prospects in Metal–Organic Framework (MOF)-Based and MOF-Derived Nanocatalysis. *Chem. Rev.* **2020**, *120*, 1438–1511. [CrossRef] [PubMed]
3. Yuan, G.; Tan, L.; Wang, P.; Wang, Y.; Wang, C.; Yan, H.; Wang, Y.-Y. MOF-COF Composite Photocatalysts: Design, Synthesis, and Mechanism. *Cryst. Growth Des.* **2022**, *22*, 893–908. [CrossRef]
4. Shu, Y.; Ye, Q.; Dai, T.; Xu, Q.; Hu, X. Encapsulation of Luminescent Guests to Construct Luminescent Metal–Organic Frameworks for Chemical Sensing. *ACS Sens.* **2021**, *6*, 641–658. [CrossRef]
5. Hassanein, K.; Cappuccino, C.; Amo-Ochoa, P.; López-Molina, J.; Maini, L.C.; Bandini, E.; Ventura, B. Multifunctional coordination polymers based on copper(I) and mercaptonicotinic ligands: Synthesis, and structural, optical and electrical characterization. *Dalton. Trans.* **2020**, *30*, 10545–10553. [CrossRef]
6. Rogovoy, M.I.; Berezin, A.S.; Samsonenko, D.G.; Artem'ev, A.V. Silver(I)-organic frameworks showing remarkable thermo-, solvato- and vapochromic phosphorescence as well as reversible solvent-driven 3D-to-0D transformations. *Inorg. Chem.* **2021**, *9*, 6680–6687. [CrossRef]
7. Artem'ev, A.V.; Davydova, M.P.; Hei, X.-Z.; Rakhmanova, M.I.; Samsonenko, D.G.; Bagryanskaya, I.Y.; Brylev, K.A.; Fedin, V.P.; Chen, J.-S.; Cotlet, M.; et al. Family of robust and strongly luminescent CuI-Based hybrid networks made of ionic and dative bonds. *Chem. Mater.* **2020**, *24*, 10708–10718. [CrossRef]

8. Troyano, J.; Zapata, E.; Perles, J.; Amo-Ochoa, P.; Fernandez-Moreira, V.; Martínez, J.I.; Zamora, F.; Delgado, S. Multifunctional copper(I) coordination polymers with aromatic mono- and ditopic thioamides. *Inorg. Chem.* **2019**, *5*, 3290–3301. [CrossRef]
9. Qian, Q.; Asinger, P.A.; Lee, M.J.; Han, G.; Rodriguez, K.M.; Lin, S.; Benedetti, F.M.; Wu, A.X.; Chi, W.S.; Smith, Z.P. MOF-Based Membranes for Gas Separations. *Chem. Rev.* **2020**, *120*, 8161–8266.
10. Zeng, M.-H.; Yin, Z.; Tan, Y.-X.; Zhang, W.-X.; He, Y.-P.; Kurmoo, M. Nanoporous Cobalt(II) MOF Exhibiting Four Magnetic Ground States and Changes in Gas Sorption upon Post-Synthetic Modification. *J. Am. Chem. Soc.* **2014**, *136*, 4680–4688. [CrossRef]
11. Horcajada, P.; Gref, R.; Baati, T.; Allan, P.K.; Maurin, G.; Couvreur, P.; Férey, G.; Morris, R.E.; Serre, C. Metal–Organic Frameworks in Biomedicine. *Chem. Rev.* **2012**, *112*, 1232–1268. [CrossRef] [PubMed]
12. Song, Y.; Fan, R.-Q.; Wang, P.; Wang, X.-M.; Gao, S.; Dua, X.; Yang, Y.-L.; Luan, T.-Z. Copper(I)-iodide based coordination polymers: Bifunctional properties related to thermochromism and PMMA-Doped polymer film materials. *J. Mater. Chem. C* **2015**, *24*, 6249–6259. [CrossRef]
13. Chakraborty, G.; Park, I.-H.; Medishetty, R.; Vittal, J.J. Two-Dimensional Metal-Organic Framework Materials: Synthesis, Structures, Properties and Applications. *Chem. Rev.* **2021**, *121*, 3751–3891. [CrossRef] [PubMed]
14. Rodenas, T.; Luz, I.; Prieto, G.; Seoane, B.; Miro, H.; Corma, A.; Kapteijn, F.; Llabrés i Xamena, F.X.; Gascon, J. Metal–organic framework nanosheets in polymer composite materials for gas separation. *Nat. Mater.* **2015**, *14*, 48–55. [CrossRef]
15. Peng, Y.; Li, Y.; Ban, Y.; Jin, H.; Jiao, W.; Liu, X.; Yang, W. Metal-organic framework nanosheets as building blocks for molecular sieving membranes. *Science* **2014**, *346*, 1356–1359. [CrossRef]
16. Duan, J.; Li, Y.; Pan, Y.; Behera, N.; Jin, W. Metal-organic framework nanosheets: An emerging family of multifunctional 2D materials. *Coordin. Chem. Rev.* **2019**, *395*, 25–45. [CrossRef]
17. Wang, H.-H.; Liu, Q.-Y.; Li, L.; Krishna, R.; Wang, Y.-L.; Peng, X.-W.; He, C.-T.; Lin, R.-B.; Chen, B. Nickel-4′-(3,5-dicarboxyphenyl)-2,2′,6′,2″-terpyridine Framework: Efficient Separation of Ethylene from Acetylene/Ethylene Mixtures with a High Productivity. *Inorg. Chem.* **2018**, *57*, 9489–9494. [CrossRef]
18. Kang, X.-M.; Wang, W.-M.; Yao, L.-H.; Ren, H.-X.; Zhao, B. Solvent-dependent variations of both structure and catalytic performance in three manganese coordination polymers. *Dalton. Trans.* **2018**, *47*, 6986–6994. [CrossRef]
19. Kang, X.-M.; Yao, L.-H.; Jiao, Z.-H.; Zhao, B. Two Stable Heterometal-MOFs as Highly Efficient and Recyclable Catalysts in the $CO_2$ Coupling Reaction with Aziridines. *Chem. Asian J.* **2019**, *14*, 3668–3674. [CrossRef]
20. Mao, H.-J.; Chen, Q.-X.; Han, B. Two Metal—Organic Coordination Polymers Based on Polypyridyl Ligands: Crystal Structures and Inhibition of Human Spinal Tumour Cells. *Aust. J. Chem.* **2018**, *71*, 902–906. [CrossRef]
21. Liu, S.-L.; Chen, Q.-W.; Zhang, Z.-W.; Chen, Q.; Wei, L.-Q.; Lin, N. Efficient heterogeneous catalyst of Fe(II)-based coordination complexes for Friedel-Crafts alkylation reaction. *J. Solid. State. Chem.* **2022**, *310*, 123045. [CrossRef]
22. Bai, N.-N.; Hou, L.; Gao, R.-C.; Liang, J.-Y.; Yang, F.; Wang, Y.-Y. Five 1D to 3D Zn(II)/Mn(II)-CPs based on dicarboxyphenyl-terpyridine ligand: Stepwise adsorptivity and magnetic properties. *Cryst. Eng. Comm.* **2017**, *19*, 4789–4796. [CrossRef]
23. Massouda, S.S.; Louk, F.R.; David, R.N.; Dartez, M.J.; Nguyn, Q.L.; Labry, N.J.; Fischer, R.C.; Mautner, F.A. Five-coordinate metal(II) complexes based pyrazolyl ligands. *Polyhedron* **2015**, *90*, 258–265. [CrossRef]
24. Almeida, K.J.d.; Murugan, N.A.; Rinkevicius, Z.; Hugosson, H.W.; Vahtras, O.; Ågren, H.; Cesar, A. Conformations, structural transitions and visible near-infrared absorption spectra of four-, five- and six-coordinated Cu(II) aqua complexes. *Phys. Chem. Chem. Phys.* **2009**, *11*, 508–519. [CrossRef] [PubMed]
25. Sheldrick, G.M. *SADABS. Program for Empirical Absorption-Correction of Area Detector Data*; University of Goöttingen: Goöttingen, Germany, 1996.
26. Sheldrick, G.M. SHELXL 2014. *Acta Crystallogr. Sect. C Struct. Chem.* **2015**, *71*, 3–8. [CrossRef] [PubMed]
27. Dolomanov, O.V.; Bourhis, L.J.; Gildea, R.J.; Howard, J.A.K.; Puschmann, H. OLEX2: A complete structure solution, refinement and analysis program. *J. Appl. Crystallogr.* **2009**, *14*, 339–341. [CrossRef]
28. Tiwari, R.K.; Kumar, J.; Behera, J.N. H-bond supported coordination polymers of transition metal sulfites with different dimensionalities. *RSC Adv.* **2015**, *5*, 78389–78395. [CrossRef]
29. Mishra, B.K.; Sathyamurthy, N. π−π Interaction in Pyridine. *J. Phys. Chem. A* **2005**, *109*, 6–8. [CrossRef]

Article

# Study on Deposition Conditions in Coupled Polysilicon CVD Furnaces by Simulations

Shengtao Zhang [1], Hao Fu [1], Guofeng Fan [2], Tie Li [3], Jindou Han [1] and Lili Zhao [1,*]

1. School of Chemistry and Chemical Engineering, Harbin Institute of Technology, Harbin 150001, China
2. Soft-Impact China (Harbin), Ltd., Harbin 150028, China
3. Harbin KY Semiconductor, Inc., Harbin 150028, China
* Correspondence: zhaolili@hit.edu.cn; Tel.: +86-188-4613-0714; Fax: +0451-840-10483

**Abstract:** Electronic-grade polysilicon is the cornerstone of the information industry. Considering the demand for this material in the semiconductor industry, any technological improvement has great potential benefits. Due to the quality requirements of electronic polysilicon, its preparation process is characterized by low raw material utilization and high cost. Simply increasing the deposition rate by increasing the chemical reaction rate will easily lead to a reduction in the proportion of dense materials. For the first time, a coupled furnace scheme is proposed to improve the utilization of raw materials while maintaining the same deposition quality. The deposition conditions on the surface of silicon rods with different base plate designs were modeled and analyzed using the software PolySim, and a design characterized by a high flow rate and the use of 9 mm and 15 mm nozzles was selected for the coupling scheme. In coupling mode, the simulation results show that the utilization of raw materials is increased by 17.5%, and the deposition rate is increased by 44.9%, while the deposition quality and uniformity remain approximately unchanged. The results show that the coupling scheme with high feed flow is beneficial for significantly improving the deposition conditions and the utilization rate of raw materials, which also provides guidance for material preparation processes with similar principles.

**Keywords:** electronic polysilicon; flow field; temperature field; boundary layer; coupled furnaces

**Citation:** Zhang, S.; Fu, H.; Fan, G.; Li, T.; Han, J.; Zhao, L. Study on Deposition Conditions in Coupled Polysilicon CVD Furnaces by Simulations. *Crystals* **2022**, *12*, 1129. https://doi.org/10.3390/cryst12081129

Academic Editor: Mingyang Chen

Received: 11 July 2022
Accepted: 8 August 2022
Published: 12 August 2022

**Publisher's Note:** MDPI stays neutral with regard to jurisdictional claims in published maps and institutional affiliations.

**Copyright:** © 2022 by the authors. Licensee MDPI, Basel, Switzerland. This article is an open access article distributed under the terms and conditions of the Creative Commons Attribution (CC BY) license (https://creativecommons.org/licenses/by/4.0/).

## 1. Introduction

With the development of the electronic information industry, semiconductor devices, integrated circuits and other high-tech fields have an increasing demand for electronic-grade (EG) high-purity polysilicon [1–3]. The modified Siemens process is the mainstream process for the production of high-purity polysilicon [4], accounting for more than 78% of the global production capacity [5]. The reduction process is the core process of the modified Siemens method. The principle of this method is that raw material with high purity, including trichlorosilane (TCS) and hydrogen ($H_2$), enter the reactor through nozzles in the base plate, and a chemical vapor deposition reaction (CVD) occurs on the surface of polysilicon rods at a high temperature around 1323 K [6,7]. As a result, the diameters of the rods increase with the growth process. Compared with dozens of solar-grade (SG) polysilicon manufacturing enterprises, the number of electronic-grade polysilicon enterprises is very small, and the production is limited [8,9]. Electronic polysilicon has a very high requirement for product purity [10]. SG polysilicon has a minimum of 6 N or 99.9999% purity, while EG polysilicon has 9 N. The atomic fractions of acceptor impurities and donor impurities are usually below 50 ppt and 150 ppt, respectively, which puts forward a very high demand for the impurity control of the whole closed-loop process [11], resulting in extremely high costs.

Energy and material consumption in the reduction stage accounts for a remarkable proportion of the whole production process [12]. Some researchers have tried to decrease production costs from different directions. Luo [13] pointed out that the well-polished substrate surface suggests an excellent radiation energy-saving capacity. Nie [14] used ceramic lining on a reactor vessel to emit thermal radiation and obtain smoother radial-dependent temperature and thermal stress distributions. However, both of them increase the furnace cost for expensive materials or processing. Sun [15] studied the influence of the reaction temperature on the yield of silicon and power consumption under certain conditions for the production of electronic-grade polysilicon, which showed that higher temperatures are good for unit consumption but need to be as uniform as possible. Some researchers have studied deposition conditions from the perspective of electricity. Nie [16] presented an electrical heating model using alternating current (AC) for the silicon rods. The influences of the location of the silicon rods, AC frequency, the radius of the rod and wall emissivity on the temperature profile and current density were studied through the application of the developed model. Du [17] found that high-frequency current is conducive to reducing the temperature gradient and put forward the concept of the mixed-frequency heating process. The deposition conditions on the surface of the silicon rod in the reactor determine the polysilicon quality and deposition rate, which is hard to improve and related to many factors, such as the uniformity of rod surface temperature, gas flow rate, flow velocity along the rods and so on. Meanwhile, complex operating parameters, including the feed gas, electronic current, reactor shape and the layout of polysilicon rods, nozzles and outlet, can significantly influence the deposition conditions [18]. Since this stage is in a closed environment and the polysilicon rod surface has a very high temperature of around 1323 K during the deposition process [10,19], gas flow in the reactor and the temperature field distribution of the silicon rod surface are difficult to directly and precisely measure and characterize [20]. The PolySim software employed in this study was used as a numerical simulation tool to understand what happens inside the CVD reactor.

In order to improve the utilization rate of raw materials and reduce costs, a coupled furnace scheme is proposed, which means connecting several furnaces in series to change the amounts of raw materials or the flow rate. This concept is expected to achieve exhaust gas recovery, improve the overall conversion rate of raw materials and reduce material and energy consumption. Due to the reduction in raw materials from the first furnace to the next furnace, the feed material quantity needs to be changed to meet requirements for consistent deposition conditions. Usually, the gas inlet nozzles are fixed on the base plate, whose distribution directly determines the injection rate and flow rate of the raw materials supplied. Therefore, it is necessary to carry out simulation studies on furnaces with different characteristics of inlet gases. In this study, the original scheme, a high inlet gas flow velocity scheme and a high inlet gas flow rate scheme were studied and compared. By analyzing the deposition conditions on the surface of the silicon rods, the high flow rate base plate design was selected for the coupling equipment. The deposition characteristics and process parameter results of the coupling equipment were also simulated, which showed good prospects for the coupled reduction furnace scheme.

## 2. Modeling Process

The CVD furnace used for modeling is a bell jar type, and its main structure is shown in Figure 1a [21]. Numbers 1 to 13 indicate the base plate, electrodes, exhaust gas outlet pipes, mixed gas inlets, mixed gas inlet pipes, base plate coolant inlet pipes, base plate coolant outlet pipes, furnace coolant inlet pipes and outlet pipes, furnace, silicon rods, coolant interlayer and observation window. The 3D model was built and meshed using PolySim software, and the physical meaning of some parameters is shown in Figure 1b. After neglecting the irrelevant details, the key structures of the reactor, such as the inner wall, base plate, gas inlets, gas outlets and polysilicon rods, were abstracted, and the numerical domain required for the modeling of a CVD reactor was obtained. The main geometric parameters of the reactor are shown in Table 1. Due to the need for resistance heating, in actual production, the tops of the two silicon rods will be connected in series through a small segment of the silicon rods in the initial stage. However, in the model, it is simplified into a small segment with an arc shape (bridge part). In order to facilitate the comparison of the physical field distribution inside the furnace chamber, a silicon rod diameter of 50 mm was preferentially set during 3D modeling.

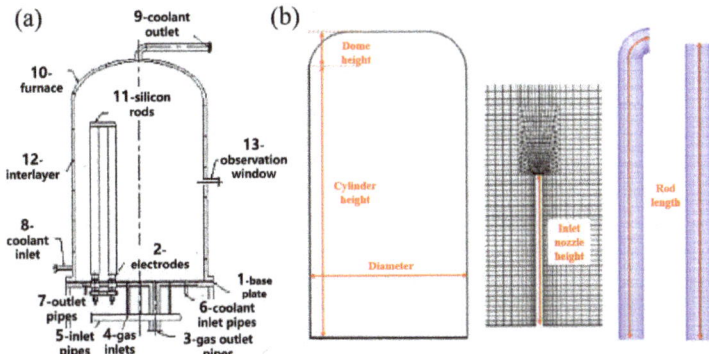

Figure 1. Schematic diagram of bell-type polysilicon reduction furnace: (**a**) Profile; (**b**) Furnace shell and silicon rod.

Table 1. Geometrical dimensions of the model of 18-rod CVD reactor.

| Categories | Parameters (mm) |
| --- | --- |
| Reactor height | 2500 |
| Chamber height | 320 |
| Reactor diameter | 1400 |
| Outlet diameter | 35 |
| Silicon rod height | 2300 |
| Typical silicon rod diameter | 50 |

The reactor chamber was discretized by a mesh with different kinds of elements (hexahedral, polyhedral and prismatic) to adapt to different characteristics of different domains. To be specific, the prismatic element is used along the polysilicon rods, which allows the extrusion of cells along the cylinders. The boundary layer uses hexahedral structure meshes to provide cell size control, and the polyhedral element is used to discretize the dome due to its complex shape. The first unit of the furnace body boundary layer and the first unit of the silicon rod boundary layer are respectively set to 2 mm and 5 mm, and the transition factor is 1.4. A schematic diagram of the furnace body and silicon rods after meshing is shown in Figure 2. Due to similar settings of mesh generation, the different models have similar grids, which contain about 7 million cells, 19 million faces and 6 million nodes. The minimum orthogonal quality and the maximum aspect ratio are about 0.32 and 32, respectively, and the quality parameters of the grid meet the requirements.

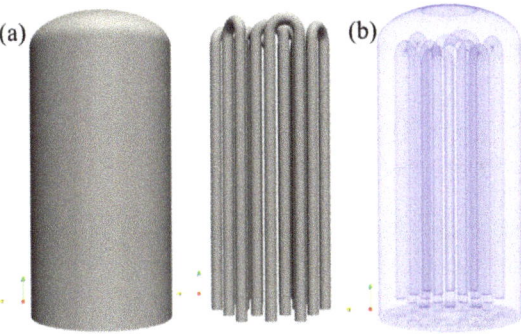

Figure 2. Schematic diagram of furnace body and silicon rods: (**a**) 3D models; (**b**) grid structure.

To explore the requirements of the coupling scheme for the reduction furnace, it is necessary to conduct modeling comparisons from the perspective of increasing the flow rate and increasing the flow rate. In view of this, different base plates with different distributions of gas inlet nozzles were designed, while the overall shape of the equipment was not changed. As shown in Figure 3, the base plate has four rings, including outlet nozzles, silicon rods, inlet nozzles and inner silicon rods from the circumference to the center. Six inlet nozzles are evenly distributed on the nozzle ring, while different schemes have different inlet nozzle arrangements. In design A (original scheme), the inlet nozzles have two diameters of 7 mm and 11 mm, and these nozzles with two diameters are arranged at intervals. In design B (high flow rate scheme), the nozzle diameters are 4.2 mm and 7 mm. Considering the total cross-sectional area of each nozzle, the feed flow rate of scheme B remains unchanged, but the feed flow velocity is increased by 2.5 times. In design C (large flow scheme), the nozzle diameters are 9 mm and 15 mm. The feed flow rate increases twofold. It should be pointed out that the above calculation is simplified, which assumes that the outlet pressure of each nozzle is the same.

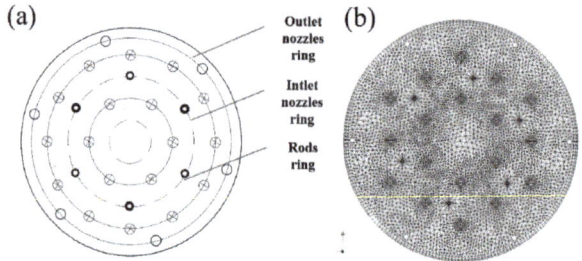

**Figure 3.** Diagram of base plate: (**a**) structure; (**b**) after meshing.

During the deposition process, large amounts of raw components for the reaction and enough electricity for heating are necessary. For controlling variables, the main process parameters and boundary conditions for modeling these designs are the same, such as electric current, the temperature of the wall, operating pressure, $H_2$/Si mole ratio and so on, as shown in Table 2. In particular, in order to be close to the actual gas composition, there is some DCS (Dichlorosilane) in the raw material supply. The reduction furnace body is a double-layer structure, and the coolant in the middle layer absorbs the heat of the furnace wall and reduces its temperature. The furnace body is generally made of stainless steel, though some manufacturers use silver-plated materials. The boundary conditions are shown in Table 3. The furnace side wall and base plate are made of the same material in this study.

**Table 2.** Process parameters.

| Schemes | Electric Current (A) | Temperature of Wall (K) | Temperature of Feed Gas (K) | Operating Pressure (kPa) | TCS Flow kg/h | DCS/TCS % | $H_2$/Si |
|---|---|---|---|---|---|---|---|
| Design A | 1480 | 370 | 393 | 600 | 550 | 3 | 3.24 |
| Design B | 1480 | 370 | 393 | 600 | 550 | 3 | 3.24 |
| Design C | 1480 | 370 | 393 | 600 | 1375 | 3 | 3.24 |

**Table 3.** Boundary conditions.

| Categories | | Parameters |
|---|---|---|
| Emissivity | Silicon rods | 0.76 |
| | Furnace wall | 0.36 |
| | Base plate | 0.36 |
| Heat Conductivity (W/(m·K)) | Furnace wall | 17 |
| | Base plate | 17 |

During the simulation, processes including turbulent flow (by k-eps model), heat transfer in the gas and inside the rods, radiation and electric current inside rods were modeled. Navier–Stokes equations were solved, together with equations for enthalpy and equations for k and epsilon, respectively. In addition, a surface-to-surface radiation model was used, also known as the view factor model. PolySim 3D was used to solve all of the above equations. The representative process time was selected when the silicon rod diameter was 50 mm so as to evaluate the growth conditions to the greatest extent. To achieve convergence, about 60,000 iterations are required. Residuals for solving equations are less than 0.0001. The heat and radiation imbalances are less than 1.5%, and the mass imbalances are less than 0.5%, which shows that the modeling results converge well. An internal visualizer of PolySim 3D was used to visualize 3D distributions of physical fields, including the temperature field, flow field and boundary layer in the reactor. The modeling results of design A were analyzed and compared with an experiment [22], and highly consistent results were obtained. In the previous study, under similar process conditions and furnace structure, the error of key parameters between the simulation results and the production results was about 4%. The accuracy meeting the requirements in design A provides reliable support for the simulation of the coupling mode.

## 3. Results and Discussion

### 3.1. Temperature Distribution

Although a temperature of around 1300 K is required for polysilicon deposition, the furnace wall maintains a relatively low temperature due to the existence of the interlayer cooling water outside the furnace. Figure 4 shows the temperature distribution inside the furnace with different designs. The temperature of the furnace wall is typically somewhere between 400–460 K, and the mid-upper part has a higher temperature, which corresponds to the higher temperature of silicon rods. In contrast, design C has a lower furnace wall temperature and better distribution uniformity than design B, which is related to the heat consumption of the larger flow of inlet gases. In order to understand the temperature distribution in the furnace cavity, four horizontal sections with different heights (from bottom to top, they are 0.5 m, 1.1 m, 1.7 m and 2.3 m, respectively) and a symmetrical vertical section (passing through two nozzles and the center of the base plate) in the reduction furnace are displayed. Compared with simply increasing the flow velocity, a larger flow rate is conducive to reducing the temperature in the furnace cavity to a certain extent, and its temperature distribution is relatively uniform. The results imply that design C can provide similar uniform deposition conditions to those of the original scheme. Although design B improves the gas inlet flow velocity, the overheated area at the top of the furnace cavity is large and concentrated in the middle and upper parts of the silicon rod, which may lead to an intense gas-phase reaction and the formation of too much silicon dust or powder. In actual production, excessive silicon dust may lead to a short circuit between electrodes or silicon rods, which increases the risk of production process termination and greatly reduces the output.

The heat inside the whole furnace comes from the resistance heating of the silicon rods; the heat is mainly concentrated on the surface of the silicon rod, and the current density in the center of the silicon rod is almost zero [23]. The surface of the silicon rods is where the reaction takes place directly, so the heat transfer state on the surface of the silicon rod deserves attention. The temperature distribution along the rod surface was obtained after the heat transfer was simulated, as shown in Figure 5. The surface temperature of the silicon rods is around 1360–1400 K, but the temperature at the upper part of the silicon rod is particularly high. In the actual process, these areas correspond to areas where a popcorn-shaped structure is most likely to be produced [6]. According to the Arrhenius formula, the reaction rate constant is exponentially related to temperature. The higher surface temperature of the silicon rod greatly increases the reaction deposition rate in this area, resulting in the phenomenon of popcorn-shaped polysilicon near the bridge part of the silicon rods in the actual process, which directly leads to a reduction in the proportion of dense or qualified materials. The rod surface temperature in design B and design C is generally higher, which means a faster deposition rate and greater preparation efficiency. Meanwhile, the temperature at the silicon rod bridge in designs B and C is not very prominent compared with that in design A, which indicates that the temperature uniformity in design B and design C is improved. The temperature measurement at a height of 2 m for the outer-ring silicon rods shows that the difference between the simulated value and the measured value is equivalent to the error range of the temperature meter, which suggests that the calculated result is reliable.

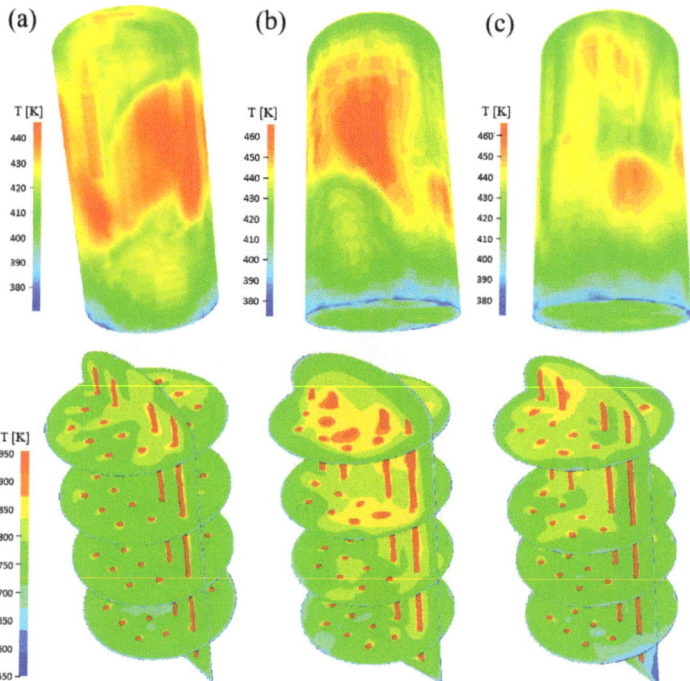

**Figure 4.** Temperature distribution of furnaces. (**a**) Design A; (**b**) design B; (**c**) design C.

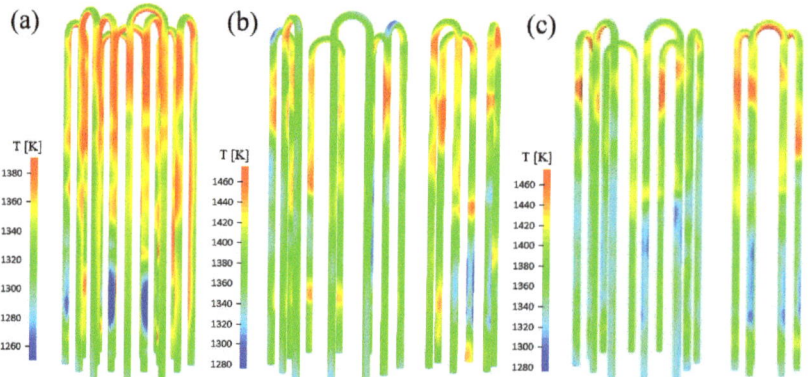

**Figure 5.** Temperature distribution along rod surface. (**a**) Design A; (**b**) design B; (**c**) design C.

A higher temperature in the deposition area will make the surface unstable and produce popcorn polysilicon. The design of the furnace structure means that the silicon rod is easily overheated at the rod bridge part, because more Joule heat is formed at this part due to the concentration of current density. Local overheating of the silicon rod will cause too intense a reaction to form unqualified polysilicon material. In addition, as the feed gas enters from the base plate inlet nozzles, the raw material is gradually heated during its rising process. At the same time, the decrease in the gas flow rate forms a gas stagnation zone, which leads to the failure of the timely replenishment of reaction gas and timely removal of tail mixed gas, and also promotes the heat concentrated in this area. Therefore, in order to improve the utilization rate of raw materials and reduce energy consumption, it is necessary to ensure the uniformity of deposition conditions on the surface of the silicon rods, reduce the overheating area and then improve the reaction rate, especially in the upper part of the reactor. In view of this, the gas flow rate and the thickness of the boundary layer on the surface of the silicon rods in different schemes are analyzed in the following content.

*3.2. Surface Velocity and Boundary Layer Thickness of Silicon Rods*

In order to find ways to relieve the excessively high temperature in the upper bridge area of the silicon rods, the gas flow field in the same section in the three schemes is visualized, as shown in Figure 6. The flow velocity in designs B and C at each cross-section is larger than that of design A, especially at the upper part of the furnace body. The area with a flow velocity above 3.5 m/s increases significantly, while the area with flow velocity below 0.5 m/s decreases significantly, which can be called a gas stagnation zone. Although the gas flow rate decreases rapidly with the rise in the vertical height, the flow velocity of the latter two designs is mostly maintained at 1–1.5 m/s above the height of 1.7 m and even reaches a high flow velocity of 2 m/s at the top-most part. As a result, the gas stagnation zone is greatly reduced, indicating that designs B and C have a relatively stable flow structure on the vertical cross-section. Meanwhile, the flow velocity on the surface of the silicon rod is more noteworthy, since it directly determines the replenishment of reaction gas and the displacement of reaction tail gas. The calculation shows that the average velocity around the silicon rod surface is 1.19 m/s, 1.80 m/s and 2.53 m/s in designs A, B and C, respectively. The increase in flow rate and flow velocity can increase the gas flow velocity on the surface of the silicon rods, which will be more conducive to reducing the volume of the overheated gas area and supplementing the reaction gas in time.

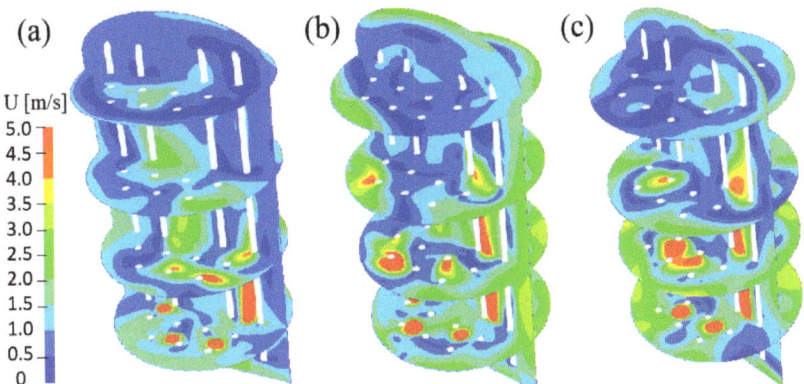

**Figure 6.** Flow field distribution. (**a**) Design A; (**b**) design B; (**c**) design C (height of horizontal sections is the same as in Figure 4).

The boundary layer thickness in different schemes is plotted to better evaluate the velocity distribution on the surface of the silicon rods, as shown in Figure 7. The boundary layer refers to the thickness of the area with a large velocity gradient near the surface of the silicon rods. From the wall of the boundary layer, the velocity in this area decreases rapidly along the tangential direction of the wall. Different schemes have similar value distributions, and larger values appear in the upper part and bridge part of the silicon rods. The difference is that the values around the silicon rod bridge part in design A and design B is greater than those in design C, even reaching 7 mm, while the values in design C are mostly between 2–5 mm. In the middle position of the outer-ring silicon rods, designs B and C have smaller values than design A, which means that the boundary layer thickness of design A is larger than that of designs B and C. In order to compare the distribution uniformity, the deviations of the parameter on the surface of the silicon rods at different times in different schemes were obtained and are shown in Table 4. These values also evaluate the uniformity of the boundary layer thickness at different thicknesses of silicon rods (different deposition stages). The boundary layer thickness deviation in design A is the lowest, with an average of 46.6%. Meanwhile, the average values of design B and design C reach 50% and 48%, respectively. These values show that the increasing flow rate and increasing flow velocity achieve smaller boundary layer thickness but also improve the non-uniformity of the boundary layer on the surface of the silicon rods, especially when the diameter of the silicon rod is small. By comparison, design C has better uniformity than design B, and its boundary layer thickness deviation is closer to that of design A, which can provide both better gas flow structure and better polysilicon deposition conditions. Therefore, design C is determined as the base plate scheme for the reduction furnaces in coupled mode.

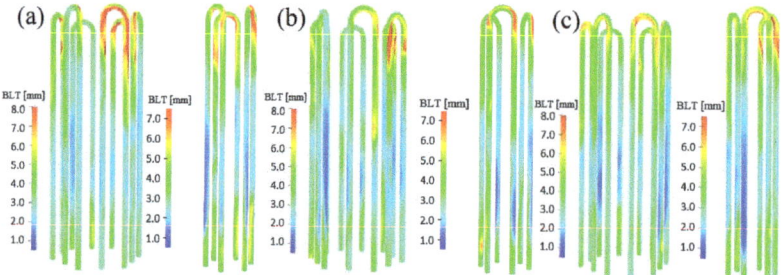

**Figure 7.** Boundary layer thickness along the rods. (**a**) Design A; (**b**) design B; (**c**) design C.

**Table 4.** Boundary layer thickness deviation (%).

| Rod Diameter | Design A | Design B | Design C |
| --- | --- | --- | --- |
| 50 mm | 47.8 | 53 | 51.8 |
| 90 mm | 45.3 | 48 | 44.1 |
| 125 mm | - | 48.9 | 48.1 |
| Average value | 46.6 | 50.0 | 48.0 |

### 3.3. Results of Coupling Scheme

In coupling mode, several reduction furnaces are connected in series to maximize the utilization of raw materials, which also provides the possibility to make full use of the heat of exhaust gas. Since the conversion rate of raw materials is generally about 10%, the influence of the difference in the contents of gas components on the composition of raw materials is ignored in the calculation, and attention is paid to the inlet gas temperature of furnace #1 and furnace #2. With the help of the calculation platform, the pipe between the coupled reduction furnaces with different silicon rod diameters was modeled. With reference to the actual process, the interlayer cooling water temperature was set to 413 K, and the corresponding silicon rod diameter inside the furnace body was 50 mm. The temperature field distribution in the pipe was obtained and is shown in Figure 8. The tail gas of furnace #1 is discharged from the base plate and transported to the inlet nozzles of furnace #2 through pipes. The outlet temperature of furnace #1 reaches 780 K. From the base plate outlet, the channel is connected to the annular pipeline via a 1 m vertical pipeline. At this time, the gas temperature reaches 622 K, and the temperature decreases by about 158 K. Before the tail gas enters furnace #2, it needs to be cooled by water cooling through a 4 m long pipe, and the temperature drops to 523 K. After the water-cooled pipe, combined with the actual distance, the outlet temperature can be maintained at around 523 K through the thermal insulation pipeline and then connected to reduction furnace #2.

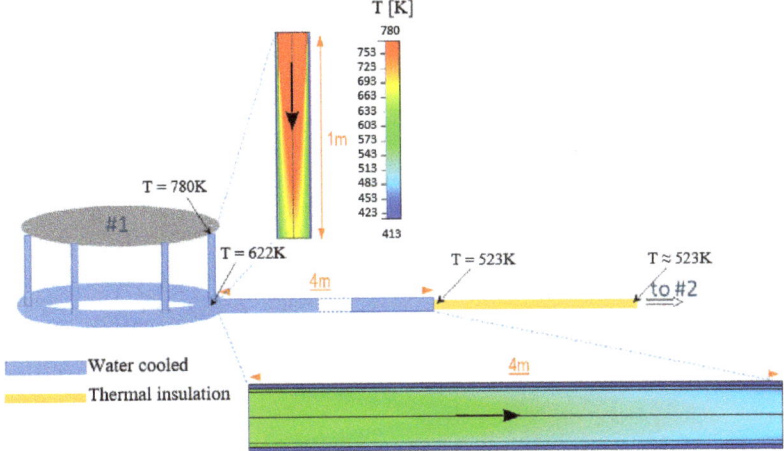

**Figure 8.** Temperature distribution of tail gas.

The gas temperatures corresponding to the different silicon rod diameters at the outlet of furnace #1 are between 534 K and 745 K, so the inlet temperature of furnace #2 is set according to these values. Further, gas temperature and pipe outlet gas temperature corresponding to different diameters of the silicon rods of coupled reduction furnace #1 and furnace #2, including 30 mm, 50 mm, 90 mm and 125 mm, were calculated, as shown in Table 5. With the increase in diameter, the gas temperature at the outlet gradually increases, and the tail gas temperature gradually rises. However, when the rod diameter reaches more than 90 mm, temperatures remain basically unchanged, and the temperature at the outlet of the pipeline also has similar trends. The pipe outlet temperature of furnace #2 is higher than that of furnace #1, ranging from 170 K to 450 K, which can be easily explained because fresh gas as raw material is heated twice in two furnaces.

**Table 5.** Temperature of tail gas (K).

| Rod Diameter | Tail Gas Temperature | Furnace #1 Pipe Outlet Temperature | Furnace #2 Pipe Outlet Temperature |
| --- | --- | --- | --- |
| 30 mm | 646 | 472 | 534 |
| 50 mm | 780 | 523 | 622 |
| 90 mm | 1025 | 582 | 749 |
| 125 mm | 1028 | 573 | 745 |

In order to calculate the process characteristics of coupling mode, design C was used in both coupled reduction furnaces #1 and #2. The internal physical fields of the two furnaces with different silicon rod diameters were modeled and calculated, and the thickness deviation of the boundary layer is shown in Table 6. The data show that the difference in the thickness deviations of the boundary layer between the two furnaces is very small, which means that the performance of the two reduction furnaces is similar in coupling mode. Design C provides approximate gas flow conditions and ensures the consistency of the deposition conditions of the two furnaces to a certain extent. At the same time, the boundary layer deviation decreases rapidly and then increases slightly with the increase in the silicon rod diameter. The consistency of deposition conditions is best when the silicon rod diameter is 90 mm, which means that the gas flow structure of design C plays the largest role when silicon rods grow to this diameter. Therefore, the base plate scheme in design C was adopted for the two reduction furnaces, i.e., furnace #1 and furnace #2. The coupling reduction process can adopt single-direction feeding, and the feeding switching step is no longer required.

**Table 6.** Boundary layer thickness deviation (%).

| Rod Diameter | Furnace #1 | Furnace #2 | Difference |
|---|---|---|---|
| 50 mm | 51.8 | 52.3 | 0.5 |
| 90 mm | 44.1 | 45.0 | 0.9 |
| 125 mm | 48.1 | 48.2 | 0.1 |

When the diameter of the silicon rods is 90 mm, the flow field and temperature field of the two reduction furnaces are shown in Figure 9. The gas flow rate near the bridge position is still high; an area of more than 2.5 m/s in the 2.3 m height section accounts for more than 50%, and the area of a gas flow rate of less than 0.5 m/s accounts for less than 5%: that is, the gas stagnation area is greatly compressed. Although the temperature in the furnace is high at this growth stage, due to the strong gas flow circulation, the temperature distribution near the silicon rod with the same section in the upper section of the overall reduction furnace is relatively uniform, and the excessive-temperature area in design A no longer exists. Due to the increase in flow velocity and the improvement in flow structure, the surface temperature distribution of the silicon rods is relatively uniform in the upper area of the silicon rods. Only the temperature near the bridge of the silicon rods in the inner ring is slightly higher than 1450 K, and the temperature distribution is mostly concentrated between 1350–1450 K. Compared with furnace #1, the overall gas temperature of furnace #2 is higher for higher inlet temperature. As the diameter of the silicon rod becomes larger, the outer-ring silicon rods have a greater shielding effect on the radiation of the inner silicon rods, and the gas near the inner-ring silicon rods is about 50 K higher than that near the outer-ring silicon rods. In summary, the temperature distribution of the silicon rod is uniform, and the temperature at the bridge is also maintained at an appropriate level, which is conducive to improving the deposition rate and reducing energy consumption while maintaining the uniformity of deposition.

The modeling of the deposition process was also carried out using PolySim software, and the quasi-steady-state method was used to approximate the silicon rod growth process. Based on the former calculated results of different deposition times, including silicon rod surface temperature, boundary layer thickness and its deviation, silicon rod surface flow velocity, inlet gas composition and other parameters, process characteristic parameters such as deposition rate were obtained using the built-in model. While maintaining similar deposition quality between the two reduction furnaces, the expected process characteristics were obtained and are shown in Table 7. The deposition rate in coupling mode reaches 20.8 kg/h, which is 44.9% higher than that of the two single reduction furnaces. The unit energy consumption is 17.2% lower than it was in the original scheme, reaching 74.2 kWh/kg. Although a single furnace in coupling mode has a lower conversion rate of raw materials of about 7%, this parameter increased by 17.5% to 14.1% from 12%. This means that the coupling scheme achieves a significant improvement in the characteristics of the deposition process, significantly reducing the reaction time from about 80 h to 53 h, and the optimization range is obvious.

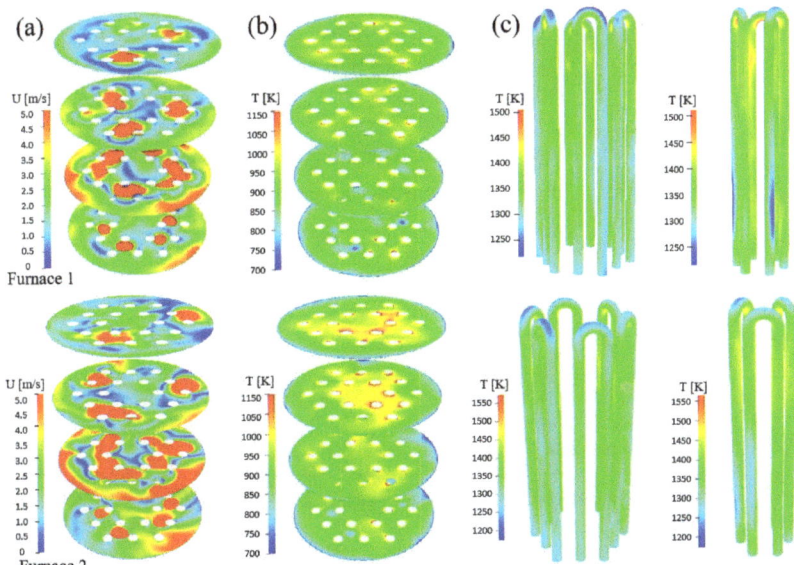

**Figure 9.** Temperature and velocity distributions. (**a**) Cross-sections of gas velocity; (**b**) cross-sections of temperature; (**c**) rod surface temperature (the height of cross-section is 0.5 m, 1.1 m, 1.7 m and 2.3 m, respectively).

**Table 7.** Process characteristics.

| Parameters | Design A | Furnace #1 | Furnace #2 | Coupling Mode | Optimization Proportion |
|---|---|---|---|---|---|
| Deposition rate (kg/h) | 14.08 | 21.4 | 19.4 | 20.4 | 44.9% |
| Unit energy consumption (kWh/kg) | 89.6 | 66.6 | 82.5 | 74.2 | 17.2% |
| Conversion rate (%) | 12 | 7.37 | 7.22 | 14.1 | 17.5% |
| Process time (h) | 79.8 | 52.6 | 52.6 | 52.6 | 34.1% |
| Total output (kg) | 1124 | 1124 | 1020 | 1072 | 4.6% |

## 4. Conclusions

For the first time, we propose a coupling device scheme for the preparation of electronic polysilicon by the modified Siemens method and carried out modeling and prediction work on the coupled furnaces based on numerical methods. The distribution of physical fields in the chamber and the uniformity of deposition conditions on the surface of the silicon rods in different base plate designs of the single furnace were analyzed. Based on a large flow rate scheme, the coupled furnaces' silicon rod deposition process was modeled, as well as connecting pipelines and structures. By increasing the flow rate by about 2 times, the coupling mode improves the utilization of raw materials by 17.5% and the deposition rate by 44.9%, while the deposition quality is approximately unchanged. This work provides a new idea for optimizing the deposition characteristics and preparation process of electronic-grade polysilicon at a low cost.

**Author Contributions:** Conceptualization, S.Z.; methodology, S.Z. and H.F.; software, J.H.; validation, G.F.; formal analysis, H.F. and G.F.; investigation, T.L.; resources, J.H.; data curation, S.Z.; writing—original draft preparation, S.Z.; writing—review and editing, L.Z.; visualization, H.F. and J.H.; supervision, L.Z. and T.L.; project administration, T.L.; funding acquisition, L.Z. All authors have read and agreed to the published version of the manuscript.

**Funding:** This research was funded by [Major Scientific and Technological Achievements Transformation Projects of Heilongjiang Province of China] grant number [CG20A008], [Natural Science Foundation of Heilongjiang Province] grant number [JQ2019E003], [2021 Harbin Science and Technology Special Plan Project] grant number [2021ZSZZGH10]. The APC was funded by [Major Scientific and Technological Achievements Transformation Projects of Heilongjiang Province of China].

**Institutional Review Board Statement:** Not applicable.

**Informed Consent Statement:** Not applicable.

**Acknowledgments:** The authors thank Soft-Impact China (Harbin), Ltd. Harbin. for its support in simulation content.

**Conflicts of Interest:** The authors declare no conflict of interest.

## References

1. Ramirez-Marquez, C.; Otero, M.V.; Vazquez-Castillo, J.A.; Martin, M.; Segovia-Hernandez, J.G. Process design and intensification for the production of solar grade silicon. *J. Clean. Prod.* **2018**, *170*, 1579–1593. [CrossRef]
2. Trinh, A.-K.; González, I.; Fournier, L.; Pelletier, R.; Sandoval, V.J.C.; Lesage, F.J. Solar thermal energy conversion to electrical power. *Appl. Therm. Eng.* **2014**, *70*, 675–686. [CrossRef]
3. Reznichenko, M. Evolution of Requirements for Solar Grade Silicon. *Procedia Eng.* **2016**, *139*, 41–46. [CrossRef]
4. Li, X.-G.; Xiao, W.-D. Model on transport phenomena and control of rod growth uniformity in siemens CVD reactor. *Comput. Chem. Eng.* **2018**, *117*, 351–358. [CrossRef]
5. Bye, G.; Ceccaroli, B. Solar grade silicon: Technology status and industrial trends. *Sol. Energy Mater. Sol. Cells* **2014**, *130*, 634–646. [CrossRef]
6. del Coso, G.; del Cañizo, C.; Luque, A. Chemical Vapor Deposition Model of Polysilicon in a Trichlorosilane and Hydrogen System. *J. Electrochem. Soc.* **2008**, *155*, 485–491. [CrossRef]
7. Fang, M.; Xiong, Y.Y.; Yuan, X.Z.; Liu, Y.W. Numerical Analysis of the Chemical Vapor Deposition of Polysilicon in a Trichlorosilane and Hydrogen System. In *International Conference on Applied Energy, Icae2014*; Yan, J., Lee, D.J., Chou, S.K., Desideri, U., Li, H., Eds.; Elsevier Ltd.: Taipei, Taiwan, 2014; pp. 1987–1991.
8. Wang, Z.; Wei, W. External cost of photovoltaic oriented silicon production: A case in China. *Energy Policy* **2017**, *107*, 437–447. [CrossRef]
9. Pizzini, S. Towards solar grade silicon: Challenges and benefits for low cost photovoltaics. *Sol. Energy Mater. Sol. Cells* **2010**, *94*, 1528–1533. [CrossRef]
10. Yadav, S.; Chattopadhyay, K.; Singh, C.V. Solar grade silicon production: A review of kinetic, thermodynamic and fluid dynamics based continuum scale modeling. *Renew. Sustain. Energy Rev.* **2017**, *78*, 1288–1314. [CrossRef]
11. Nie, Z.; Ramachandran, P.A.; Hou, Y. Optimization of effective parameters on Siemens reactor to achieve potential maximum deposition radius: An energy consumption analysis and numerical simulation. *Int. J. Heat Mass Transf.* **2018**, *117*, 1083–1098. [CrossRef]
12. Vallerio, M.; Claessens, D.; Logist, F.; Van Impe, J. Multi-Objective and Robust Optimal Control of a CVD Reactor for Polysilicon Production. In *Computer Aided Chemical Engineering*; Klemes, J.J., Varbanov, P.S., Liew, P.Y., Eds.; Elsevier: Amsterdam, The Netherlands, 2014; pp. 571–576.
13. Luo, X.; Li, S.; Li, G.; Xie, Y.-C.; Zhang, H.; Huang, R.-Z.; Li, C.-J. Cold Spray (CS) Deposition of a Durable Silver Coating with High Infrared Reflectivity for Radiation Energy Saving in the Polysilicon CVD Reactor. *Surf. Coat. Technol.* **2021**, *409*, 126841. [CrossRef]
14. Nie, Z.; Wang, Y.; Wang, C.; Guo, Q.; Hou, Y.; Ramachandran, P.A.; Xie, G. Mathematical Model and Energy Efficiency Analysis of Siemens Reactor with a Quartz Ceramic Lining. *Appl. Therm. Eng.* **2021**, *199*, 117522. [CrossRef]
15. Sun, Q.; Chen, H.; Duan, C.; Wan, Y. Effect of Reaction Temperature and Mixture Ratio on Polysilicon Production Process. In *International Conference on Advanced Materials, Processing and Testing Technology*; Trans Tech Publications, Ltd.: Guangzhou, China, 2021.
16. Nie, Z.; Zhou, Y.; Deng, J.; Wen, S.; Hou, Y. Thermal and Electrical Behavior of Silicon Rod with Varying Radius in a 24-Rod Siemens Reactor Considering Skin Effect and Wall Emissivity. *Int. J. Heat Mass Transf.* **2017**, *111*, 1142–1156. [CrossRef]
17. Du, P.; Zhou, Y.; Ramachandran, P.A.; Xie, G.; Hou, Y. Numerical Investigations of Heat Transfer and Skin Effect Characteristics within Rods Located in a 48-Rod Siemens Reactor Heated by AC. *Appl. Therm. Eng.* **2021**, *193*, 116972. [CrossRef]
18. An, L.-S.; Liu, C.-J.; Liu, Y.-W. Optimization of operating parameters in polysilicon chemical vapor deposition reactor with response surface methodology. *J. Cryst. Growth* **2018**, *489*, 11–19. [CrossRef]
19. Ramos, A.; Rodríguez, A.; del Cañizo, C.; Valdehita, J.; Zamorano, J.C.; Luque, A. Heat losses in a CVD reactor for polysilicon production: Comprehensive model and experimental validation. *J. Cryst. Growth* **2014**, *402*, 138–146. [CrossRef]
20. Tang, G.; Chen, C.; Cai, Y.; Zong, B.; Cai, Y.; Wang, T. Numerical Simulations of a 96-rod Polysilicon CVD Reactor. *J. Cryst. Growth* **2018**, *489*, 68–71.
21. Xin, Y.; Shao-fen, W.; Da-zhou, Y. Three-dimensional Numerical Simulation and Optimization of Polysilicon Reduction Furnace Based on Fluent. *Energy Sav. Nonferrous Metall.* **2011**, *27*, 48–52+56.

22. Shimin, L.; Shengtao, Z.; Yinfeng, H.; Min, H. Research on improvement of flow structure of electronic polysilicon reduction furnace based on numerical simulation. *J. Synth. Cryst.* **2019**, *48*, 545–549.
23. Yu, C.; Zhou, Y.M.; Du, P.; Zhao, L.; Zhao, D.; Wang, P.J.; Tian, L.; Xie, G.; Hou, Y.Q. Alternating Current Heating Model of Rods Located in Siemens Reactor. *Fuel Cells* **2021**, *21*, 11–17. [CrossRef]

MDPI
St. Alban-Anlage 66
4052 Basel
Switzerland
www.mdpi.com

*Crystals* Editorial Office
E-mail: crystals@mdpi.com
www.mdpi.com/journal/crystals

Disclaimer/Publisher's Note: The statements, opinions and data contained in all publications are solely those of the individual author(s) and contributor(s) and not of MDPI and/or the editor(s). MDPI and/or the editor(s) disclaim responsibility for any injury to people or property resulting from any ideas, methods, instructions or products referred to in the content.